HANDBOOK
on the
IEE WIRING REGULATIONS

Revised Third Edition

Trevor E. Marks

ARTCS, CEng, MIEE, MInstMC, FInstD

Copies can be obtained from:

William Ernest Publishing
136A Breck Hill Road Woodthorpe
Nottingham NG3 5JP England

Copyright © T. E. Marks 1985, 1988, 1992
All rights reserved by the author
The content and material contained in this publication must not be circulated, copied in whole or in part, reproduced or transmitted by any method or means, including electrical, mechanical, electronic, photographic, photocopying, recording, or otherwise stored in a retrieval system without the prior written permission of the author.

First published 1985
Reprinted: 1985, 1986, 1987

Revised Second Edition published 1988
Reprinted: 1988, 1989, 1990, 1991

Revised Third Edition published 1992

Published by :
William Ernest Publishing
(Division of William Ernest Ltd)

British Library Cataloguing in Publication Data

Marks, Trevor E. (Trevor Ernest), *1931-*
  Handbook on the IEE wiring regulations. - 3rd ed
  I. Title
  621.31924
  ISBN 0-9510156-3-X

DISCLAIMER
Although the contents have been checked, because of the large quantity of calculations involved no responsibility for loss to any person, company or organisation for whatever reason will be accepted by the author or publishers; this book is provided for guidance only.

Printed and bound in Great Britain by S.R.P. Ltd Exeter

Book Contents

Preface to the Third Edition … iv

Copyright … ii

Disclaimer … ii

Introduction … vi

Acknowledgements … vii

Prefaces to First and Second Editions … v

PART ONE
Regulations in general use
Contents … 1

PART TWO
Explanations and worked examples of the Regulations
Contents … 161

PART THREE
Installation tables
Contents … 281

PART FOUR
Design tables, fuse, mcb, and mccb characteristics
Contents … 377

Preface to Third Edition

It is now some three years since the handbook was last revised. During this period there have been many changes to British and IEC Standards; in addition, the IEE Wiring Regulations have again been amended by the introduction of the 16th Edition. These changes have necessitated the publication of a new edition of the book.

Opportunity has been taken to include more information whilst revising the book which now contains 80 pages more than the first edition.

In many of the articles I wrote for the technical press, I included tables giving the maximum length of conductors to comply with the regulations; these have formed the basis of the 'INST' tables, which give the maximum length that cable from 1 mm^2 up to 16 mm^2 can be sized to comply with the regulations for different sized loads.

A further improvement to earlier editions is the re-arrangement of the tables that give the impedance of the armouring of armoured cables, so that they give the phase earth loop impedance at the design temperature, as well as the impedance of the cable armour.

Additionally, the table giving the resistance, reactance, and impedance of XLPE cables has been expanded to cover all sizes from 1 mm^2 up to 300 mm^2.

The table provided to save calculating the sine of the power factor angle (for voltage drop calculations) has been changed; originally it gave the conversion from cosines to sines; it now gives the conversion of power factor to sine, starting at 0.1 power factor, and finishing at 0.99.

A table of temperature conversion factors has also been included in part 4, and more mcb and mccb characteristics have been included.

I have felt, for some time now, that the list of symbols and definitions were in the wrong part of the book; accordingly, they will now be found at the beginning of Part 1.

The book has again been printed on coloured paper, for easy identification of each part. As before, the additional tables that are required by the engineer and contractor, when designing an installation, have been placed in Part 4.

Part 3 has also been completely rewritten to bring it up-to-date. The tables giving the current-carrying capacity of cables have been rearranged, so that tables for similar types of cables are together: for instance, all armoured cables are grouped in one location. As before, tables giving the size of conductor required when grouped with other cables have been provided, since these assist the electrician and contractor, and enable the engineer and consultant to carry out spot checks when inspecting projects. Part 3 also contains the tables giving the maximum length of a circuit with a given size of cable, to comply with the regulations. Tables are provided for 0.4 and 5 seconds disconnection times, as well as tables for socket outlet circuits, and voltage drop in lighting circuits.

Since the handbook is used by many colleges and teachers for the City and Guilds installation and BTEC electrical installation students, as well as by engineers and electricians, Part 2 has been completely revised, and now includes chapters on the assessment of the installation, as well as selecting protective devices; as before, numerous examples are given of how to carry out the calculations.

Part 1 follows the same format as that used in previous editions, but has been completely revised in line with the 16th Edition of the IEE Wiring Regulations. As before, it still lists the regulations in an understandable language, under subject titles.

T.E.M. December 1991

Introduction

This handbook has been produced in small format so that it can be carried in the pocket or toolbox. It is not intended as a substitute for the Wiring Regulations issued by the Institution of Electrical Engineers, but to provide a pocket sized reference book. Used in this context, the handbook can be used by electricians, technicians, students, and engineers.

The handbook includes only those regulations which can be expected to be in use every day. Bearing this in mind, if the exact wording of a regulation is required, then the latest Edition of the Wiring Regulations should be consulted. To assist in this respect, the regulation numbers have been placed in the left hand margin below the text to which it refers.

The handbook is in four parts. Part 1 gives the regulations under subject headings, in an easily understandable language, and to obviate having to look in different sections for applicable regulations, all the appropriate regulations are included under the subject title. Certain regulations are therefore repeated under different subject headings, because they are associated with the subject concerned: for example, reference to Regulation 525-01 will be found under each of the subject headings of Selection of equipment, Diversity, Selection of cables, Safety extra-low voltage, and Functional extra-low voltage.

Part 2 gives explanations of the regulations, with numerous examples of the calculations required, and should therefore be of particular help to the electrician, teacher, lecturer, student and engineer.

Part 3 comprises installation tables, with explanations and examples of how to use them, eliminating the need to carry out complicated calculations for small contracts, the calculations being confined to addition, subtraction, multiplication and division. The aim is to assist the contractor and electrician, although engineers may find them useful for checking final circuits, or seeing the effect grouping has on cable size, without having to do the calculations. They will also find the earth loop impedance tables to be useful when testing an installation.

Part 4 contains tables giving information which is needed when designing an installation, and has been provided to assist the contractor and engineer.

The opinions and interpretations expressed in the handbook are the author's own, and do not purport to represent the official view of the Institution of Electrical Engineers.

December 1991   Trevor E. Marks

Preface of earlier editions

FIRST EDITION

The first copy of this handbook was produced to be used on the William Ernest Ltd Fifteenth Edition Wiring Regulations training course for supervisors and electricians, and permission was given to Culham Laboratories to print 50 copies of the Handbook for use by their electricians.

They have subsequently been given permission to print 50 copies of this revised edition.

The amendments to the Fifteenth Edition published in May 1984 by the Institution of Electrical Engineers generally affected the text of the Handbook, and, in particular, the installation tables.

The Handbook has therefore been revised to take into consideration the May 1984 Amendments, and the whole of Part 3 Installation Tables has been rewritten.

Revising the handbook has also enabled improvements to be made, one such improvement being to rearrange the table in Part 3, by placing tables of design data at the rear of the installation tables.

T. E. M - July 1984

SECOND EDITION

It is now some three years since the handbook was first published, during which time the regulations have been amended, the amendments of 1987 affecting the handbook most, particularly the tables giving the cable sizes for grouped cables and circuits.

It was for this reason, and the fact that the 16th Edition will not be published for at least two years, that it was decided to publish a revised edition, advantage being taken of the revision to include more information.

As before, each part of the book has been printed on coloured paper for easy identification. Part 4 contains tables required when designing an installation, such as the CRG tables which give the current-carrying capacity required for conductors for all the various types of grouping for protective devices up to 250 amperes. This should prove useful to designers, since it enables them to see the effect of grouping cables without having to do any calculations. Part 4 also contains a table of sines and cosines for those who want to take advantage of the calculations for determining the voltage drop in a conductor when the load has a power factor.

Part 3 follows the same format as that used in the first edition, but has been completely rewritten, Again, the designer as well as the contractor and electrician may find the CSG tables helpful, since they give the actual size of cable required for various types of grouping; thus a quick comparison can be made on cable sizes with different grouping arrangements. As before, a chapter with worked examples is given to illustrate how the tables should be used, and where possible, instructions have been included with the tables.

Bearing in mind that the handbook is used by colleges and students as well as by contractors and engineers, Part 2 has been completely rewritten, and gives more detail and worked examples, which show how to use the various formulae given in the regulations, as well as how to carry out the necessary calculations.

Part 1 is basically the same as the first edition by following the same format, but revised to bring it up to date with the amendments. It still lists the regulations in an understandable language under subject titles, but the subjects have now been put into alphabetical order, for although a comprehensive 'Contents - Index' has been provided, it was felt that the new arrangement would assist in locating a particular subject.

I have received many letters and phone calls from readers advising me of how useful they found the first edition; my only hope is that they find this edition even more useful.

T.E.M - March 1988

Acknowledgments

This handbook includes material reproduced by kind permission of the IEE from the IEE Wiring Regulations. Such material may not be copied or reproduced without the permission in writing from the IEE. The IEE Regulations are amended and re-issued from time to time, and it is desirable that reference should always be made to the latest edition.

Extracts from IEC Standards have been used with the permission of the International Electrotechnical Commission, which retain the copyright.

Extracts from British Standards are quoted by kind permission of the British Standards Institution.

I would also like to thank

G.E.C. Alsthom Installation Equipment Ltd., for permission to reproduce characteristics from their publication IEF/483, and for providing the specially constructed T type fuse characteristics used in Part 4;

Merlin Gerin Ltd., for providing the specially constructed mcb and mccb characteristics used in Part 4;

Hawker Fusegear Ltd;

BICC Pyrotenax Ltd;

BICC Cables Ltd;

Delta Crompton Cables Ltd;

Power Centre Trunking Ltd;

Walsall Conduits Ltd;

Wylex Ltd;

William Ernest Training for permission to use material from their electrical safety training courses;

The many equipment and cable manufacturers for providing the information to enable me to construct the design tables in Part 4.

My thanks are also due to Gilbert Payne (who was with UKAEA at that time), without whose encouragement the first edition of the handbook would never have been written, and finally my wife, who not only had the patience to proof read each edition, but also helped in checking the 100 tables in the latest edition.

December 1991   Trevor E. Marks

Are you complying with Section 2 (2C) of the Health & Safety at Work etc. Act ?

Play safe -

increase your

safety

Use our Professional Service for Professionals

Save money - courses given on your own premises

- Two day beginners' course on the IEE Wiring Regulations.
- Two day course for technicians, engineers and contractors, covering the application of the Wiring Regulations.
- Two day course on inspection and testing to the Wiring Regulations.

- Two day course on inspection and testing to comply with the Electricity at Work Regulations.
- Designing installations to comply with the Electricity at Work Regulations.
- Three day course on electrical safety for civil and mechanical engineers.

Courses can be designed to suit your particular requirements.

Special arrangements available for other training organisations.

WILLIAM ERNEST TRAINING
136A Breck Hill Road, Woodthorpe, Nottingham NG3 5JP
FAX: 0602 693 436

PART ONE

REGULATIONS UNDER SUBJECT TITLES

SPACE FOR NOTES AND AMENDMENTS:

Contents - index

1. **Symbols** ---------- 10
2. **Definitions** ---------- 13
3. **Assessment of the installation** ---------- 17
   - Arrangement of circuits ---------- 18
   - Compatibility ---------- 18
   - Environmental conditions ---------- 18
   - Maintain ability ---------- 18
   - Maximum demand ---------- 17
   - Purpose of installation ---------- 17
   - Standby or safety services ---------- 17
   - Structure of the installation ---------- 18
   - Supply characteristics ---------- 17
4. **Agricultural and Horticultural installations** ---------- 19
   - Cable installation ---------- 22
   - Calculations ---------- 22
   - Electric fence controllers ---------- 23
   - Equipment outside the equipotential zone ---------- 21
   - Fire, Protection against ---------- 23
   - Installation and materials ---------- 23
   - Protection by safety extra-low voltage ---------- 19
   - Socket outlets ---------- 19
   - Supplementary equipotential bonding ---------- 22
   - TN system ---------- 19
   - Use of residual current devices ---------- 21
   - Where regulations are applicable ---------- 19
   - Where the system is IT ---------- 22
   - Where the system is TT ---------- 21

Automatic disconnection see indirect contact ---------- 105

5. **Bath and shower rooms** ---------- 24
   - Cable and equipment installation ---------- 27
   - Disconnection time ---------- 26
   - Earthing ---------- 27
   - Electric heating ---------- 25
   - Electrical equipment ---------- 25
   - Equipment or wiring not allowed ---------- 24
   - Instantaneous showers ---------- 27
   - Lampholders ---------- 24
   - Protective measures not allowed ---------- 25
   - Size of supplementary bonds ---------- 25
   - Socket outlets ---------- 24
   - Supplementary bonding ---------- 25
   - Supplementary bonding of a fixed appliance ---------- 26
   - Switches ---------- 24
   - Where regulations are applicable ---------- 27

Bending radius of cables see selection and erection of cables ---------- 128
Bonding see equipotential bonding ---------- 60
Cable see selection and erection of cables ---------- 121

6. **Circuit protective conductors** ---------- 28
   - Accessibility of joints ---------- 31
   - Caravan sites ---------- 32

1

Circuit protective conductor continued:
- Data processing equipment —33
- Factory built equipment —31
- General requirements —28
- Items not allowed —31
- Marking of protective conductors —28
- Protective conductor colour —32
- Protective conductors compulsory —34
- Radial circuits —29
- Residual current devices —32
- Ring circuits —29
- Size of conductor —29
- Items which can be used as protective conductors —31

7. **Conduit** —35
- Ambient temperature —36
- Category 1, 2 and 3 circuits —39
- Colour code —36
- Conduit as a protective conductor —38
- Conduit bends —35
- Conduit fixings —37
- Conduit used for mechanical protection —39
- Conduit wiring —39
- Corrosion and moisture —35
- Damage during installation —39
- Earthing of equipment with high leakage current —39
- Fire barriers —36
- Flexible conduit —38
- Material —35
- Overhead wiring —37
- Prefabricated systems —36
- Socket outlets —38
- Terminations and junctions —36
- Underground wiring —37
- Wiring capacity of conduit —373

8. **Construction site temporary installations** —40
- Cable installation —44
- Calculations —44
- Distribution and distribution equipment —43
- Extent of requirements for construction site —40
- Protection against indirect contact —41
- Regulations that are not applicable to construction sites —43
- Socket outlets —43
- Types of construction site —40
- Use of residual current devices —42
- Voltage of equipment —40
- Where not applicable —40
- Where the system is IT —42
- Where the system is TT —42

9. **Data processing equipment** —45
- Low noise earthing for safety —47

Derating factors   see Part 2 —180

10. **Diversity** —48
- Allowances for diversity at distribution boards (Table 2) —49 /50
- Distribution —51
- Diversity —48

Diversity continued:
- Final circuits --- 48
- Maximum demand --- 48
- Typical final circuit loads --- 48
- Voltage drop --- 51

11. Earthing --- **52**
- Buried earthing conductors -- 55
- Compliance with other regulations ------------------------------ 54
- Earth electrode -- 54
- Earth electrode connections -------------------------------------- 54
- Earth electrode material -- 54
- Earthing general requirements ----------------------------------- 52
- Fundamental requirements -- 52
- Installation with more than one supply ----------------------- 55
- Lightning protection --- 53
- Main earthing terminal --- 53
- Size of earthing conductor -- 55

Electric shock See protection against direct contact and indirect contact ------------ 103 - 105

12. Emergency switching -- **56**
- Accessibility --- 56
- Colour, labelling and position of switch handle -------------- 58
- Devices not allowed -- 59
- Emergency stopping -- 57
- Emergency switching -- 56
- Exclusions -- 57
- Fireman's emergency switch -------------------------------------- 57
- How installed -- 56
- Operational safety -- 57
- Requirements for emergency switches ------------------------- 58
- Requirements for fireman's switches --------------------------- 58
- Where required --- 56

13. Equipotential bonding --- **60**
- Basic measure of protection -------------------------------------- 60
- Bath and shower rooms -- 63
- Bonding exposed to extraneous conductive parts ------------ 62
- Bonding two exposed conductive parts together ------------- 62
- Bonding two extraneous conductive parts to an exposed conductive part -------- 62
- Bonding two extraneous conductive parts together --------- 62
- Caravans and motor caravans ------------------------------------ 64
- Conductors not allowed --- 60
- Contact with non-electrical services ---------------------------- 64
- Extraneous conductive parts used for bonding -------------- 62
- General --- 62
- Instantaneous showers -- 64
- Main equipotential bonding --------------------------------------- 60
- Object of bonding --- 60
- Size of main equipotential bonding conductors ------------- 60
- Supplementary bonding - exposed to extraneous conductive parts ------------ 61
- Supplies to other buildings --------------------------------------- 60
- When bonding not required --------------------------------------- 64
- Where bonding connections required -------------------------- 61

Fireman's emergency switch See Emergency switching --------------------------- 56

14.	**Functional extra-low voltage**	66
-	Barriers or enclosures	67
-	Extra- low voltage	66
-	Functional extra-low voltage	66
-	Protection against direct and indirect contact	66
-	Protection against indirect contact	67
-	Socket outlets	68
-	Voltage drop	68
15.	**Identification and notices**	69
-	Areas for skilled and instructed persons	72
-	Caravans	72
-	Colour identification	70
-	Diagrams and charts	69
-	Earthing label	69
-	Emergency switches	72
-	Fireman's emergency switch	71
-	Identification of equipment	69
-	Isolation	71
-	Mechanical maintenance	72
-	Residual current devices	72
-	Socket outlet label	71
-	Street lighting	72
-	Voltage exceeding 250 V	69
16.	**Index of protection**	74
-	First characteristic numeral	74
-	Second characteristic numeral	74
-	Third characteristic numeral	75
-	Using the IP code	75
17.	**Inspection and Testing**	76
-	Alterations and additions to an installation	80
-	Applicable installations	76
-	Completion certificate	81
-	Continuity of protective conductors	77
-	Continuity of ring circuit conductors	78
-	Earth electrode resistance	80
-	Earth fault loop impedance	80
-	Frequency of testing	81
-	General requirements	76
-	Insulation applied on site	78
-	Insulation resistance	78
-	Introduction	76
-	Non-conducting floors and walls	79
-	Periodic inspection and testing	81
-	Polarity	80
-	Protection by barriers	79
-	Protection by electrical separation	79
-	Requirement for testing	76
-	Residual current device	80
-	SELV	79
-	Sequence of tests	77
-	Testing notice	81
-	Visual inspection	77

18.	Isolation	82
-	Accessibility	86
-	Capacitors and inductors	85
-	Circuits to be isolated	83
-	Discharge lighting	83
-	General requirements	82
-	Identification	85
-	Isolating switchgear	84
-	Isolation at the origin	84
-	Motors	83
-	Object of isolation	82
-	Prevention of re-energisation	83
-	Remote isolators	83
-	Requirements for isolators	85
-	What is an isolator	82
-	What is isolation	82

Labelling equipment See Identification and notices --- 69

Low voltage
- Functional extra-low voltage --- 66
- Reduced voltage systems --- 112
- SELV --- 116

19.	Mechanical maintenance	87
-	Accessibility	88
-	Identification	87
-	Position of device	87
-	Provision of devices	88
-	Provisions for safety	87
-	Requirements for devices	88
-	Switching off for mechanical maintenance	87
20.	Overcurrent protection	89
-	Assessment of general characteristics	89
-	Automatic re-closure	89
-	Circuit breakers operated by unskilled persons	92
-	Co-ordination of overload and short circuit protection	91
-	Disconnection times	92
-	Discrimination	91
-	Energised fuse replacement	91
-	Fuse replacement by unskilled persons	92
-	General requirements	89
-	Identification	91
-	Lampholders	93
-	Overcurrent detection	91
- 21	Fault currents	94
-	Conductor protection	94
-	Conductors in parallel	96
-	Discrimination	96
-	Fault current capacity of conductors	95
-	Fault current protective device	95
-	Limitation of mechanical and thermal effects	94
-	Omission of fault current protection	95
-	Position of fault current protective device	94
-	Prospective fault current	95
-	Values of k	96 /397
-	Withstand capacity	94

- 22 Overloads ---97
  - Conductor protection ---97
  - Conductor size ---97
  - Conductors in parallel ---98
  - Derating semi-enclosed fuses ---98
  - Device fusing current ---97
  - Device size ---97
  - Devices that comply with requirements for fusing current ---98
  - Discrimination ---99
  - Duration of overload ---97
  - Exception to normal overload position ---98
  - Identification of overload devices ---99
  - Limitation of temperature rise ---97
  - Omission of overload devices ---98
  - Overload replacement ---99
  - Position of overload devices ---98
  - Withstand capacity ---97

Overhead wiring between buildings See selection and erection of cables ---126

23. **Protection against heat and fire** ---100
  - Fire barriers ---102
  - Flammable liquid ---101
  - Hazardous areas ---100
  - Heat barriers in vertical ducts etc. ---100
  - High temperature equipment ---100
  - Location of equipment ---100
  - Protection against burns ---101
  - Prevention of overheating ---102
  - Protection against arcing ---100
  - Reducing the risk of fire ---102
  - Terminations ---101

24. **Protection against direct contact** ---103
  - Method 1 - Insulation ---103
  - Method 2 - Barriers or enclosures ---103
  - Method 3 - Protection by obstacles ---104
  - Method 4 - Placing out of reach ---104
  - Methods allowed ---103
  - Protection by residual current device ---104

25. **Protection against indirect contact** ---105
  - Agricultural installations ---109
  - Automatic disconnection when using RCDs in a TN system ---106
  - Basic measure of protection ---105
  - Basic rule for protection ---105
  - Bathrooms ---109
  - Caravans and motor caravans ---110
  - Compliance with Tables 41A to 41D ---108
  - Detrimental influences ---111
  - Equipment outside the equipotential zone ---107
  - Equipotential zone ---107
  - Exceptions for protection against indirect contact ---110
  - General requirements ---105
  - Maintainability ---111
  - Normal body resistance ---108
  - Protection in TN system ---108
  - Protection in TT systems ---108
  - Protection when voltage to earth is 220 V to 270 V in a TN system ---105

Protection against indirect contact continued:
- protection at different voltages to earth in a TN system ---106
- Protective conductors ---110
- Protective devices allowed in a TN system ---108
- Protective devices allowed in TT systems ---108
- Reduced voltage systems ---109
- Supplementary bonding ---111
- ZS for different voltages between 220 and 277 V ---107

26. Reduced voltage systems ---112
- Maximum voltage ---112
- Plugs and socket outlets ---112
- Protection against direct contact ---112
- Protection against indirect contact ---112
- Reduced voltage system ---112
- Source of supply ---112

27. Residual current devices ---113
- Caravan site installation ---114
- Direct contact ---114
- Discrimination ---115
- Electrical equipment outside the equipotential zone ---114
- Fundamental requirement for safety ---113
- Indirect contact ---114
- IT system installations ---113
- Maximum phase earth loop impedance allowed ---113
- PEN conductors ---113
- Protection not provided ---115
- Reduced voltage systems ---114
- Selection and erection of residual current devices ---115
- Socket outlets in TT system ---113
- TT system installations ---113

28. SELV ---116
- Extra-low voltage ---116
- Other extra-low voltage systems ---118
- Plugs and sockets ---117
- Protection against direct and indirect contact ---116
- Protection against indirect contact ---118
- Protection against touching live parts (direct contact) ---118
- Segregation of conductors ---117
- Segregation of live and exposed parts ---117
- SELV sources ---116
- Voltage drop ---118

29. Sauna heaters ---119
- Electrical equipment ---119
- Installation ---120
- Temperature Zones ---119
- Type of appliance ---119
- Wiring ---120

30. Selection and erection of cables ---121
- Bending radii ---128 / 130
- Cable colours ---135
- Cable installation ---129
- Cables installed in walls, floors and ceilings ---130
- Cable selection ---122
- Category 1, 2 and 3 circuits ---132
- Contact with non-electrical services ---133

Selection and erection of cable continued:
- Current-carrying capacity --- 123
- Eddy currents --- 130
- Electrical equipment --- 121
- Fire barriers --- 129
- Flexible cords --- 136
- General --- 121
- Hazardous areas --- 134
- Joints and terminations --- 133
- Lift shafts --- 130
- Mechanical damage --- 127
- Moisture and corrosion --- 134
- Overhead lines between buildings --- 126
- Rotating machines --- 127
- Selection of cables --- 121
- Spacing of supports for cables --- 124
- Standard of cables --- 123
- Temperature --- 130
- Vermin and sunlight --- 132
- Voltage drop --- 135

31. Selection and erection of equipment --- 137
- Accessibility and maintainability --- 140
- Agricultural installations --- 142
- Building structures --- 145
- Cables --- 145
- Ceiling roses --- 144
- Compatibility and external influences --- 139
- Contact with non-electrical services --- 141
- Discharge lighting --- 144
- Dusty atmospheres --- 143
- General --- 137
- Immersion heaters --- 144
- Instantaneous water heaters --- 145
- Joints and terminations --- 142
- Lampholders and luminaires --- 145
- Location of equipment --- 137
- Moisture and corrosion --- 141
- Motors --- 143
- Outside installations --- 143
- Socket outlets --- 144
- Standard of equipment --- 137
- Suitability of supply --- 139
- Switchboards and switchgear --- 144
- Transformers --- 143
- Voltage drop --- 140

Short-circuit protection --- 94

32. Socket outlet circuits --- 146
- Bathrooms --- 149
- BS 196 socket outlets --- 147
- BS 1363 13 A socket outlets --- 146
- BS 4343 16 A socket outlets --- 147
- Cable couplers --- 152
- Caravan site installations --- 150
- Determination of cable size --- 146
- Equipment not allowed --- 149
- Equipment to be used outside --- 149

Socket outlet circuits continued:
- FELV systems --- 150
- Floor area allowed per socket outlet circuit --- 152
- General --- 146
- High earth leakage current equipment --- 152
- Protective conductors --- 150
- Reduced voltage systems --- 150
- Requirements for plug and socket outlets --- 151
- SELV systems --- 150
- Shock protection when the voltage is 240 V --- 148
- Spurs off the ring --- 147
- Switching --- 150
- TT installations --- 151

33. **Switching** --- 153
    - Accessibility --- 155
    - Appliances --- 154
    - Bathrooms --- 154
    - Discharge lighting --- 155
    - Fireman's switch --- 155
    - General --- 153
    - Heating and cooking appliances --- 154
    - Motors --- 154
    - Plugs and socket outlets --- 155
    - Semi conductor devices --- 155
    - Single pole switches --- 153
    - Switches prohibited --- 155
    - Transformers --- 155

Swimming pools --- 256

34. **Trunking** --- 156
    - Busbar trunking --- 157
    - Cables in trunking --- 158
    - Category 1, 2 and 3 circuits --- 158
    - Corrosion and exposure to weather --- 156
    - Fire barriers --- 156
    - Heat barriers --- 156
    - Material --- 156
    - Spacing of supports for trunking --- 157
    - Termination of trunking --- 158
    - Trunking as a protective conductor --- 157
    - Trunking supports --- 156
    - Socket outlets in trunking --- 158
    - Wiring capacity of trunking --- 374

Voltage drop
- See Selection and erection of cables --- 136
- Selection and erection of equipment --- 140

Wiring capacity
- Conduit see Table CCC 1 in Part 3 --- 373
- Trunking see Table CCT 1 in Part 3 --- 374

1. Symbols

Introduction
Most of the following symbols will be found in the IEE Wiring Regulations. The others have been added so that there is less chance of making a mistake when carrying out calculations. For instance, the symbol I_p is used to denote prospective short-circuit current; but since this symbol is used in the regulations irrespective of whether three phase symmetrical faults or phase neutral faults are being considered, this handbook uses the notation that I_p represents a three phase symmetrical fault and I_{pn} represents a phase to neutral fault. Using this notation keeps track of the type of short circuit being considered.

Conductors
PEN	Protective Earthed Neutral.
cpc	Circuit Protective Conductor.
S	Cross-sectional area of live conductor.
S_p	Cross-sectional area of protective conductor.

Current
I_a	Current causing automatic operation of a protective device in a specified time.
I_b	Design or load current.
I_n	Nominal rating or setting of protective device.
I_z	Current-carrying capacity of the circuit conductor.
I_2	Current causing the effective operation of an overload device (with overload current).
I_f	Earth fault current = $U_o \div Z_S$.
I_p	Prospective short circuit current. Also used to indicate a 3 phase symmetrical short circuit current in this book.
I_{pp}	Prospective short circuit current between two phases
I_{pn}	Prospective short circuit current between a phase and neutral conductor.
$I_{\Delta a}$	Earth leakage current.
$I_{\Delta n}$	Rated residual operating current of an RCD.
I_t	The tabulated current required for a conductor.
I_{tab}	The actual tabulated current for a conductor given in the tables.
mA	Milliamp (0.001 A)

Devices
FBA	Factory built assembly.
gG	General purpose fuse having the full range breaking capacity and also suitable for motor protection
gM	A full range breaking capacity fuse suitable only for motor protection.
mcb	Miniature circuit breaker.
mccb	Moulded case circuit breaker.
RCD	Residual current device.
rccb	Residual current circuit breaker.

Factors (correction)

G	Factor for grouping (Cg).
A	Factor for ambient temperature (Ca).
T	Factor for contact with thermal insulation (Ci).
S	Factor for protective device not having a fusing factor of 1.45 (example semi-enclosed rewirable fuse to BS 3036) (C4).
C_t	Factor for correcting resistance for operating temperature of conductor.

Impedance and resistance

$R_1 (Z_1)$	Resistance (impedance) of phase conductor from the origin to the end of a final circuit.
$R_2 (Z_2)$	Resistance (impedance) of the protective conductor from the origin to the end of the final circuit.
R_a	The sum of resistances of earth electrodes.
R_b	The resistance of the earth electrode for an exposed conductive part.
R_A	The sum of the resistance of the earth electrode and of the protective conductors connecting it to exposed conductive parts
Z_E	Impedance of the phase/earth loop external to the installation, or of the system up to the point under consideration.
Z_{inst}	Phase earth loop impedance in the installation (i.e excluding external earth loop impedance).
Z_s	Phase earth loop impedance from the source of energy to the end of a final circuit. Called the system impedance, and is equal to $U_o \div I_f$ or $Z_s = Z_E + Z_{inst}$, or $Z_s = Z_E + R_1 + R_2$, or $Z_s = Z_E + Z_1 + Z_2$.
Z_p	Impedance (or resistance) of one phase.
Z_{pn}	Impedance (or resistance) of phase and neutral.
Z_{pp}	Impedance (or resistance) of two phases.

Systems

T	Source of energy directly connected with earth, or installation directly connected with earth.
N	Installation's exposed conductive parts connected to the source earth.
C	Protective conductor connecting the installation's exposed conductive parts to the source earth, the protective conductor being combined with the neutral.
S	Protective conductor connecting the installation's exposed conductive parts to the source earth by a separate conductor.
I	The source of energy is either not connected with earth, or is connected to earth through a high impedance.
TN-C	Combined neutral and protective conductor system.
TN-S	Separate neutral and protective conductor system.
TN-C-S	Neutral and protective conductor combined from the source to the origin of the installation, and separate within the installation.
TT	System earthed at source, but installation is earthed locally, (i.e no protective conductor from source to origin).
IT	Source isolated from earth or through an impedance, installation earthed locally.

Voltage

V_L	Line voltage.
V_{ph}	Phase voltage.
U_o	Nominal voltage to earth.
mV	millivolts (0.001 V)

Miscellaneous

t_a	Ambient temperature.
t_f	Final operating temperature of a conductor.
t_p	Maximum permitted operating temperature of a conductor in °C
t	Time (eg. disconnection time of protective device).
k	Factor used in adiabatic equation.
IP	Index of Protection (see BS 5490).
PME	Protective Multiple Earthing.
FELV	Functional Extra Low Voltage.

Mathematical

<	Less than.
≤	Less than or equal to.
>	More than.
≥	More than or equal to.
≃	Approximately equal to.

Multiples and submultiples of units

10^6	mega	1 000 000	10^0		1
10^3	kilo	1 000	10^{-1}	deci	0.1
10^2	hecto	100	10^{-2}	centi	0.01
10^1	deca	10	10^{-3}	milli	0.001
10^0		1	10^{-6}	micro	0.000 001

2. Definitions

Before delving into the regulations in detail, it is wise to consider the terminology. Learning the more important definitions will enable a better understanding of the regulations. The following definitions are the basic ones which should be learnt and include terminology such as 'Energy let-through', with which the plant engineer, electrician and electrical contractor, is less likely to be familiar.

AMBIENT TEMPERATURE
The normal temperature of the air or other medium in the location where equipment is to be used.

ARM'S REACH
The limit to which a person can reach without assistance, within a zone of accessibility, as outlined in the following diagram.

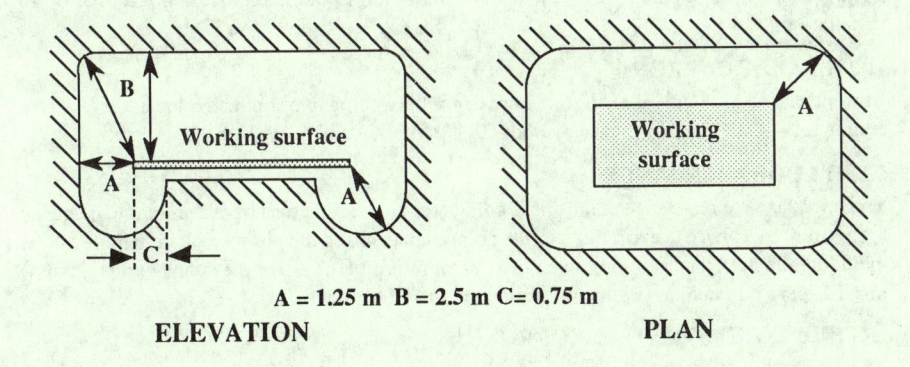

$A = 1.25$ m $B = 2.5$ m $C = 0.75$ m

ELEVATION   PLAN

AUTOMATIC DISCONNECTION
The protective device operates automatically when a fault occurs in the circuit it protects.

BONDING CONDUCTOR
A conductor used for equipotential bonding

BREAKING CAPACITY
A value of short-circuit current that a protective device is capable of breaking under defined conditions.

CIRCUIT BREAKER (LINKED)
A circuit breaker arranged to make and break all the poles at the same time, or in a pre-determined sequence.

CIRCUIT PROTECTIVE CONDUCTOR
A conductor connecting exposed conductive parts to the main earthing terminal.

CLASS I EQUIPMENT
Equipment having only basic insulation, and required to be earthed via a protective conductor.

CLASS II EQUIPMENT
Equipment that does not rely solely on the basic insulation for protection against electric shock i.e., the equipment is either all insulated or double insulated. The equipment does provide for or requires the connection of a protective conductor. The installation may still require a protective conductor at accessories such as, double insulated ceiling roses.

CURRENT-CARRYING CAPACITY
The maximum current which equipment or cable can carry without exceeding the designed operating temperature of the equipment or cable, with a given ambient temperature.

DIRECT CONTACT
Contact with normally energised phase or neutral parts that may result in electric shock.

EARTHING
Connecting exposed conductive parts of an installation to the main earthing terminal of an installation.

EARTH FAULT CURRENT
Current caused by a fault of negligible impedance between a live conductor and earth, or a live conductor and an earthed exposed conductive part.

EARTH LOOP IMPEDANCE
The impedance or resistance of the phase conductor and earth return from the source of energy to the point of a phase to earth fault, or the point of measuring the phase earth loop impedance; including the impedance of the source. The earth return being either the actual earth (ground) in a TT and IT system, or protective condctors in a TN system.

EARTHED EQUIPOTENTIAL ZONE
An area in which the exposed and extraneous conductive parts are bonded together, so that the magnitude and duration of the voltage appearing between them when an electrical fault occurs, will not lead to a fatal electric shock.

ELECTRICAL EQUIPMENT
Abbreviated to equipment, and when mentioned in a regulation includes all items used in an electrical installation, including wiring materials, accessories and appliances.

ENERGY LET-THROUGH
The total energy let-through a protective device before the device finally interrupts the fault current flowing in the circuit concerned.

EQUIPOTENTIAL BONDING
An electrical connection between exposed conductive parts and extraneous conductive parts which puts them at approximately the same potential.

EXPOSED CONDUCTIVE PART
Metalwork of electrical equipment which can be touched and which is not a live part but may become live under fault conditions.

EXTRANEOUS CONDUCTIVE PART
Metalwork that is not part of the electrical installation, and which is liable to introduce a potential, generally earth potential.

EXTRA-LOW VOLTAGE
A voltage not exceeding 50 V a.c., or 120 V d.c between conductors, or between any conductor and earth.

INDIRECT CONTACT
Contact with exposed conductive parts (metalwork surrounding live parts), and which may result in electric shock under fault conditions.

LIVE PART
A phase or neutral conductor, or part intended to be energised in normal use.

LOW VOLTAGE
A voltage exceeding 50 V a.c., but not exceeding 1000 V a.c., between conductors, or 600 V a.c between any conductor and earth.
A voltage exceeding 120 V d.c., but not exceeding 1500 V d.c between conductors, or 900 V d.c. between any conductor and earth.

MAKING CAPACITY
The peak short-circuit current that a protective device is capable of making onto under defined conditions.

OVERCURRENT
A current exceeding the nominal current rating of a circuit. The nominal current rating of conductors is their current-carrying capacity.

OVERLOAD CURRENT
A current exceeding the nominal current rating of an electrically sound circuit or equipment.

PROSPECTIVE FAULT CURRENT
The current that could result from a fault of negligible impedance between live conductors having a difference in potential between them, or between a live conductor and an exposed conductive part, or earth.

PROTECTIVE CONDUCTOR
A conductor used as a means of protection against electric shock by connecting extraneous conductive parts together, or to exposed conductive parts, or connecting any of them to the main earth terminal.

PROTECTIVE DEVICE
In the text of the Handbook, it is a device installed in a circuit to protect the circuit by disconnecting the supply to it if there is an overload, short circuit, or a fault to earth.

REDUCED VOLTAGE SYSTEM
An electrical supply in which the voltage between phases is 110 volts, and the phase to earth volts does not exceed 63.5 volts for three-phase, or 55 volts for single-phase

RESIDUAL CURRENT
The vector sum of the instantaneous current flowing in all live conductors of a circuit at a point in the installation. If the circuit is balanced, the residual current is zero.

RESIDUAL CURRENT DEVICE
A current-operated earth-leakage circuit breaker. Abbreviated to RCD.

SHORT-CIRCUIT CURRENT
A current caused by a fault of negligible impedance between either phase conductors, or phase and neutral conductors.

SIMULTANEOUSLY ACCESSIBLE
Equipment or parts which can be touched at the same time by a person, or where applicable, by livestock.

SUPPLIER
A person or company who supplies electrical energy, including the person or company owning the transmission lines and cables used to deliver the electrical energy.

SWITCH
A mechanical switching device capable of making, breaking and carrying its rated current. Where specified, it may also be capable of operating under overload conditions. It may also be capable of making, but not breaking, short-circuit and earth fault currents.

SWITCH (LINKED)
A switch arranged to make and break all the poles at the same time or in a predetermined sequence

WITHSTAND CAPACITY
The maximum current flowing through it that a device can withstand without being damaged, or causing a danger to arise.

3. Assessment of the installation

Before an alteration, an extension or a new installation is started, an assessment has to be made of the characteristics of the installation. This involves determining the purpose for which the installation is required, the maximum demand of the installation, the characteristics of the supply, the distribution and circuit arrangements, external influences, compatibility with other equipment and services, and the frequency and quality of the maintenance it will receive.

Purpose of the installation
The first assessment is of the purpose for which the installation is intended. This will include such items as: the type of building structure, whether the building will be used for handicapped people, are there any hazardous areas, are corrosive substances being used?
300-01

Maximum demand
The first requirement is to work out the total electrical load that will be connected to the installation. Diversity may be taken into account when assessing the maximum demand of the installation. When assessing the maximum demand, starting currents or in-rush currents of equipment can be ignored.
311-01

Supply characteristics
The next step is to determine the number and types of live conductors. This to some extent will depend upon the purpose for which the installation is intended, and the electricity supplier.

The next requirement is to determine the type of system of which the installation will form part. Will it be TN-C, TN-C-S, TN-S , TT or IT?
312-01, 312-02, 312-03

The supply company must also be asked for the supply characteristics, to enable the protection within the installation to be designed. These characteristics are still required, even if the supply is from a private source such as a generator. The characteristics required can be listed as follows:
1. nominal voltage,
2. nature of current and frequency -   (a.c., or d.c., 50 Hz or 60 Hz etc.),
3. the single-phase and three-phase prospective short -circuit current at the origin of the installation,
4. type and rating of the overcurrent protective device at the origin of the installation,
5. the earth fault loop impedance external to the origin of the installation.

313-01

Standby or safety services
Where the installation concerns the provision of standby or safety services, the characteristics of the source of any standby power supplies shall be determined.

Additionally the capacity, reliability, and rating of such sources shall be determined, along with the time required to change-over in the event of an interruption in the normal supply source.

Where the installation receives its normal supply from a supplier, then arrangements shall be made with the supplier concerning the switching arrangements, such that both sources of power are either prevented from operating in parallel, or are arranged to operate in parallel.
313-02

Structure of the installation

Separate circuits have to be provided for those parts of the installation that need to be separately controlled to avoid danger. This is to ensure that other circuits remain energised when there is a fault on one circuit. This means making certain that there is discrimination between protective devices. Consideration has to be given to what would happen when a protective device operates. For instance, what would happen if all the lights went out due to a fault in a power circuit? Could there be an accident due to the lights going out? Each final circuit has to be connected to a separate way in a distribution board, so that it is electrically separate from every other final circuit. This means that where the final circuits are single-phase, and are taken from a three-phase and neutral distribution board, each circuit must have its own neutral conductor.
300-01

Arrangement of circuits

Every installation has to be broken down into circuits, so that the installation can be operated and maintained safely; this also allows inspection and testing to be carried out safely. Each final circuit in an installation shall be connected to a separate way in a distribution board, making it electrically separate from every other circuit, so as to prevent danger. This means that where single-phase circuits are derived from a three-phase supply, a separate neutral shall be installed for each single-phase circuit.

The number of circuits required shall be planned, so that the circuits and installation will comply with the regulations for current-carrying capacity, overcurrent protection, and isolation.
314-01

Environmental conditions

The environmental conditions applicable to the installation have to be taken into account when selecting equipment and wiring materials. Consideration shall be given to the ambient temperature, mechanical stresses, corrosive substances, presence of water and moisture, and to any other conditions which can foreseeably be seen to affect the installation now and in the future.
Chapter 32

Compatibility

Consideration has to be given to what effects the installation is likely to have on other equipment and services. For instance, installing mains cables in close proximity to telemetry cables can affect the signals carried by those cables. Starting currents of large motors can cause a dip in voltage which will affect electronic control circuits. The supplier of the electricity will also need to be informed of any equipment which is likely to affect their supplies, as this would affect other customers connected to the same service.
331-01

Maintainability

When selecting the protective devices for the installation, consideration has to be given to the frequency and quality of maintenance that the installation will receive throughout its intended life.

This is required so that the protective measures taken to provide safety and the reliability of the equipment selected, remain effective throughout the intended life of the installation. In addition, consideration has to be given to any periodic inspection, testing, maintenance and repairs that may be necessary on the installation during its intended life.
341-01

4. Agricultural and horticultural installations

Where applicable
With the exception of dwellings used solely for human habitation, the regulations concerning agricultural and horticultural installations apply to the whole site, including both outdoor and indoor installations.
605-01

Protection by safety extra-low voltage
Where SELV is used to protect against both direct and indirect contact in areas accessible to livestock, the maximum value of extra low voltage of 50 V shall be reduced to a value appropriate to the type of livestock using the areas concerned. (Note 12 V is sufficient to kill some types of animal)
605-02-01

No matter how low the voltage used for the SELV circuit is, protection against direct contact shall be provided by either barriers or enclosures giving a minimum protection of IP2X, or by insulation that will withstand 500 V a.c., r.m.s. for 1 minute.
605-02-02

Socket outlets
Socket outlet circuits shall be protected by an RCD manufactured to British Standards. The operating current shall be 30 mA, and it will have a maximum operating time of 40 msec with a residual current of 150 mA. This means that Regulation 471-16-01 for portable equipment outdoors is not applicable.
605-03 , 605-09-03

Protection against indirect contact by equipotential bonding, and automatic disconnection of the supply within the equipotential zone

TN system
The limiting values of earth fault loop impedance for installations forming part of a TN system are only applicable where the exposed and extraneous conductive parts are within the equipotential zone created by the main equipotential bonding.
605-09-01

Protection against indirect contact shall be by either or both of the following methods:
1) overcurrent protective device.
2) residual current device.

Either method gives protection by automatic disconnection.
413-02-07

The characteristic of each overcurrent protective device, and the earth fault loop impedance of each circuit, has to be such that :

$$Z_S \leq \frac{U_o}{I_a}$$

Where Z_S is the earth fault loop impedance of the circuit, U_o is the nominal voltage to earth, and I_a is the current causing the protective device to disconnect within a specified time.

The specified disconnection time in locations in which livestock is kept is:
1. In accordance with Table 605A for final circuits feeding:
   (a) portable equipment intended for manual movement during use or,
   (b) hand-held Class I equipment.
2. 5 seconds for:
   (a) a distribution circuit or,
   (b) a final circuit supplying only stationary equipment or,
   (c) a final circuit not used for hand-held Class 1 equipment or,
   (d) a circuit feeding equipment which is not going to be moved manually.
3. 5 seconds for circuits complying with the reduced low voltage regulations.

Maximum disconnection times for TN systems (605A)

U_o	t (seconds)
120	0.35
220 to 277	0.2
400, 480	0.05
580	0.02

605-04, 605-05-01, 605-05-02, 605A, 605-05-06

Where the voltage to earth U_o is 240 V, and a protective device is used to satisfy the disconnection times in Table 605A, the maximum value of Z_S for each type of protective device is given in Table ZS 9 in Part 3.

605-05-03, 605-05-04, 605-05-07

For reduced low voltage installations, where the voltage between phases is 110 V, the maximum values of earth loop impedance are given in Table ZS 8.

471-15-06

Where the voltage to earth U_o is 240 V, and the disconnection time allowed is 5 seconds, the maximum values of Z_S allowed are given in ZS1 to ZS7 in Part 3.

605-05-06

The limitation of electric shock by using Table 41C is not allowed.

605-05-05

Where the disconnection time allowed for a final circuit is 5 seconds, and the actual disconnection time of the circuit exceeds that given in Table 605A, whilst another final circuit which requires disconnecting within the time specified in Table 605A, is connected to the same distribution board or distribution circuit, then one of the following conditions shall be fulfilled:
1. the protective conductor impedance from the distribution board to the main earth bar for equipotential bonding shall not exceed $25\ V \times Z_S \div U_o$, where Z_S is the earth loop impedance for 5 seconds disconnection time. Where U_o is 240 V, the formula becomes $0.104 \times Z_S$.
2. equipotential bonding shall be carried out at the distribution board to the same types of extraneous conductive parts as those to which the main equipotential bonding is connected; the size of the bonding conductors shall be the same as that of the main equipotential bonding.

605-05-06

AGRICULTURAL AND HORTICULTURAL INSTALLATIONS

Where the disconnection times detailed above cannot be met by using an overcurrent protective device, then one of the following remedies shall be provided:
1. local supplementary equipotential bonding shall be installed, connecting together the exposed conductive parts of the circuit concerned (including the earthing terminal of socket outlets) and extraneous conductive parts. The resistance of the supplementary bonding conductor shall be less than or equal to 25 V ÷ Ia, where I_a is the minimum current needed by an overcurrent device to disconnect the circuit in 5 seconds, or is the residual operating current of an RCD.
2. Protection shall be provided by an RCD complying with $Z_S \times I_{\Delta n} \leq 25$ V.

605-05-08, 605-05-09

Where exposed conductive parts are simultaneously accessible, they shall be connected to the same earthing system. They can be connected individually, collectively, or in groups.
413-02-03

Every exposed conductive part of the installation shall be connected by protective conductors to the main earthing terminal of the source of the supply
413-02-06

Use of residual current devices (RCDs)

If the overcurrent protective devices cannot disconnect the low voltage circuits within the time specified in Table 605A, or disconnect fixed equipment circuits and reduced low voltage circuits within 5 seconds, then protection shall be provided by a residual current device, such that $Z_S \times I_{\Delta n} \leq 25$ V.
605-05-08

Where RCDs are used for automatic disconnection, exposed conductive parts need not be connected to the TN system's protective conductors, providing they are connected to an earth electrode such that $R_A \times I_a \leq 25$ V
413-02-17

Equipment outside the equipotential zone

Where a circuit which feeds fixed equipment with exposed conductive parts is outside the earthed equipotential zone, and where it can be touched by a person directly in contact with the ground, the disconnection time allowed for the circuit must comply with Table 605A
605-09-02

Where the system is TT

Where exposed conductive parts are protected by a single protective device, they shall be connected via a protective conductor to the main earthing terminal, and then to a common earth electrode. Where several protective devices are in series, the exposed conductive parts may be connected to a separate earth electrode associated with each protective device.

The protective device can be either an overcurrent device, or a residual current device, the preferred protection being a residual current device.
413-02-18, 413-02-19

Each circuit must comply with the formula $R_A \times I_a \leq 25$ V, where R_A is the resistance of the protective conductor from the exposed conductive part to the earth electrode, plus the resistance of the earth electrode.

The value I_a is the current that will cause disconnection of the overcurrent protective device within 5 seconds, or is the residual operating current $I_{\Delta n}$ of a residual current protective device.
605-06

Where the system is TT, and protection is by equipotential bonding and automatic disconnection, every socket outlet circuit shall be protected by an RCD complying with $R_A \times I_{\Delta n} \leq 25$ V
471-08-06, 605-05-09, 413-02-16

Where the system is IT
Where the system is IT, Regulations 413-02-21 to 413-02-25 are applicable, except that the voltage in regulation 413-02-23 is changed from 50 V to 25 V
605-07-01

Regulation 413-02-26 is applicable, except that Table 41E is changed to 605E, and I_a is the current which either disconnects the circuit within the time specified in Table 605E, or within 5 seconds, when this disconnection time is permitted.

Voltage between phase and neutral U_o Voltage between phases U volts	Neutral not distributed t (seconds)	Neutral distributed t (seconds)
120 to 240 volts	0.4	1
220/380 to 277/480	0.2	0.5
400/690	0.06	0.2
580/1000	0.02	0.08

605-07-02, 605E

Supplementary equipotential bonding
The minimum size of supplementary bonding conductors shall be the larger of either:
1) the size determined by the formula $Ia \times R \leq 25$ V, or
2) the size in accordance with the general regulations for supplementary bonding.
605-08-01

In areas where livestock is kept, supplementary equipotential bonding shall be installed. The supplementary equipotential bonding shall connect together all exposed and extraneous conductive parts (including non-insulating floors) which can be touched by livestock.
Where a metallic grid is laid in non-insulating floors for the purpose of supplementary bonding, it shall be connected to the protective conductors of the installation..
605-08-02, 605-08-03

Cable installation
Cables shall be sized in accordance with the overload, short-circuit, and protective conductor regulations.
600-01, 600-02

Calculations
The temperature rise of a conductor due to fault current shall be taken into account when checking that the circuit is protected against indirect contact.
413-02-05

AGRICULTURAL AND HORTICULTURAL INSTALLATIONS

Protection against fire
With the exception of equipment essential to the welfare of livestock, an RCD having a maximum residual operating current of 500 mA shall be installed for the supply to equipment.
605-10-01

To minimise the risk of burns to livestock and of fire, heating appliances shall be installed so as to maintain a suitable distance from combustible material and livestock. Where the heaters are radiant heaters, the suitable clearance distance shall be as recommended by the manufacturer, but in any event not less than 500 mm.
605-10-02

Installation and materials
Electrical equipment shall, so far as is practicable, be of Class II construction in areas accessible to livestock in and around agricultural buildings. The electrical equipment can be constructed of or protected by insulating materials. Electrical equipment shall be suitable for the environmental conditions applicable to the installation, such as water, dust, corrosion, flora, fauna, solar radiation, wind, and mechanical stresses.

The fixed wiring shall be inaccessible to livestock in areas where livestock are kept, and any cables that are liable to attack by vermin shall be suitably protected.

Devices installed for emergency switching or emergency stopping shall be installed so that they are inaccessible to livestock. The operation of such safety devices must not be impeded by livestock, due consideration being given to the conditions that could arise if the livestock panicked.
605-11 , 605-12 , 605-13

Electric fence controllers
Electric fence controllers shall comply with the following British Standards:
BS 2632 Mains-operated electric fence controllers.
BS 6369 Battery-operated electric fence controllers suitable for connection to the supply mains.
605-14-01

The installation of mains operated fence controllers, so far as is reasonably practicable, shall be free from risk of mechanical damage or unauthorised interference; they shall not be installed on a pole carrying an overhead power or telephone line, unless the overhead power line is insulated, and is the supply for the fence controller.
605-14-02, 605-14-03

An earth electrode that is connected to fence controller shall be kept completely separate from the earthing system of any other circuit, and its resistance area must not overlap the resistance area of any electrode used for protective earthing.
605-14-04

The electric fence controller, the electric fence or conductor shall be installed so that it will not be liable to come into contact with any other equipment or conductor; not more than one controller shall be connected to each electric fence or similar system of conductors.
605-14-05, 605-14-06

5. Bath and shower rooms

Socket outlets
Unless the circuit is derived from a SELV safety source with a nominal voltage not exceeding 12 V r.m.s. a.c., or d.c., which safety source is out of reach of a person in a bath or shower, socket outlets shall not be installed in a room containing a fixed bath or shower, and there shall be no provision for connecting portable equipment.

Where a socket outlet is installed from a 12 V SELV source, the socket outlet must not have any accessible metal parts, and shall be insulated or protected with barriers or enclosures against direct contact.

Where a shower is installed in a room that is not a bathroom, sockets shall be at least 2.5 metres from the shower cubicle.

Providing the supply in a shaver socket has been manufactured to BS 3535, it can be installed in a bathroom for the use of electric shavers; in this case the protective conductor of the final circuit feeding the shaver socket shall be connected to the earth terminal of the shaver socket. The shaver socket can be independent or part of a luminaire.
601-03, 601-09, 601-10

Lampholders
If a lampholder is within 2.5 metres of a fixed bath or shower it shall be constructed of, or shrouded in insulating material, and have an insulating shroud. Alternatively, luminaires that are totally enclosed may be used.
601-11

Switches
Switches and control devices shall be mounted so that they are inaccessible to a person using a fixed bath or shower.

The only switches or controls that are allowed to be accessible in a room with a fixed bath or shower are: shaver sockets manufactured in accordance with BS 3535, the switches or controls of instantaneous water heaters manufactured to BS 3456, insulating cords of cord-operated switches which comply with BS 3676, switches which form part of a circuit derived from a SELV safety source with a nominal voltage not exceeding 12 V r.m.s. a.c., or d.c., and switches or controls operated by mechanical actuators with insulating linkages.

The safety source for the SELV circuit shall be out of reach of a person using the bath or shower, the switches must not have any accessible metal parts, and shall be insulated or protected with barriers or enclosures against direct contact.
601-08, 601-03,

Equipment or wiring not allowed
No electrical equipment shall be installed in the interior of a bath tub or shower basin.

Installations carried out on the surface shall not use metal conduit, metal trunking, a cable with an exposed metallic sheath, or an exposed earthing or bonding conductor.
601-02, 601-07

Protective measures not allowed
Protection against electric shock, by means of obstacles, by placing out of reach, by non-conducting location, or earth free local equipotential bonding, is not allowed.
601-05, 601-06

Electrical equipment
Stationary appliances which have heating elements, including silica glass sheathed elements that can be touched, must not be installed within reach of a person using a fixed bath or shower.
　　Where electrical equipment is installed in the area below a bath, access to that area must only be available by using a key or tool, and supplementary bonding shall be installed unless the equipment is supplied from a SELV source.
601-12-01, 601-04-03

Electric heating
Heating units embedded in the floor and intended for heating the room may be installed, providing they are either covered by a metallic grid, or by an earthed metallic sheath, which is connected to the local equipotential bonding conductors.
601-12-02

Supplementary bonding
Except for equipment supplied from a SELV circuit, supplementary equipotential bonding shall be provided between all metal parts that are simultaneously accessible in a room containing a fixed bath or shower. This includes all exposed and extraneous conductive parts, such as the metalwork of electrical equipment, metal waste pipes, metal baths or shower trays, metal water pipes, central heating pipes, or exposed structural metalwork. It includes electrical equipment that is installed in the space below a bath, even though it is only accessible by use of a key or tool, but equipment supplied from a SELV circuit is excluded.
601-04-02, 601-04-03

Size of supplementary bonds
The minimum size of supplementary bonding conductors shall be:
1. **For exposed conductive parts bonded together:**
   If sheathed or otherwise mechanically protected, not less than the conductance of the smallest cpc connected to the exposed conductive parts. If not mechanically protected, the minimum size shall be 4 mm^2.
2. **For exposed conductive parts bonded to extraneous conductive parts:**
   If sheathed or otherwise mechanically protected, not less than half the conductance of the cpc connected to the exposed conductive part. If not mechanically protected, the minimum size shall be 4 mm^2.
3. **For extraneous conductive parts bonded together**
   If sheathed or otherwise mechanically protected, not less than 2.5 mm^2. If not mechanically protected, 4 mm^2. If any of the extraneous conductive parts are bonded to an exposed conductive part, the conductor size shall be determined in accordance with paragraph 'b'.

547-03-01, 547-03-02
547-03-03, 547-03-04

BATH AND SHOWER ROOMS

Exposed conductive part to exposed conductive part

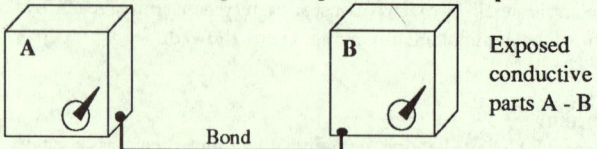

Same conductance as the smallest cpc feeding A or B, providing that it is mechanically protected: otherwise minimum size is 4 mm².

Exposed conductive part C to extraneous conductive part

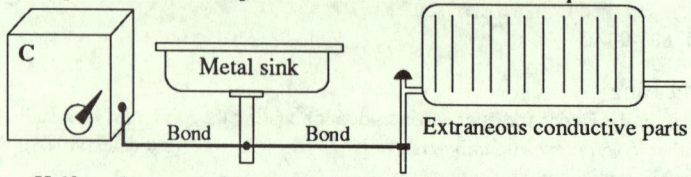

Half conductance of the cpc in 'C' if sheathed or mechanically protected: otherwise minimum size required is 4 mm².

Extraneous conductive part to extraneous conductive part

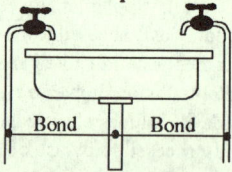

Minimum size 2.5 mm² if sheathed or otherwise mechanically protected: 4 mm² if not mechanically protected.

Figure 3 - Minimum size of supplementary bonding conductors

Supplementary bonding a fixed appliance

Where a fixed appliance has to be included in the supplementary bonding, and that appliance is supplied from a flex outlet type accessory through a short length of flexible cable containing a cpc, then the supplementary bonding connection can be made at the earth terminal in the flex outlet accessory.
547-03-05

Disconnection time

Except for SELV circuits, where electrical equipment is simultaneously accessible to the metalwork of other electrical equipment, or to extraneous conductive parts, the electrical equipment shall be disconnected by the protective device in 0.4 seconds in the event of an earth fault in the equipment.
601-04-01

BATH AND SHOWER ROOMS 27

Earthing
Where exposed conductive parts may become charged, they shall be connected to earth in such a manner as to discharge the electrical energy without danger. Either overcurrent or residual current protective devices shall be used to protect the circuits against the persistence of an earth fault current. Where exposed conductive parts are connected with earth and are accessible to extraneous conductive parts, the extraneous conductive parts shall be connected to the main earthing terminal of the installation.
130-04

Instantaneous showers
Every instantaneous shower heater must contain an automatic device for preventing a dangerous rise in temperature.
554-04

The metal water pipe through which the water supply to the heater is provided shall be solidly and mechanically connected to all metal parts (other than live parts) of the heater or boiler. There must also be an effective electrical connection between the metal water pipe and earth independent of the circuit protective conductor.
554-05-02

The supply to the heater shall be through a double pole linked switch; the cables from the switch must connect directly to the heater, and not through a plug and socket.

The switch shall be separate from and within easy reach of the heater. Alternatively, the switch can be incorporated in the heater, providing it complies with BS 3456.

Where the heater is installed in a room containing a fixed bath or shower, the switch shall be out of reach of a person using the bath or shower.
554-05-03, 601-08-01

A switch or control device is not considered accessible to a person using a bath or shower, if it is remote and operated by an insulating cord, or by insulated mechanical actuators, or if the controls comply with BS 3456.
601-08-01

Where a shower tray is made from a metallic material, or where it is made from concrete, it shall be bonded to the metal water pipe in the shower heater. In the case of a concrete shower tray, the reinforcing can be used for the bonding connection.
INF

Where regulations are applicable
The regulations for bath and shower rooms apply to bath and shower tubs, including their surroundings.
601-01

Cable and equipment installation
Except where modified by the above regulations, the installation shall comply with all the other regulations. Cables shall be sized in accordance with the overload, fault current and protective conductor regulations.
600-01, 600-02

6. Circuit protective conductors

General requirements
Where a voltage may appear on exposed conductive parts, then the exposed conductive parts shall be earthed so that the electrical energy is discharged without danger.
130-04-01

Where exposed conductive parts are earthed, the circuits shall be protected by either overcurrent protective devices, or residual current devices, to prevent the persistence of dangerous earth leakage currents.
   Where the phase earth loop impedance is too large to allow the prompt disconnection of the circuit by overcurrent devices because the prospective earth fault current is insufficient, then a residual current protective device shall be used for circuit protection.
130-04-02, 130-04-03

Where exposed conductive parts are earthed, and are simultaneously accessible with extraneous conductive parts of other services, the extraneous conductive parts shall be connected to the main earthing terminal of the installation.
130-04-04

The only protective device allowed in an earthed neutral is a linked circuit breaker, or switch, where the related phase conductors are arranged so as to be disconnected at the same time.
130-05-02

Protective conductors can be used jointly or separately for protective or functional purposes, providing the functional purpose does not affect the protective measures of the conductor.
542-01-06

Protective conductors shall be protected against mechanical and chemical deterioration, and electrodynamic effects.
543-03-01

Where a protective conductor is installed through several installations, each one of which has its own earthing arrangement, then either of the following requirements shall be complied with:
1. the protective conductor shall be earthed within one installation only, and insulated from the earthing arrangement in the other installations, or
2. be capable of carrying the maximum fault current likely to flow through it from any of the installations.

Where the protective conductor is part of a cable, it shall only be earthed in the installation containing the protective device for the circuit.
542-01-09

Marking of protective conductors
Protective conductors up to 6 mm^2 shall be insulated throughout, the same as conduit wiring cables to BS 6004 or BS 7211, unless they form part of a multicore cable or cable enclosure, such as conduit or trunking.
543-03-02

The un-insulated protective conductor in a sheathed cable shall be protected with green/yellow insulating sleeving complying with BS 2848, where the sheath is removed at joints and terminations, if the size of the protective conductor is less than 6 mm^2.
543-03-02, 514-03-01

When a bare conductor or busbar is used as a protective conductor, it shall be marked by equal green and yellow stripes each stripe being between 15 and 100 mm wide. The stripes shall be placed close together, and be positioned either throughout the length of the conductor, or in each compartment and at each accessible position. Where adhesive tape is used for marking, it shall be coloured green/yellow.
514-03-01

Ring circuits
Every final ring circuit protective conductor, unless formed by conduit, trunking, armouring or the metal sheath of a cable, shall also be installed in the form of a ring, and both ends shall be connected to the same earth terminal at the origin of the circuit.
543-02-09

When conduit, trunking, ducting, armouring or the metal sheath of a cable is used as the protective conductor for socket outlet circuits, the earthing terminal of each socket outlet shall be connected by a separate insulated protective conductor to an earth terminal in the associated socket outlet box or enclosure.
543-02-07

Radial circuits
When conduit, trunking, ducting, armouring or the metal sheath of a cable is used as the protective conductor for radial circuits, the earthing terminal of each outlet accessory shall be connected by a separate insulated protective conductor to an earth terminal in the associated outlet box or enclosure.
543-02-07

Size of conductor
The size of a circuit protective conductor shall either be calculated in accordance with 543-01-03, or determined from table 54G
543-01-01

Where the phase conductor has been sized for protection against short-circuit currents, and the earth fault current is likely to be less than the short-circuit current, then the size of the protective conductor shall be determined by calculation.
543-01-01

Where the circuit protective conductor is not conduit, trunking, ducting, armouring or the sheath of a cable, or enclosed in a cable, then the minimum size of conductor must be equivalent to 2.5 mm^2 copper, and be protected against mechanical, chemical deterioration, and electrodynamic effects.
543-01-01, 543-03-01

When one protective conductor is used for several circuits, its cross-sectional area shall be either -
1. selected in accordance with Table 54G by using the largest phase conductor of the group of circuits concerned, or

CIRCUIT PROTECTIVE CONDUCTORS

2. be selected by determining the fault current and disconnection time for each circuit, and then using the values of the circuit with the most unfavourable value of fault current and disconnection time, to size the conductor, by using the formula of Regulation 543-01-03.

543-01-02

Where the size of the protective conductor is determined by calculation, the following formula shall be used.

$$S\ mm^2 = \frac{\sqrt{I^2 t}}{k}$$

The value of 'k' is obtained from tables 54B to 54F in the Regulations, or from Table K1 & K1A in Part 4. The most important item to watch when selecting 'k' is the initial and final temperature of the conductor. $S\ mm^2$ is the minimum size of protective conductor required, so in most cases, the next largest standard conductor size will have to be chosen. 'I' is the phase to earth fault current in amperes, and 't' is the actual time taken in seconds for the protective device to disconnect the circuit.

543-01-03

Where the disconnection time 't' is less than 0.1 seconds, then $I^2 t$ from the manufacturer's $I^2 t$ characteristics shall be less than $k^2 S^2$ for the protective conductor.

434-03-03

Table 54G can be used instead of calculating the minimum cross sectional area of the protective conductor. A modified table 54G based on 'k' factors, giving values of Sp for copper, aluminium and steel is contained in MISC1 in Part 4, as M54G.

Cross-sectional area of phase conductor (S) in sq.mm.	Minimum cross-sectional area of protective conductor (S_P) in sq. mm.	
	Size required when c.p.c. is made from the same material as phase conductor	Size required when c.p.c. is made from a different material to phase conductor
S not exceeding 16	S	$\dfrac{k_1 S}{k_2}$
S from 16 to 35	16	$\dfrac{k_1 16}{k_2}$
S larger than 35	$\dfrac{S}{2}$	$\dfrac{k_1 S}{k_2 2}$

where S is the cross-sectional area of the phase conductor, S_P is the cross sectional area of the cpc, k_1 is the 'k' value for the phase conductor, and k_2 is the 'k' value for the cpc material.

Table 54G, 543-01-04

Table 54G gives the minimum sizes of protective conductor required, so in most cases, the nearest larger standard size of cable will have to be installed.

543-01-04

Items which can be used as protective conductors
A circuit protective conductor can be one of the following:
1. a single core cable.
2. an insulated or bare conductor contained in a cable.
3. the metal sheath or armour of a cable.
4. metal conduit, trunking, ducting or any other type of metallic enclosure.
5. an insulated or bare conductor in a common enclosure with insulated live conductors.

The protective conductor shall be copper where its size does not exceed 10 mm^2.
543-02-02, 543-02-03

Where the sheath of an micc cable, conduit, trunking, ducting or the metal sheath of a cable is used as a protective conductor it shall:
1. have its electrical continuity ensured by its construction or by its connection.
2. be protected against mechanical, chemical, and electrochemical deterioration.
3. have a cross-sectional area not less than that given by the formula of 543-01-03, or Table 54G, or by testing in accordance with BS 5486 Part 1.

543-02-05

Factory built equipment
Where the case, frame, or enclosure of busbar trunking, switchgear, control gear and distribution boards etc., is used as a protective conductor, it shall:
1. have its electrical continuity ensured, by its construction or by its connection.
2. be protected against mechanical, chemical and electrochemical deterioration.
3. have a cross-sectional area not less than that given by the formula of 543-01-03, or Table 54G, or by testing in accordance with BS 5486 Part 1, and
4. allow the connection of other protective conductors at every pre-determined tap-off point.

543-02-04

Accessibility of joints
With the exception of cables buried underground, conduit, trunking or ducting, all joints in protective conductors shall be accessible for inspection, unless the joint is made by welding, soldering, brazing or compression tool.
543-03-03, 526-04-01

Joints that can be disconnected for test purposes are allowed in protective conductors.
543-03-04

Items not allowed
In general, protective conductors must not have a switch or isolator installed in them, but a multipole switching device, or plug in device is allowed, providing the protective conductor circuit is opened last, and closed no later than the associated live conductors.

There is the special case, where an installation is supplied from more than one source of energy which requires independent earthing, and it is necessary to make sure that not more than one means of earthing is connected at any one time. In this case, a switch may be installed between the neutral point and the means of earthing, providing the switch is linked so as to switch the live conductors at the same time as the earthing conductor.
543-03-04, 460-01-04

Aluminium or copper-clad aluminium shall not be used as a bonding conductor.
547-01

The plain colours of yellow or green shall not be used in flexible cables or cords. Neither shall any bi-colour, other than green/yellow, be used.
514-07-02

In fixed wiring, the single colour green shall not be used.
514-06-02

Plain slip or pin grip sockets shall not be used for conduit joints when they are used as a protective conductor.
  Joints in conduit shall be screwed or made with substantial mechanical clamps, so that they are electrically continuous.
543-03-06

Flexible or pliable conduit is not allowed to be used as a protective conductor. Neither is a gas or oil pipe allowed to be used as a protective conductor.
543-02-01

Except for switchboards and busbar trunking, the exposed metalwork of equipment cannot be used as part of the circuit protective conductor of other equipment.
543-02-08

Extraneous conductive parts shall not be used as a circuit protective conductor.
543-02-06

Residual current devices

Where an RCD is used for protection, only the live conductors are to pass through the magnetic circuit, the protective conductor being outside that circuit.
531-02-02

Protective conductor colour

The colour combination green/yellow shall only be used for protective conductors.
514-03-01

Caravan sites

Where the installation for mobile touring caravans forms part of a TN-C-S system, the protective conductor of each socket outlet shall be connected to an earth electrode. The socket outlet circuit shall be protected by a residual current device, and $R_A I_a$ must not exceed 50 volts, where R_A is the total resistance of the earth electrode and protective conductors up to the socket outlet, and I_a is the residual operating current $I_{\Delta n}$.
608-13-05, 413-02-18,
413-02-19, 413-02-20

Where a residual current device is used to protect a mobile touring caravan socket outlet, or a group of six socket outlets, the earthing terminal of the socket outlets must not be connected to the PME terminal
608-13-05

Data processing equipment complying with BS 7002
Where the earth leakage current from equipment complying with BS 7002 does not exceed 3.5 mA, no special precautions are necessary.
607-02-02
The earthing arrangements for installations supplying equipment such as industrial control equipment, micro-electronics or computers, having an earth leakage current in excess of 3.5 mA, shall comply with the following regulations:
607-01
Where an installation is supplied through a residual current device, and there is more than one item of equipment having an earth leakage current in excess of 3.5 mA, the total leakage current must not exceed 25% of the tripping current of the residual current device.
607-02-03
Where the total leakage current exceeds 25% of the tripping current of the residual current device, the equipment shall be supplied through a double wound transformer or other device, in which the input and output circuits are electrically separate. The protective conductor shall connect the exposed conductive part of the equipment to the secondary winding of the transformer. The protective conductor shall comply with one of the arrangements outlined in Regulation 607-02-07.
607-02-03
Where an item of stationary equipment has an earth leakage current between 3.5 mA and 10 mA, it shall be permanently connected to the fixed wiring, or connected by means of a BS 4343 socket-outlet.
607-02-04
Where an item of stationary equipment has an earth leakage current exceeding 10 mA in normal use, it shall be permanently connected to the fixed wiring of the installation. The equipment can be connected through a BS 4343 socket-outlet, providing the protective conductor in the flexible connecting cable is supplemented by a separate additional 4 mm^2 conductor, and an additional contact in the plug and socket-outlet.
607-02-05
Where a final circuit feeds a number of socket outlets, in an area where it can be expected that the total earth leakage current from the equipment connected to the socket-outlets will exceed 10 mA, the circuit shall be provided with a high integrity protective connection complying with items 1 to 6 of Regulation 607-02-07.

Alternatively, a ring circuit can be used, providing there are no spurs taken from the ring circuit, the minimum size of protective conductor used is 1.5 mm^2, and the ends of the protective conductor are connected into separate earth terminals at the source of the ring circuit.
607-02-06
Where the fixed wiring of every final circuit is installed in a location intended to accommodate several items of stationary equipment in which the total earth leakage current exceeds 10 mA, the circuit shall be provided with a high integrity protective connection complying with one or more of the following:
1. a single protective conductor with a minimum cross-sectional area of 10 mm^2.
2. a separate duplicated protective conductor, having independent mechanically and electrically sound terminations, protected from mechanical damage and vibration, and

CIRCUIT PROTECTIVE CONDUCTORS

      made by compression-sockets, soldering, brazing, welding, or mechanical clamps, each protective conductor having a minimum cross sectional area of 4 mm^2.

3.   duplicate protective conductors with live conductors in a multicore cable, providing that the total cross sectional area of all of the conductors in the cable is not less than 10 mm^2. One of the protective conductors can be formed by the metallic armour, sheath or braid of the cable, providing its electrical continuity can be assured, and providing it has a minimum cross-sectional area calculated in accordance with Regulation 543-01.
4.   an earth monitored protective conductor, the circuit being disconnected in the event of a failure in that protective conductor; the circuit shall comply with the automatic disconnection and equipotential bonding regulations.
5.   duplicate protective conductors formed by trunking and conduit, sized in accordance with Regulation 543-01-03, and having at least a 2.5 mm^2 conductor installed in the same enclosure and connected in parallel with it,
6.   equipment being connected to the supply through a double wound transformer having electrically separate primary and secondary circuits. The circuit protective conductor shall be connected to the exposed conductive parts of the equipment, and to a point on the secondary side of the transformer. The protective conductors for the circuit shall be installed in accordance with one of the arrangements (1) to (5) above.

607-02-07

Where the installation forms part of a TT system and the stationary items of equipment have an earth leakage current exceeding 3.5 mA, the product of the total earth leakage current and twice the resistance of the installation earth electrodes shall not exceed 50. Where this arrangement cannot be achieved, the equipment shall be supplied through a double wound transformer described in (6) of Regulation 607-02-07.

607-03

Where the installation forms part of an IT system, equipment having a high earth leakage current shall not be directly connected.

607-04

Each exposed conductive part of data processing equipment, together with the metallic enclosures of Class II and Class III equipment, and any FELV circuit earthed for functional reasons, shall be connected to the main earthing terminal.

    A protective conductor which is only being used for a functional purpose, and not as a protective conductor, does not need to comply with Section 543 of the Regulations.

607-05

The manufacturer of the equipment shall be consulted when a 'clean" (low noise) earth is specified, in order to confirm that the arrangements made are satisfactory and suitable for the functional purposes.

545-01

The requirements of protective measures shall take precedence over function purposes when protective and functional earthing purposes are combined.

546-01, 542-01-06

Protective conductors compulsory

With the exception of double insulated lampholders, a cpc shall be installed to every point in the wiring, and to each accessory in an installation protected by automatic disconnection of the supply.

471-08-08

7. Conduit

Material
Conduit and conduit fittings used should have been manufactured in accordance with British Standards:
BS 31 Specification steel conduit and fittings for electric wiring.
BS 731 Flexible steel conduit and adaptors.
BS 4568 Part 1 Steel conduit bends and couplers metric version of BS 31.
BS 4568 Part 2 Plain and threaded fittings for use with conduit in Part 1.
BS 4607 Insulating conduits, fittings and components.
BS 6053 Outside diameters of conduits and threads for conduits and fittings.
BS 6099 Part 1 Specification of general requirements for conduit.
BS 6099 Part 2 Section 2.2 Specification of rigid plain conduits of insulating material
511-01-01, 521-04

Corrosion and moisture
Protection shall be provided against corrosion that may occur in damp situations, where metallic conduit is installed in contact with building materials containing magnesium chloride, corrosive salts, lime, and acidic woods, such as oak.

Protection includes coating the conduit, or preventing contact by separating with plastics or other suitable material.

Special care shall be taken when installing aluminium conduit in damp situations, particularly with fixing clips. Generally, contact between dissimilar metals shall be avoided, especially where brass with a high copper content is involved, since dissimilar metals can set up electrolytic action in damp situations. (Galvanised steel would however be suitable for fixing clips).

Metal conduit and its fixings shall be corrosion-resistant when exposed to the weather, or installed in damp situations, and must not be in contact with dissimilar metals likely to set up electrolytic action.
INF, 522-05-01, 522-03-02

Drainage outlets are to be provided in unsealed conduit systems where moisture may collect., e.g., through condensation, and protection against mechanical damage shall be provided where it is subject to waves.
522-03-02, 523-03-03

Conduit bends.
The radius of every conduit bend shall be not less than 2.5 times the overall diameter of the conduit. Where cables other than conduit wiring cables are installed in conduit, the radius of conduit bends shall be increased from the minimum stated above, as determined by the cable radius required from the manufacturers (See Table 3 under Selection of Cables).
522-08-03

Solid elbows or tees are only allowed immediately behind an accessible outlet box, inspection fitting or luminaire. Where a conduit run does not exceed 10 metres between outlet points, and the sum of all bends between the outlets is equivalent to only one 90° bend or less, then a solid elbow may be installed, up to 500 mm from an accessible outlet box.
INF

Terminations and junctions
Conduit ends shall be free from burrs. Terminations in equipment not fitted with spout entries shall be treated to prevent damage to the cables. (A female bush on the end of the conduit is a suitable means of achieving this).
522-08-01

Where cables are connected to a conduit system, substantial boxes with ample capacity shall be installed.
522-08-02

Every outlet for cables from a conduit system to a duct, or ducting system, shall be so made that it is mechanically strong, and cables must not be damaged whilst being installed.
522-08-01, 522-08-02

The termination of conduit, non-sheathed cables, or sheathed cables that have had the outer sheath removed, shall be enclosed in fire resistant material that complies with British Standards, unless they are terminated in a box, accessory or luminaire complying with British Standards.
   The building structure may form part of the enclosure referred to above.
526-03-03, 526-03-02

Ambient temperature
Conduits shall be suitable for the maximum and minimum values of ambient temperature to which they will be exposed in normal service.
Where non-metallic or plastics boxes are in contact with a luminaire, they shall be suitable for the suspended load, and the temperature to which they will be subjected.
522-01-01

Long straight lengths of rigid plastic conduit must allow for expansion and contraction; using slip joints in plastic conduit is a means of achieving this.
522-12-01

Colour code
Where conduit is to be distinguished from other services, it shall be coloured orange.
514-02

Prefabricated systems
Where conduit systems are prefabricated and not wired in situ, allowance shall be made for tolerances in the buildings dimensions. Precautions against damage to such systems shall be taken, so that the conduits or cable ends are not damaged during erection, or during building operations.
522-06-01, 522-08, 522-12-01.

Fire barriers
Where conduits pass through floors, walls, partitions or roofs, the opening through such walls or floors etc., shall be sealed to the same degree of fire resistance as that of the material through which it is passing.
527-02-01

Where conduits pass through walls and floors etc., that have a specified degree of fire resistance, the hole round the conduit in the wall or floor etc., shall be sealed to the same degree of fire resistance.

Additionally, the conduit shall be internally sealed to maintain the degree of fire resistance of the material through which the conduit is passing.

Where the conduit system has a maximum internal cross-sectional area of 710 mm^2, and is a non-flame propagating system, it need not be internally sealed. (With the exception of 38 mm conduit, all conduits have an internal area less than 710 mm^2.)
527-02-02

Conduit fixings

The conduit system has to be selected and erected to minimise any damage occurring to the sheath and insulation of the conductors during the installation, use and maintenance, of the system.
522-08-01

The conduit system has to be so supported that the conductors in the conduit are not exposed to undue mechanical strain, and so that there is no undue mechanical strain on the terminations of the conductors. Consideration shall be given to the conductor's own weight.
522-08-05

Cables that are installed in a vertical conduit shall, where necessary, be provided with additional support, so that the cables are not damaged by their own weight.
522-08-04

Distance between fixings for conduit

Conduit size in mm	Maximum distance between fixings in metres					
	Metal		Plastic		Pliable	
	Horizontal	Vertical	Horizontal	Vertical	Horizontal	Vertical
16	0.75	1.0	0.75	1.0	0.3	0.5
20	1.75	2.0	1.5	1.75	0.4	0.6
25	1.75	2.0	1.5	1.75	0.4	0.6
32 & 40	2.00	2.25	1.75	2.0	0.6	0.8
Over 40	2.25	2.5	2.0	2.0	0.8	1.0

INF

Overhead wiring

Cables sheathed with pvc, or having an oil-resisting, flame-retardant, or h.o.f.r sheath for overhead wiring between separate buildings, may be installed in heavy gauge steel conduit providing the area is inaccessible to vehicular traffic.

Where conduit is used to span between buildings above ground, the maximum length of span allowed is 3 metres. No joints are allowed in its span, and the minimum size of conduit shall be 20 mm., and the minimum height allowed above ground 3 metres.
INF

Underground wiring

Flat pvc insulated and sheathed cables can be installed in conduit underground.
INF, 522-06-03

Conduit used as protective conductor
Plain slip or pin grip sockets shall not be used for conduit joints. Joints shall be screwed or made with substantial mechanical clamps.
543-03-06

Where metal conduit is used as a protective conductor, its cross-sectional area shall be determined either by application of the formula of 543-01-03, or Table 54G.

543-01-03 formula $S \text{ mm}^2 = \dfrac{\sqrt{I^2 t}}{k}$

Table M54G

Cross sectional area of copper conductor	Area required for steel conduit
S ≤ 16	2.6S
S > 35	1.3S

Area of steel conduit is given in MISC 1 Part 4. The area required for the conduit given above is based on the worst factor given in the tables. In practice, conduit will always be large enough to act as a protective conductor.
543-02-05

Where a conduit is used as a protective conductor, its electrical continuity has to be assured, and has to be protected against mechanical, chemical or electrochemical deterioration.
543-02-05, 543-02-04

Where a conduit is used as the circuit protective conductor for more than one circuit, it shall be sized either by;
1) calculation, using the formula 543-01-03 and the most onerous value of fault current and disconnection time encountered in each of the various circuits, or
2) selected by using Table 54G, with the cross-sectional area of the largest cable enclosed in the conduit.

543-01-02

When conduit is used as a protective conductor, it shall be protected against mechanical, chemical and electrodynamic effects.
543-03-01

The joints in conduit when used as a circuit protective conductor do not need to be accessible.
543-03-03

Conduit shall not be used as a PEN conductor.
546-02-07, 543-02-10

Where a protective conductor is used for both protective and functional purposes, the requirements for the protective measures shall prevail.
546-01

Flexible conduit
Flexible or pliable conduit shall not be used as a protective conductor.
543-02-01

Conduit wiring
When conduit is used as a protective conductor, the earthing terminal of each accessory and socket outlet shall be connected by a protective conductor to an earthing terminal fixed in the associated box or other enclosure.
543-02-07

Cables to be installed in underground conduits shall be sheathed or armoured.
522-06-03, 522-08-01

Earthing of equipment with high earth leakage current
Where a final circuit supplies an item of stationary equipment having an earth leakage current exceeding 10 mA in normal service, or where a final circuit supplies a number of socket outlets in an area where it is expected or known that several items of equipment installed in that area will have a total earth leakage current exceeding 10 mA in normal service, a high integrity protective connection shall be made by duplicate protective conductors. This can comprise a metal conduit sized in accordance with Regulation 543-01-03, and a conductor having a cross sectional area not less than 2.5 mm^2 installed in the same conduit and connected in parallel with the conduit.
607-02-07, 607-02-06

Category 1,2, and 3 circuits.
Category 2 circuits must not be installed in the same conduit as Category 1 circuits, unless the Category 2 circuits are insulated to the highest Category 1 voltage present.
Category 3 circuits must not be installed in the same conduit as Category 1 circuits.
528-01-04, 528-01-05

Where controls or accessories are mounted in or on a common box, switch plate or block, category 1 and 2 cables and terminations shall be segregated from each other by a partition. If the partition is metal, it shall be earthed.
528-01-07

Conduit used for mechanical protection
Short lengths of metal conduit not exceeding 150 mm in length which are inaccessible, do not require protection against indirect contact.
471-13-04 (v)

Conduit can be used for protecting cables to be installed outdoors on walls, providing it is suitably protected against corrosion.
522-05-01

Damage during installation
The number of cables drawn into a conduit shall be such that no damage occurs to the cable or conduit during the installation of the cables.
522-08

Wiring capacity of conduit
See Table CCC 1 (Cable Capacity Conduit) in Part 3 .

8. Construction site temporary installations

Extent of the requirements for construction sites
Except where modified by the following regulations, the general regulations are applicable to construction sites. Thus the regulations for protection against direct contact, overload protection, short-circuit protection, and the sizing of circuit protective conductors, isolation etc., are still applicable.
600-01, 600-02

Types of construction site
Construction sites fall into different categories depending upon the type of work being carried out; the following are the names of different types of site which can be considered construction sites.
1. new building construction, which would include housing sites, factory sites etc.
2. extensions to existing buildings and the demolition of existing buildings.
3. alterations and repairs to existing buildings
4. works of engineering construction , such as the Thames barrier, Channel tunnel.
5. earthworks, such as excavating for roads, reservoirs, railway cuttings.
6. work of a similar nature to the above where electrical, mechanical, civil or building work is being carried out.

All these types of works require to compliance with the special regulations for construction sites.
604-01

Where not applicable
The special requirements for construction sites are not applicable to the management buildings on a site. Management buildings can comprise, site offices, cloakrooms, site meeting rooms, canteens, restaurants, dormitories and toilets. These buildings and areas are considered to be fixed installations to which the normal regulations apply.

The fixed installations on a construction site are limited to those mentioned above, and to the main switchgear for the site in which the principal protective devices are contained. Any installation other than the management buildings, on the load side of the main switchgear, are considered to be moveable
604-01

Voltage of equipment
Equipment shall be identified for the particular supply from which it is energised, and shall be suitable for the supply voltage and frequency available. The equipment shall only contain voltages derived from the same source of supply, unless they are control or signalling circuits, or from a standby supply. On large sites, high voltage distribution is not precluded.

The maximum voltage that can be used in various locations is as follows:
1. 25 V single-phase SELV for use with portable hand lamps in damp and confined locations.
2. 50 V single-phase, centre tapped to earth, for portable hand lamps in damp and confined locations.
3. 110 V single-phase reduced low voltage, centre tapped to earth for portable hand lamps in general use, portable handheld tools, and local lighting up to 2 kW.

4. 110 V three-phase reduced low voltage, centre tapped to earth, for single-phase portable hand lamps in general use, portable handheld tools, and local lighting up to 2 kW, as well as small single-phase and three-phase mobile plant up to 3.75 kW.
5. 240 V single-phase for fixed floodlighting.
6. 415 V three-phase for fixed and movable equipment larger than 3.75 kW.

Where safety and standby supplies are connected, devices shall be provided to prevent the interconnection of different supplies.
604-02-01, 604-02-02, 604-11-06

Protection against indirect contact

Protection against indirect contact shall be by either or both of the following methods:
1. overcurrent protective device.
2. residual current device.

Either method gives protection by automatic disconnection.
413-02-07
Where protection is by means of automatic disconnection of the supply then :

$$Z_S \leq \frac{U_o}{I_a}$$

where Z_S is the earth fault loop impedance of each circuit, U_o is the nominal voltage to earth, and I_a is the current causing the protective device for the circuit to disconnect the circuit within the following times for TN systems:

Voltage to earth U_o volts	Disconnection time 't' seconds
120 V	0.35
From 220 to 277 V	0.2
400 to 480 V	0.05
580 V	0.02

604-04-01, Table 604A
The above maximum disconnection times are applicable to all movable installations, whether they are directly connected to the supply, or connected through a plug and socket. The exception, however, is reduced low voltage systems installed in accordance with the reduced low voltage regulations, where the disconnection time allowed is 5 seconds.
604-04-01, 604-04-02, 604-04-06
Where the voltage to earth U_o is 240 V, the earth loop impedances corresponding to a disconnection time of 0.2 seconds for all types of protective devices are given in Table ZS 9 Part 3.

For reduced low voltage installations, where the voltage between phases is 110 V, the maximum values of earth loop impedance are given in Table ZS 8.
604-04-03, 604-04-04, 604-04-07
Where exposed conductive parts are simultaneously accessible, they shall be connected to the same earthing system. They can be connected individually, collectively, or in groups.
413-02-03

Every exposed conductive part of the installation shall be connected by protective conductors to the main earthing terminal of the source of the supply
413-02-06

Use of residual current devices (RCDs)

If the overcurrent protective devices cannot disconnect the low voltage circuits within the time specified in Table 604A, or reduced low voltage circuits within 5 seconds, then protection shall be provided by a residual current device, such that $Z_S \times I_{\Delta n} \leq 25$ V.
604-04-08

Where RCDs are used for automatic disconnection, exposed conductive parts need not be connected to the TN system protective conductors, providing they are connected to an earth electrode, such that $R_A \times I_a \leq 25$ V, where R_A is the resistance of the cpc and earth electrode.
413-02-17

Where the system is TT

Where exposed conductive parts are protected by a single protective device, they shall be connected via a protective conductor to the main earthing terminal, and then to a common earth electrode. Where several protective devices are in series, the exposed conductive parts may be connected to a separate earth electrode associated with each protective device.

The protective device can either be an overcurrent device or a residual current device, the preferred protection being a residual current device.
413-02-18, 413-02-19

Each circuit must comply with the formula $R_A \times I_a \leq 25$ V, where R_A is the resistance of the protective conductor from the exposed conductive part to the earth electrode, plus the resistance of the earth electrode. The value I_a is the current that will cause disconnection of the overcurrent protective device within 5 seconds, or is the residual operating current $I_{\Delta n}$ of a residual current protective device.
604-05

Where the system is IT

An IT system should not be used unless there is no other system to provide the supply. If an IT system is used, then permanent earth fault monitoring shall be provided.
604-03-01

Where the system is IT, Regulations 413-02-21 to 413-02-25 are applicable, except that the voltage in Regulation 413-02-23 is changed from 50 V to 25 V
604-06-01

Regulation 413-02-26 is applicable, except that table 41E is changed to the following:

Voltage between phase and neutral U_o Voltage between phases U volts	Neutral not distributed t (seconds)	Neutral distributed t (seconds)
120 to 240 volts	0.4	1
220/380 to 277/480	0.2	0.5
400/690	0.06	0.2
580/1000	0.02	0.08

604-06-02

Regulations that are not applicable to construction sites
The group of regulations concerned with installations within the equipotential zone, or circuits feeding fixed equipment or socket outlets outside the equipotential zone, are not applicable to a construction site. This means that the following regulations are not applicable: 471-08-02 to 05.
604-08-01, 604-08-02, 604-08-03

Socket outlets
Every socket outlet shall be protected by one or more of the following:
1. reduced low voltage system using automatic disconnection.
2. a 30 mA residual current device manufactured to BS 4293, having an operating time of 40 ms with a residual current of 150 mA.
3. a SELV supply complying with the SELV regulations.
4. electrical separation complying with the regulations for electrical separation., each socket outlet being supplied from a separate transformer.

604-08-03, 604-08-04

Where the system is TT, and where protection is by automatic disconnection, every socket outlet circuit shall be protected by an RCD complying with $R_A \times I_{\Delta n} \leq 25$ V
471-08-06

Every socket outlet shall comply with BS 4343, and be incorporated into a unit manufactured in accordance with BS 4363 and BS 5486 Part 4.
604-12-01, 604-12-02

Cable couplers must comply with BS 4343, and luminaire supporting couplers are not allowed
604-12-03, 604-13

Distribution and distribution equipment
An assembly comprising the main control gear and principal protective devices shall be installed at the origin of each installation. Each supply and distribution assembly shall incorporate a means of isolating and switching the incoming supply, the isolator being suitable for locking in the off position.

Emergency switching shall be provided on the supply to equipment where it may be necessary to disconnect all live conductors in order to remove a hazard.
604-11-01, 604-11-02, 604-11-03, 604-11-04

The general regulations concerning isolation are applicable to construction sites.
600-01, 600-02

Every circuit supplying equipment shall be fed from a distribution assembly complying with BS 4363 and BS 5486 Part 4. Each distribution assembly shall comprise:
1. overcurrent protective devices.
2. devices affording protection against indirect contact

If required, socket outlets can be incorporated in the distribution unit.
604-09-01, 604-11-05

Except for assemblies manufactured to BS 4363 and BS 5486 Part 4, equipment shall be protected to at least IP44.
604-09-02

Cable installation
Adequate protection shall be given to cables installed across site roads and walkways, and the cables shall be so installed that no strain is placed on the terminations of conductors, unless the termination has been designed to withstand such strain.
604-10-01, 604-10-02

Cables shall be sized in accordance with the overload, short-circuit, and protective conductor regulations.
600-01, 600-02

Calculations
The temperature rise of a conductor due to fault current shall be taken into account when checking that the circuit is protected against indirect contact.
413-02-05

Tables 41C in the regulations, or PCZ 8 in Part 3 are not applicable to construction sites, since these tables limit the touch voltage by limiting the impedance of the protective conductor.
604-04-05

9. Data processing equipment

The manufacturer of the equipment shall be consulted when a low noise earth is specified for a particular item of equipment, to confirm that the arrangements detailed in this Chapter will be suitable for functional purposes.

607-02-01

Where the earth leakage current from equipment complying with BS 7002 does not exceed 3.5 mA, no special precautions are necessary.

607-02-02

The earthing arrangements for installations supplying equipment such as industrial control equipment, microelectronics, or computers, having an earth leakage current in excess of 3.5 mA, shall comply with the following regulations:

607-01

Where an installation is supplied through a residual current device, and where there is more than one item of equipment having an earth leakage current in excess of 3.5 mA, the total leakage current must not exceed 25% of the tripping current of the residual current device.

607-02-03

Where the total leakage current exceeds 25% of the tripping current of the residual current device, the equipment shall be supplied through a double wound transformer, or other device, in which the input and output circuits are electrically separate. The protective conductor shall connect the exposed conductive part of the equipment to the secondary winding of the transformer. The protective conductor shall comply with one of the arrangements outlined in Regulation 607-02-07.

607-02-03

Where an item of stationary equipment has an earth leakage current between 3.5 mA and 10 mA, it shall be permanently connected to the fixed wiring, or connected by means of a BS 4343 socket-outlet.

607-02-04

Where an item of stationary equipment has an earth leakage current exceeding 10 mA in normal use, it shall be permanently connected to the fixed wiring of the installation. The equipment can be connected through a BS 4343 socket-outlet, providing the protective conductor in the flexible connecting cable is supplemented by a separate additional 4 mm^2 conductor, and an additional contact in the plug and socket-outlet.

607-02-05

Where a final circuit feeds a number of socket outlets, in an area where it can be expected that the total earth leakage current from the equipment connected to the socket-outlets will exceed 10 mA, the circuit shall be provided with a high integrity protective connection complying with items 1 to 6 of Regulation 607-02-07.

Alternatively, a ring circuit can be used, providing there are no spurs taken from the ring circuit, the minimum size of protective conductor used is 1.5 mm^2, and the ends of the protective conductor are connected into separate earth terminals at the source of the ring circuit.

607-02-06

DATA PROCESSING EQUIPMENT

Where the fixed wiring of every final circuit is installed in a location intended to accommodate several items of stationary equipment, and where the total earth leakage current exceeds 10 mA, the circuit shall be provided with a high integrity protective connection complying with one or more of the following:

1. a single protective conductor with a minimum cross-sectional area of 10 mm^2.
2. a separate duplicated protective conductor, having independent mechanically and electrically sound terminations, protected from mechanical damage and vibration and made by compression-sockets, soldering, brazing, welding, or mechanical clamps, each protective conductor having a minimum cross sectional area of 4 mm^2.
3. duplicate protective conductors with live conductors in a multicore cable, providing that the total cross sectional area of all of the conductors in the cable is not less than 10 mm^2. One of the protective conductors can be formed by the metallic armour, sheath or braid of the cable, providing its electrical continuity can be assured, and providing it has a minimum cross-sectional area calculated in accordance with Regulation 543-01.
4. an earth monitored protective conductor, the circuit being disconnected in the event of a failure in the protective conductor. The circuit shall comply with the automatic disconnection and equipotential bonding regulations.
5. duplicate protective conductors formed by trunking and conduit sized in accordance with Regulation 543-01-03, and having at least a 2.5 mm^2 conductor installed in the same enclosure and connected in parallel with it.
6. equipment being connected to the supply through a double wound transformer having electrically separate primary and secondary circuits. The circuit protective conductor shall be connected to the exposed conductive parts of the equipment, and to a point on the secondary side of the transformer. The protective conductors for the circuit shall be installed in accordance with one of the arrangements 1 to 5 above.

All the above protective conductors shall comply with the regulations for protection against indirect contact and protective conductors; the only exception is where Regulation 607-02-05 applies.
607-02-07

Where the installation forms part of a TT system, and the stationary items of equipment have an earth leakage current exceeding 3.5 mA, the product of the total earth leakage current and twice the resistance of the installation earth electrodes shall not exceed 50. Where this arrangement cannot be achieved, the equipment shall be supplied through a double wound transformer described in (6) of Regulation 607-02-07.
607-03

Where the installation forms part of an IT system, equipment having a high earth leakage current shall not be directly connected.
607-04

Each exposed conductive part of data processing equipment, together with the metallic enclosures of Class II and Class III equipment and any FELV circuit earthed for functional reasons, shall be connected to the main earthing terminal.

A protective conductor which is only being used for a functional purpose and not as a protective conductor, does not need to comply with Section 543 of the regulations.
607-05

Low noise earthing: safety
The manufacturer of the equipment shall be consulted when a 'clean' (low noise) earth is specified, to confirm that the arrangements made are satisfactory, and suitable for the functional purposes.
545-01

The requirements of protective measures shall take precedence over functional purposes when protective and functional earthing purposes are combined.
546-01

The metallic enclosures of Class II and Class III equipment and the exposed conductive part of data processing equipment shall be connected to the main earthing terminal.

When a FELV circuit is earthed for functional reasons, it, too, shall be connected to the main earthing terminal.

Compliance with Section 543 is not required when an earth conductor is only used for a functional purpose.
607-05

10. Diversity

Maximum demand
The maximum demand of the installation has to be worked out in amperes.
311-01

Diversity
When working out the maximum demand current for an installation or circuit, diversity may be taken into account.
311-01

Where the standard circuit arrangements are used for socket outlets, no allowance for diversity is to be used, because diversity has already been taken into account.
INF

Final circuits
The full-load current of each piece of equipment connected to the circuit is taken, after allowing for any appropriate diversity. The total of all the individual loads on the circuit are then added together to obtain the total load on the circuit. Table 1 can be used to work out the load of each piece of equipment.

Typical final circuit loads

TABLE 1	
2 A socket outlets	Minimum 0.5 A.
Socket outlets exceeding 2 A.	Rated current I_n
Tungsten lighting	Connected load with minimum allowance of 100 Watts per lampholder
Household cooker	10 A plus 30% of the remaining cooker load plus 5 A if socket is in cooker unit.
Equipment not exceeding 5 VA (Clocks etc.)	Ignore.
Discharge lighting	Rated lamp watts × 1.8 divided by the voltage, providing the circuit is power factor corrected to 0.85 lagging
All other stationary equipment	Full load current (F.l..c.)

INF

TABLE 2 DIVERSITY

ALLOWANCES FOR DIVERSITY AT A DISTRIBUTION BOARD FOR SIZING FEEDER CABLE

Type of circuit connected at distribution board	Diversity to be applied depending upon the type of premises		
	Individual household dwelling & individual dwellings in a large complex	Small business premises Small shops and stores. Very small factories.	Small commercial premises including hotels, boarding houses & guest houses
Heating excluding thermal storage heaters & cookers	Full load up to 10A plus 50% of connected load in excess of 10A	100% of full load of largest appliance plus 75% full load of remaining appliances	100% full load of largest appliance +80%full load 2nd largest appliance +60% full load remaining appliances
Lighting	66% of connected load	90% of connected load	75% of connected load
Cooking appliances	10A + 30% of the remaining cooker load + 5A if socket outlet in cooker control	100% full load of largest appliance plus 80% of the full load of the second largest appliance plus 60% of the full load of any remaining appliances.	
Final circuits inst- alled in accordance with Appendix 5	100% full load of largest circuit plus 40% full load of every other circuit	100% of full load current of largest circuit plus 50% of full load current of every other circuit	
Instantaneous water heaters	100% full load of largest appliance plus 100% full load of second largest appliance plus 25% full load of any remaining appliances		

This table is provided as a guide for those occasions when it is not possible to work out diversity from experience.

TABLE 2 DIVERSITY continued
ALLOWANCES FOR DIVERSITY AT A DISTRIBUTION BOARD FOR SIZING FEEDER CABLE

Type of circuit connected at distribution board	Diversity to be applied depending upon the type of premises		
	Individual household dwelling & individual dwellings in a large complex	Small business premises Small shops and stores. Very small factories.	Small commercial premises including hotels, boarding houses & guest houses
Socket outlets and stationary equipment not installed in accordance with Appendix 5	100% full load current of the largest point of utilisation + 40% of full load current of every other point of utilisation.	100% full load current of the largest point of utilisation + 75% of full load current of every other point of utilisation current of every other point of utilisation.	100% full load current largest point of utilisation + 75% full load current for main rooms in constant use + 40 full load
Motors excluding lift motors		100% full load current of the largest motor + 80% full load current of 2nd largest motor + 60% full load current of remaining motors	100% full load current largest motor plus 50% full load current of the remaining motors.
Thermostatically controlled water heaters Off peak heating		DIVERSITY NOT ALLOWED	

This table is provided as a guide for those occasions when it is not possible to work out diversity from experience.

Distribution
The full-load current (without diversity) of each final circuit connected to the distribution board is taken, and the diversities given in Table 2 are then applied to obtain the total current demand on the distribution board.
Table 2 must not be applied to circuits whose load has been reduced by applying Table 1.

The designer of the installation is allowed to apply diversity to a group of circuits that have already had a diversity factor applied to them, but the designer is not allowed to use Table 2 to determine the current demand of that group of circuits.

A suitably qualified electrical engineer may use other methods of determining maximum demand.

Distribution boards shall be capable of carrying the connected load, and not just the current after diversity has been taken into account.
INF

Voltage drop
When calculating voltage drop in a circuit, or in a feeder cable to a group of circuits, the current demand can be used after diversity has been taken into account.
INF

11. Earthing

Fundamental requirements
Where a voltage may appear on exposed conductive parts, due to the insulation of live conductors becoming defective, or due to a fault in equipment, then the exposed conductive parts shall be earthed so that the electrical energy is discharged without danger.
130-04-01

The arrangement of circuits shall be such that the persistence of dangerous earth leakage currents is prevented.
130-04-02

Where exposed conductive parts are earthed, the circuits shall be protected either by overcurrent protective devices, or residual current devices, to prevent the persistence of dangerous earth leakage currents. Where the phase earth loop impedance is too large to cause the prompt disconnection of the circuit by overcurrent devices, because the prospective earth fault current is insufficient, then residual current protective devices shall be used for the circuit's protection.
130-04-03

Where exposed conductive parts are earthed, and are simultaneously accessible to extraneous conductive parts of other services, the extraneous conductive parts shall be connected to the main earthing terminal of the installation.
130-04-04

The only protective device allowed in an earthed neutral is a linked circuit breaker, where the related phase conductors are arranged to be disconnected at the same time.
130-05-02

Earthing general requirements
An assessment shall be made to make certain:
1. any earth leakage currents do not have a harmful effect upon other circuits or equipment,
2. whether additional connections with earth are required independent of the installation's main earth, to prevent interference with the correct operation of any equipment.

331-01

The supplier (of electricity) shall not connect an installation unless they are satisfied that it complies with the Electricity Supply Regulations 1988, amended 1989.
Electricity Supply Regulations 1988, amended

Before commencing an installation the type of earthing arrangement shall be determined: i.e. the type of system which the installation and supply will comprise.
312-03

Depending upon the requirements of the installation, joint or separate earthing arrangements may be used both for protection or functional purposes, such as equipment that requires an earth for its operation, (quick start fluorescent fittings). Any functional use must not affect the conductor's protective qualities.
542-01-06, 545-01-01,607

The earthing of the installation shall be:
1. Continuously effective,
2. Able to carry earth fault currents and earth leakage currents without danger, particularly from thermal, thermomechanical, and electromechanical stresses,
3. Adequately robust, or have additional suitable mechanical protection,
4. Arranged so that the risk of damage to other metallic parts through electrolysis is avoided.

542-01-07, 542-01-08

Where a protective conductor is installed through more than one installation, each one of which has its own earthing arrangement, then either of the following requirements shall be complied with:
1. the protective conductor shall be capable of carrying the maximum fault current likely to flow through it from any of the installations, or
2. the protective conductor shall be earthed within one installation only, and insulated from the earthing arrangements of the other installations; in this case, where the protective conductor is part of a cable, it shall only be earthed in the installation containing the protective device for the cable circuit.

542-01-09

Where the earthing system of an installation is subdivided, then each subdivision must comply with the earthing regulations.
541-01-02

Lightning protection installed

Where lightning protection is installed, the bonding of the lightning protection system to the main earth bar of the electrical installation, shall comply with BS 6651, but the bonding conductor's cross-sectional area must not exceed that of the main earthing conductor.
541-01-03

Main earthing terminal

The consumer's main earthing terminal shall be connected to the earthed point of the source of electrical energy for a TN-S system, by the supply undertaking to the neutral of the source of electrical energy for a TN-C-S system, and connected by an earthing conductor to an earth electrode for a TT or IT system.
542-01-02, 542-01-03, 542-01-04

A main earthing terminal or bar shall be provided for every installation, to connect the circuit protective conductors, main equipotential bonding conductors, any functional earthing conductors, and the lightning bonding conductor to the main earthing conductor.
542-04-01

The main earthing terminal shall be accessible, to enable the earthing conductor to be disconnected for test purposes, The joint shall be mechanically strong, reliably maintain electrical continuity, and be capable of disconnection only by means of a tool.
542-04-02

Provision shall be made for the connection of PEN conductors to the main earthing terminal, when the system is TN-C.
542-01-05

Earth electrodes
The following types of earth electrodes may be used:
1. earth rods or pipes, (providing the pipes are not gas or water services),
2. earth tapes or wires,
3. earth plates,
4. underground structural metalwork embedded in foundations,
5. except for pre-stressed concrete, welded metal reinforcing of concrete embedded in the ground,
6. metallic sheaths or other coverings of cables, providing;
   a) they are not liable to deterioration through excessive corrosion, and
   b) the cable sheath is in effective contact with earth, and the consent of the cable owner has been obtained to use the cable as an earth electrode, as long as arrangements are made with the cable owner to notify the consumer of any change that is made to the cable, which will affect its suitability as an earth electrode.

542-02-01, 542-02-04, 542-02-05

Earth electrode material
The design used, and the material from which earth electrodes are made, shall be able to withstand damage due to corrosion. The earthing arrangements shall also take into account any increase in earth resistance caused by corrosion. The earth electrode should be buried at a depth such that the earth electrode resistance is not increased by the soil drying or freezing.
542-02-02, 542-02-03

Compliance with other regulations
Every earthing conductor shall also comply with the regulations for protective conductors.
542-03-01, 541-01-01

Earth electrode connections
Connections to earth electrodes shall not be made with aluminium, or copperclad aluminium conductors.
542-03-02

Connections to earth electrodes shall be soundly made, so that they are electrically and mechanically satisfactory; they shall have a label permanently attached in a visible position, durably marked in legible type 4.75 mm high, with the words 'Safety electrical connection - do not remove'. The connection must also be suitably protected against corrosion.
542-03-03

EARTHING

Size of earthing conductor
The minimum size of earthing conductors shall be determined either by the formula

$$S \text{ mm}^2 = \frac{\sqrt{I^2 t}}{k}$$

or by using Table 54G, the requirements being outlined in the following table. The calculated minimum sizes are further limited in minimum size when they are buried in the ground.

TABLE 54G

SIZE OF MAIN EARTHING CONDUCTOR REQUIRED	
Area of phase conductor into installation main switch S	Area of main earthing conductor being the same material as the phase conductor S_p
S not exceeding 16 mm^2 S from 16 mm^2 to 35 mm^2 S exceeding 35 mm^2	S 16 mm^2 0.5 S

543-01-03, 543-01-04, 542-03-01

Buried earthing conductors
Where cables are buried in the ground they shall be subject to the following minimum cross-sectional areas:

MINIMUM CROSS-SECTIONAL AREAS OF BURIED EARTHING CONDUCTOR

	Protected against mechanical damage	Not protected against mechanical damage
Protected against corrosion by a sheath	543-01-03 formula or Table 54G	16 mm^2 copper 16 mm^2 coated steel
Not protected against corrosion	25 mm^2 copper 50 mm^2 steel	25 mm^2 copper 50 mm^2 steel

542-03-01

Installation with more than one supply
Where there is more than one source of supply to an installation, and where one of the sources of electricity requires its earth to be independent, a switch may be inserted between the neutral point and earth, providing the switch is linked, so that it connects and disconnects the live conductors at the same time as the earth connection.
460-01-04

12. Emergency switching

Emergency switching
Emergency switching is used to remove any hazard to persons, livestock, or property, by rapidly cutting off the electrical supply.
One device can be used to perform the functions of:
1. emergency switching,
2. isolation,
3. switching off for mechanical maintenance,
4. functional switching.

Where one device is used for one or more of the above functions, it shall be verified that the device complies with the regulations for each function it performs.
476-01-01

Where required
Emergency switching shall be provided for every part of an installation where it is necessary to prevent or remove a hazard by disconnecting the supply rapidly.
463-01-01

Every machine driven by electricity, and which may give rise to danger, shall be provided with an emergency stopping device.
476-03-02

How installed
Where possible, the emergency switch should act directly on the supply conductors, and be so arranged that a single initiative action will cut off the supply.
463-01-02

An emergency switch shall be installed in a readily accessible position where the hazard might occur, and if appropriate, at any other position from which the device for emergency switching may need to be operated. Additionally, further devices shall be provided where additional emergency switching may be required.
537-04-04

Accessibility
Where emergency switching is provided for machines driven by electricity, and which may give rise to danger, it shall be readily accessible and easily operated by the person in charge of the machine. Where there is more than one device for manually stopping the machine, the restarting of which could cause danger, provision shall be made to prevent the unintentional or inadvertent restarting of the machine.
476-03-02

The intended use of the premises shall be taken into account, so that access to the emergency switch is not impeded when a foreseeable emergency condition arises.
476-03-01

Emergency switches shall be readily accessible, and suitably durably labelled.
463-01-04

Operational safety
The arrangement of emergency switches must not interfere with the complete operation necessary to remove a hazard, or create another hazard when operated.
463-01-03

Emergency switches shall only be operated by skilled or instructed persons, where the inappropriate operation of an emergency switch could give rise to a greater danger.
476-03-03

The release of an emergency switching device shall not re-energise the equipment concerned.
537-04-05

Emergency stopping
Where movements caused by electrical means may give rise to danger, emergency stopping shall be provided.
463-01-05

Emergency stopping may include energising equipment to effect an emergency stop, e.g. d.c. injection braking.
537-04-02

Fireman's emergency switch
Exterior electrical installations and internal discharge lighting installations, operating at a voltage exceeding 1000 V a.c., or 1500 V d.c., shall be provided with a fireman's emergency switch in the low voltage circuit. When the voltage exceeds extra low voltage, a fireman's emergency switch shall also be provided if :
1. the installation is installed in a shopping mall,
2. the installation is installed in an arcade,
3. the installation is installed in a covered market.

A temporary installation in a permanent building used for exhibitions is not an exterior installation.
476-03-05

Exclusions
Portable discharge lighting luminaires or signs, not exceeding 100 watts, fed from an accessible local socket outlet.
476-03-05

Requirements for fireman's switches

Fireman's switches shall comply with the requirements of the local fire authority, together with the following:
1. the switch shall be mounted exterior to the building and adjacent to the equipment when the installation is exterior to the building. Alternatively, a notice is to be mounted adjacent to the equipment, giving the position of the fireman's switch, with a notice mounted at the switch, making it clearly distinguishable,
2. the switch shall be mounted in the main entrance for interior installations, or at a position agreed with the local fire authority,
3. the switch shall be mounted not more than 2.75 metres from the ground, in a conspicuous position, accessible to firemen, unless agreed differently with the local fire authority,
4. where more than one fireman's switch is installed on any one building, each switch shall be clearly marked to indicate the part of the installation it controls,
5. where practicable, one master fireman's switch shall be mounted on the building to control all the exterior installations operating at a voltage higher than 1000 V a.c., or 1500 V d.c. A separate master fireman's switch shall be installed in the building to control all the interior discharge lighting.

476-03-06, 476-03-07

Colour, labelling and position of switch handle

Every fireman's emergency switch shall:
1. be coloured red,
2. have a 150 mm x 100 mm minimum size nameplate mounted adjacent to it, with the words 'FIREMAN'S SWITCH' in 12.7 mm high minimum size characters marked on it. Where necessary, the size of the nameplate and lettering shall be increased in size, so that it is clearly legible from a distance appropriate to the site conditions,
3. have the OFF position at the top, and clearly indicate the ON and OFF positions in lettering that is clearly legible by a person on the ground,
4. be prevented from being inadvertently returned to the ON position, and arranged to facilitate operation by a fireman.

537-04-06

Requirements for emergency switches

The handles or push buttons of emergency switching devices shall be clearly identifiable, and preferably coloured red.

537-04-04

Where practicable, emergency switches shall be manually operated and directly interrupt the main circuit.

537-04-03

Circuit breakers or contactors operated by remote emergency switches shall open on de-energisation of the coils, or be opened by another suitable method.

537-04-03

Emergency switching shall comprise either:
1. a single device that directly cuts off the incoming supply, or
2. a group of devices, operated by a single initiation, which will remove the hazard by cutting off the appropriate supply.

Emergency stopping may include the supply remaining on for electrical breaking facilities.
537-04-02

The emergency device used for interrupting the supply shall be capable of cutting off the full load current of the relevant parts of the installation, taking into account stalled motor conditions.
537-04-01

An emergency switching device shall be capable of being restrained in the OFF or STOP position, or have a latching mechanism keeping the device in the OFF position.
537-04-05

The release of an emergency switching device shall not re-energise the equipment concerned.,
537-04-05

An emergency switching device which automatically resets, is permitted where both the operation of the emergency device, and the means of re-energising the equipment, are under the control of one person.
537-04-05

Devices not allowed
Plugs and socket outlets shall not be selected as devices for emergency switching.
537-04-02

13. Equipotential bonding

Basic measure of protection
One of the basic measures of protection against indirect contact is equipotential bonding and the other measure is automatic disconnection of the supply. Its application shall be based on the requirements of the system of which the installation forms a part.
413-01-01(i), 413-02-01

Object of bonding
Connecting the extraneous conductive parts within the installation to the main earthing terminal is intended to create a zone in which the magnitude of touch voltages appearing between exposed and extraneous conductive parts are kept to a minimum.
INF

Main equipotential bonding
Extraneous conductive parts that extend throughout the installation shall be connected to the main earth terminal at the origin of an installation by main equipotential bonding conductors, including the following items:
1. main water pipes,
2. gas installation pipes,
3. main service pipes and ducting,
4. risers of central heating and air conditioning systems,
5. exposed metallic parts of the building structure,
6. the lightning conductor system,
7. metallic sheath of telecommunications cable, subject to the owner's or operator's consent,
8. any extraneous conductive part which is in direct contact with earth.

413-02-02, PME REGS

Supplies to other buildings
Where an installation also feeds other buildings, main equipotential bonding shall be carried out in each building.
413-02-02

Conductors not allowed
Aluminium or copperclad aluminium shall not be used for bonding connections
547-01

Size of main equipotential bonding conductors
The cross-sectional area of main equipotential bonding conductors shall be at least half the cross-sectional area of the main earthing conductor, the minimum size allowed being 6 mm^2.

The cross-sectional area of the bonding conductor need not exceed 25 mm^2, if made from copper. A material other than copper can be used, providing it has an equivalent conductance to copper.

EQUIPOTENTIAL BONDING

If the supply to the installation is PME, the size of the main bonding conductor is determined by the electricity supplier; in this case the minimum size bonding conductor required is as follows:

Copper equivalent cross sectional area of the supply neutral conductor	Minimum copper equivalent cross-sectional area of main equipotential bonding conductor
Up to 35 mm^2	10 mm^2
over 35 mm^2 up to 50 mm^2	16 mm^2
over 50 mm^2 up to 95 mm^2	25 mm^2
over 95 mm^2 up to 150 mm^2	35 mm^2
over 150 mm^2	50 mm^2

547-02-01, PME Regs

Where bonding connections are required

The main equipotential bonding connection to water and gas services shall be on the consumer's side of any insulating insert in the pipework, and as near as possible to the point at which the service enters the premises. The connection shall be made on the consumer's side of any meter, and between the meter connection and any branch pipework. Where possible the bonding connection shall be made within 600 mm of the meter.

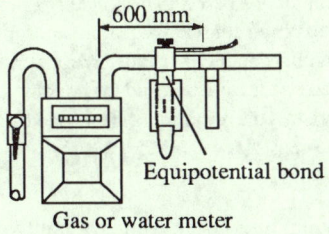

Figure 1 - Bonding with meters

547-02-02

Supplementary bonding exposed to extraneous conductive parts

With the exception of special situations, such as bathrooms, additional protection is only required if the automatic disconnection times specified for the circuit cannot be met.

Where the automatic disconnection times cannot be fulfilled, then either local supplementary equipotential bonding, or protection by an RCD, has to be used.

413-02-15

Where local supplementary equipotential bonding is to be installed, it shall connect together the exposed conductive parts of the circuit concerned, including the earth terminal in socket outlets, and the extraneous conductive parts.

The resistance of the supplementary bonding conductor shall not exceed 50 V divided by the minimum current that will disconnect the circuit to the exposed conductive part within five seconds, and providing the bonding conductor is sheathed or otherwise mechanically protected, its conductance shall not be less than half that of the cpc connected to the exposed conductive part. Where the bonding conductor is not mechanically protected, its minimum cross-sectional area shall be 4 mm^2.

413-02-15, 547-03-02, 413-02-27, 413-02-28

Where a residual current operated device is used instead of equipotential bonding, then $Z_S \times I_{\Delta n} \leq$ 50 V. The resistance of the bonding conductor, subject to the minimum sizes specified in the preceding paragraph, shall not exceed $50\text{ V} \div I_{\Delta n}$.
413-02-16, 413-02-28

Bonding two extraneous conductive parts together
A supplementary bonding conductor bonding two extraneous conductive parts together shall be subject to the following minimum sizes:
1. not less than 2.5 mm^2, if sheathed or otherwise mechanically protected,
2. a minimum size of 4 mm^2, if not mechanically protected

547-03-03

Bonding two exposed conductive parts together
Where two exposed conductive parts are to be bonded together, the supplementary bonding conductor, subject to the above minimum sizes, shall have a conductance not less than the smallest cpc installed to the exposed conductive parts. See Figure 2.
547-03-01

Bonding two extraneous parts to an exposed conductive part
Where an extraneous conductive part is to be bonded to another extraneous conductive part, which in turn, is bonded to an exposed conductive part, then, subject to the minimum sizes given above, the supplementary bonding conductor shall have a conductance equal to half that of the cpc connected to the exposed conductive part and the bonding conductor whose resistance R shall not exceed $50\text{V} \div I_a$, where I_a is the minimum current needed to disconnect the circuit to the exposed conductive part within 5 secs. See Figure 2.
547-03-03

Bonding exposed to extraneous conductive parts
Where an exposed conductive part is to be bonded to an extraneous conductive part, the supplementary bonding conductor, subject to the above minimum sizes, shall be half the size of the cpc connected to the exposed conductive part, and the bonding conductor resistance R shall not exceed $50\text{V} \div I_a$, where I_a is the minimum current needed to disconnect the circuit to the exposed conductive part within 5 secs. See Figure 2.
547-03-02

General
Supplementary bonding shall be provided by permanent and reliable conductive parts, or by supplementary conductors, or by both. Where a fixed appliance is to be connected to a supplementary bonding conductor, and the appliance is connected to an accessory by a short length of flexible cable, the protective conductor in the flexible cable can be the bonding conductor from the appliance to the earth terminal in the accessory.
547-03-04, 547-03-05

Extraneous conductive parts used for bonding
With the exception of gas pipes and oil pipes, structural metalwork and extraneous conductive parts can be used as protective conductors, providing the construction and connection maintains the

electrical continuity, and protects against mechanical, chemical or electrochemical deterioration, and the cross-sectional area is at least equal to that calculated from 543-01-03, or Table 54G.
543-02-06

Exposed conductive part to exposed conductive part

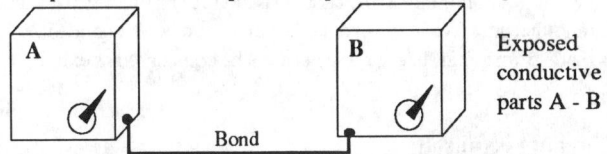

Exposed conductive parts A - B

Same conductance as the smallest cpc feeding A or B, providing that it is mechanically protected: otherwise minimum size is 4 mm².

Exposed conductive part C to extraneous conductive part

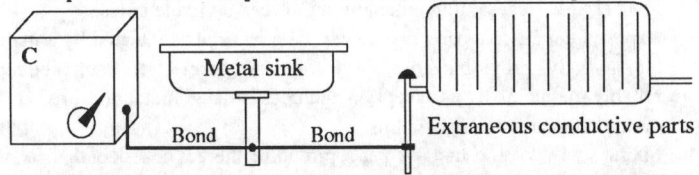

Extraneous conductive parts

Half conductance of the cpc in 'C' if sheathed or mechanically protected: otherwise minimum size required is 4 mm².

Extraneous conductive part to extraneous conductive part

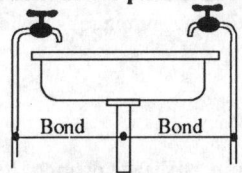

Minimum size 2.5 mm² if sheathed or otherwise mechanically protected: 4 mm² if not mechanically protected.

Figure 3 - Minimum size of supplementary bonding conductors

Bath and shower rooms

With the exception of equipment fed from a SELV circuit, local supplementary bonding shall be carried out in a room with a fixed bath or shower, even when the following conditions apply:
1. when there are no exposed conductive parts in the bath or shower room,
2. when the extraneous conductive parts are connected back to the main equipotential bonding conductors with metal to metal joints of negligible impedance.

Local supplementary bonding shall be carried out between simultaneously accessible exposed and extraneous conductive parts, and between simultaneously accessible extraneous conductive parts.
601-04-02

Electrical equipment installed in the space below the bath shall only be accessible by using a key or tool, and shall be included in the supplementary equipotential bonding in the bathroom.
601-04-03

Caravans and motor caravans

The minimum cross-sectional area of equipotential bonding conductors shall be 4.0 mm^2. These shall be used to bond extraneous conductive parts to the protective conductor of the caravan installation. Furthermore, bonding shall be applied at more than one place if the construction of the caravan does not provide continuity. If the caravan is made of insulating material, then isolated metal parts which cannot become live in the event of a fault, need not be bonded. Where the structure of the caravan is made from metal sheets, these need not be considered as extraneous conductive parts.
608-03-04

When bonding not required

The following list gives those situations where bonding need not be carried out.
1. Overhead line insulator wall brackets, or any metal connected to them, providing they are out of arm's reach.
2. Inaccessible steel reinforcement in reinforced concrete poles.
3. Exposed conductive parts that cannot be gripped or contacted by a major surface of the human body, providing a protective conductor connection cannot be readily made or be reliably maintained; this exception includes isolated metal bolts, rivets, up to 50 mm x 50 mm. nameplates and cable clips.
4. Fixing screws of non-metallic parts, providing there is no risk of them contacting live parts.
5. Short lengths of metal conduit or similar items which are inaccessible, and do not exceed 150 mm in length.
6. Metal enclosures used mechanically to protect equipment complying with the regulations for Class II (double insulated) equipment.
7. inaccessible, unearthed street lighting (furniture) supplied from an overhead line

471-13-04

Contact with non-electrical services

Metal conduit, trunking, ducting and the metal sheaths or armour of cables operating at low voltage, which might come into contact with other metalwork, shall either be effectively bonded to the other metalwork, or effectively segregated from it, irrespective of any equipotential bonding carried out at the origin of the installation.
528-02-05

Instantaneous showers

The metal water pipe through which the water supply to the heater is provided shall be solidly and mechanically connected to all metal parts (other than live parts) of the heater or boiler. There must also be an effective electrical connection between the metal water pipe and the main earthing terminal, independently of the circuit protective conductor.
554-05-02

The supply to the heater shall be through a double pole linked switch. The switch shall be separate from and within easy reach of the heater. Alternatively the switch can be incorporated in the heater. The electrical connection from the boiler or heater shall be made directly to the switch, and not through a plug and socket.

Where the heater is installed in a room containing a fixed bath or shower, the switch shall be out of reach of a person using the bath or shower; an insulated cord operated pull switch would be acceptable.

The installer of the boiler or heater shall confirm that the neutral cannot be independently opened by any single pole switch etc., from the origin of the installation up to the boiler, before the boiler or heater is connected.

554-05-03, 554-05-04

Where a shower tray is made from a metallic material, or where it is made from concrete, it shall be bonded to the metal water pipe in the shower heater; in the case of a concrete shower tray, the reinforcing can be used for the bonding connection.

INF

14. Functional extra-low voltage

Extra-low voltage
Extra-low voltage does not exceed 50 V r.m.s., a.c., or 120 V ripple free d.c., between conductors, or between conductors and earth
Definition.

Functional extra-low voltage (abbreviated to FELV)
FELV must not exceed 50 V r.m.s., a.c., or 120 V ripple-free d.c.
411-01-01

If an extra-low voltage supply is used, but does not comply with all the regulations for SELV, then to provide protection against electric shock, the appropriate requirements of Regulations 471-14-02 to 471-14-06 shall be complied with.
471-14-01, 411-03

The fact that the voltage is extra-low voltage shall not be used as the only means of protection against electric shock.
411-01-02

Protection against direct & indirect contact
Where the regulations for SELV are complied with, except that live or exposed conductive parts are connected to earth, or to the protective conductors of other systems, then protection against direct contact and indirect contact is afforded providing:
  1. enclosures give protection against the insertion of a solid body that exceeds 12 mm in diameter, and up to 80 mm long. (IP2X)
  2. the insulation withstands a test voltage of 500 V r.m.s., a.c., for one minute.

If an equivalent degree of safety is provided by equipment conforming to a relevant British Standard is installed or used, the above requirement shall not exclude its installation being carried out without supplementary protection.
471-14-02

Where the extra low-voltage system does not comply with the regulations for SELV, other than by live or exposed conductive parts being in contact with earth or protective conductors of other systems, then protection against direct contact shall be provided either by one or more of the following:
  1. barriers or enclosures complying with the regulations 412-03-01 to 04.
  2. insulation corresponding to the minimum test voltage required for the primary circuit.

Where the extra-low voltage circuit supplies equipment whose insulation does not comply with the minimum test voltage required for the primary circuit, then any insulation which is accessible shall be reinforced during erection to withstand a test voltage of 1500 V r.m.s., a.c., for one minute.
471-14-03

Protection against indirect contact

Where the primary circuit of the source of extra-low voltage is protected by automatic disconnection of the supply, then exposed conductive parts of the extra-low voltage system shall be connected to the primary circuit protective conductor.
471-14-04

Where the primary circuit of the source of supply for the FELV is protected by electrical separation, then the exposed conductive parts of the FELV circuit shall be connected to the primary circuit's non-earthed protective conductor.
471-14-05

Barriers or enclosures

Live parts shall be in enclosures or behind barriers giving protection against direct contact, giving minimum protection against the insertion of a solid body exceeding 12 mm in diameter and up to 80 mm in length. (IP2X). Where an opening larger than this is necessary for the replacement of parts, or to avoid interference with the functioning of the equipment, then both the following requirements shall be met.
1. Unintentional contact with live parts shall be prevented.
2. Provision shall be made so that a person is sure to be aware that live parts can be touched through the opening, and that they should not be touched.

A larger opening than IP2X is only allowed for equipment or accessories complying with British Standards, where it is impracticable to contain the opening to IP2X, due to the function of the equipment, e.g. a lampholder. Where such a deviation is used, the opening shall be as small as possible.
412-03-01, 471-05-02

Where horizontal top surfaces of barriers or enclosures are readily accessible, then it must not be possible to make contact with live parts with small tools, wires or other thin objects which are more than 1 mm thick (IP4X).
412-03-02

Barriers or enclosures shall be suitable for the normal working conditions to which they will be subjected in service, and be firmly secured in place, to provide stability and durability in order to maintain the degree of separation and protection required.
412-03-03

Where it is necessary to remove a barrier or to open an enclosure, then one of the following precautions has to be taken:
1. opening an enclosure or removing a barrier must only be possible by using a key or tool, or
2. it must only be possible to open an enclosure or remove a barrier after the supply is disconnected, and it must only be possible to restore the supply after the barrier has been replaced, or the unit reclosed, or
3. an intermediate barrier giving a degree of protection of at least IP2X is installed to prevent contact with live parts.

The above does not apply to BS 67 ceiling roses, BS 3676 cord operated switches, BS 5042 B.C. lampholders or to BS 6776 E.S. lampholders.
412-03-04

Socket outlets
Socket outlets and luminaire supporting couplers used for functional extra-low voltage shall not admit plugs for different systems (voltages or SELV) used in the same premises.
471-14-06

Voltage drop
The voltage at the terminals of fixed equipment must not be less than the minimum value specified in the British Standard relevant to the fixed equipment. Where no British Standard is available for the fixed equipment, the voltage drop within the installation must not exceed a value appropriate to the safe functioning of equipment.

Where the supply is given in accordance with the Electricity Supply Regulations 1988 & 1990, the previous requirements are satisfied if the voltage drop from the origin of the installation up to the terminals of the fixed equipment does not exceed 4% of the nominal voltage of the supply.
525-01

15. Identification and notices

Identification of equipment
Unless there is no possibility of confusion, switchgear and controlgear shall be labelled, and where danger could arise because it is remote from the operator, a suitable indicator complying with BS 1710 shall be placed at the operator's position.
514-01

Protective devices for circuits, including those in a distribution board, shall be identified so that the circuits protected may be easily recognised.
514-08

The nominal current rating appropriate to the circuit shall be on, or adjacent to, every fuse or circuit breaker.
533-01-01

Diagrams and charts
Diagrams, charts, or tables, or their equivalent, have to be provided, showing the size and type of conductors, and the points served by a circuit, together with the location of the circuit's protective devices, isolators, and switches, and the information necessary to identify these items, any symbols used complying with BS 3939.
A description of the method used to limit shock voltages from indirect contact has to be given.
514-09

Voltage exceeding 250 V
Where it is not expected that equipment or an enclosure contains a voltage exceeding 250 volts, the equipment shall be labelled with the maximum voltage present, the label being clearly visible before access to the equipment is possible.
Where a voltage in excess of 250 volts exists between two simultaneously accessible live parts in separate enclosures, a label shall be provided, warning of the maximum voltage present, which label shall be clearly visible before access to live parts can be made.
514-10

Earthing label
At the point of connection of every earthing conductor to an earth electrode, or a bonding conductor to an extraneous conductive part, a permanent label shall be fixed. The label shall be durably marked, with letters not less than 14 point, containing the words 'Safety Electrical Connection - Do Not Remove'.
514-13-01

On completion of an installation, a notice having a minimum character size of 14 point, and made from durable material so that the notice will remain readable throughout the life of the installation, shall be fixed as close to the origin of the installation as possible. (Usually the incoming main switch.) The notice stating that the installation shall be periodically inspected and tested, giving the dates of the last inspection, and the recommended date of the next inspection, is illustrated below.

> **IMPORTANT**
> This installation should be periodically inspected and tested, and a report on its condition obtained, as prescribed in the Regulations for Electrical Installations issued by the Institution of Electrical engineers.
> Date of last inspection ....
> Recommended date of next inspection ....

This notice need not be installed for highway power supply cables, where the installation is subject to a programmed Inspection and Testing procedure.
514-12-01, 611-04-04

Where earth-free local equipotential bonding or protection by electrical separation is used, a warning notice shall be installed in a prominent position adjacent to every access point to the location concerned. The warning notice shall be durably marked with 14 point characters, and contain the following warning:

> The equipotential protective bonding conductors associated with the electrical installation in this location
> **MUST NOT BE CONNECTED TO EARTH**
> Equipment having exposed-conductive parts connected to earth must not be brought into this location.

514-13-02

Colour identification

Where electrical services such as conduit are required to be distinguished from other services or pipelines, the colour used for identification of the electrical services shall be orange in compliance with BS 1710.
514-02

The colours reserved exclusively for identification of protective conductors are green and yellow, and it must not be used for any other purpose. One of the two colours shall cover between 30% and 70% of the surface, the other colour covering the remaining surface.

Where a bare conductor or busbar is used as a protective conductor, it shall be identified with green and yellow stripes, each being between 15 mm and 100 mm wide, and placed close together. The stripes shall either be installed throughout the length of the conductor, or placed in each compartment, and at each accessible position. Where adhesive tape is used for this purpose it shall be bicoloured green and yellow.
514-03

The single colour green shall not be used.
514-06-02

IDENTIFICATION AND NOTICES

Every single core and multicore non-flexible cable shall be identified throughout its length, red, yellow, or blue indicating the phase conductors, and black the neutral conductors. Paper-insulated cables, or pvc insulated armoured auxiliary cables, can have cores numbered in accordance with British Standards. Generally 1, 2 and 3 are used for phase conductors, and 0 for the neutral.

Micc cables can be identified at their terminations with tapes, sleeves, or discs of the appropriate colour.

Any colour used for identification of a switchboard busbar, shall comply with the colours given for cables.

A bare conductor shall be identified where necessary by tape, disc, or sleeve of the same colour as that used to identify cable cores.
514-06-01, 514-06-04, 514-06-03

Red, yellow, and blue may be used as phase colours for cables up to final circuit distribution boards, but single-phase circuits from a final distribution board shall be coloured red and black.
Table 51A

Every core of a flexible cable or cord shall be identified throughout its length, Brown for phase, Blue for neutral, Green/yellow for the protective conductor.
514-07-02

Fireman's emergency switch

Where a fireman's emergency switch is not located adjacent to the equipment, a notice shall be placed adjacent to the equipment, indicating where the switch is located. A further notice shall be placed at the switch, so that it is clear which equipment that switch operates.
476-03-07

Where more than one switch is installed on any one building, each switch shall be clearly marked to indicate which installation it controls.
476-03-07

Every fireman's switch shall be labelled with a nameplate 150 mm x 100 mm, containing the letters 'FIREMAN'S SWITCH'. The size of these words shall be appropriate to the circumstances, but in any event, not less than 36 point characters shall be used.
537-04-06

Isolation

Where an installation is supplied from more than one source, a durable warning notice shall be permanently fixed in such a position that any person wanting to operate any of the main switches is warned that all switches shall be operated to isolate the complete installation.
460-01-02

Where a single device does not isolate all the live parts in an enclosure or equipment, a durable warning notice shall be permanently fixed, warning that additional isolators require to be operated before access is made to live parts.
514-11, 461-01-05

All devices used for isolators shall be clearly identifiable, by durable marking, to indicate the installation or circuits which they isolate.
461-01-07

Mechanical maintenance
Devices used for switching off for mechanical maintenance shall be readily identifiable by durable labelling where necessary.
462-01-02

Emergency switches
Devices used for emergency switching shall be durably marked.
463-01-04

Residual current devices
A notice using indelible characters, not smaller than 14 point, shall be fixed in a prominent position at or near the origin of the installation when the installation incorporates a residual current device. The notice shall read:

> 'This installation, or part of it, is protected by a device which automatically switches off the supply if an earth fault develops. Test quarterly, by pressing the button marked "T" or "Test". The device should switch off the supply, and should then be switched on to restore the supply. If the device does not switch off the supply when the button is pressed, seek expert advice.'

514-12-02

Areas for skilled and instructed persons
Warning signs shall be provided clearly and visibly to indicate areas reserved for skilled and instructed persons.
471-13-03

Street lighting
Except where the method of installation prohibits the marking of underground cable, it shall be marked with cable covers or marking tape.

For the purposes of identification, ducting, marker tape, or cable tiles used with highway power supply cable, shall either be marked or colour coded, so that they are distinct from other services.
611-04-02, 611-04-03, 522-06-03

Caravans
A label manufactured from durable material, with easily legible characters, shall be provided next to the main switch in every caravan, advising the user what to do before connecting the electrical supply to the caravan, and what to do before disconnecting the supply from the caravan on leaving the caravan site.

INSTRUCTIONS FOR ELECTRICITY SUPPLY

To connect

1. Before connecting the caravan installation to the mains supply, check that -
   a) the supply available at the caravan pitch supply point is suitable for the caravan electrical installation and its appliances, and
   b) the caravan main switch is in the OFF position.
2. Open the cover to the appliance inlet provided at the caravan supply point, and insert the connector of the supply flexible cable.
3. Raise the cover of the electricity outlet provided on the pitch supply point, and insert the plug of the supply cable.

   THE CARAVAN SUPPLY FLEXIBLE CABLE MUST BE FULLY UNCOILED TO AVOID DAMAGE BY OVERHEATING.
4. Switch on the caravan main switch.
5. Check the operation of residual current devices, if any, fitted in the caravan by depressing the test button.

   IN CASE OF DOUBT, OR IF AFTER CARRYING OUT THE ABOVE PROCEDURE THE SUPPLY DOES NOT BECOME AVAILABLE, OR IF THE SUPPLY FAILS, CONSULT THE CARAVAN PARK OPERATOR OR HIS AGENT, OR A QUALIFIED ELECTRICIAN.

To disconnect

6. Switch off at the caravan main isolating switch, switch off at the pitch supply point, and unplug both ends of the cable.

Periodical Inspection

Preferably not less than once every three years, and more frequently if the vehicle is used for more than normal average mileage for such vehicles, the caravan electrical installation and supply cable shall be inspected and tested, and a report on their condition obtained, as prescribed in the Regulations for Electrical Installations published by the Institution of Electrical Engineers.

608-07-05

A notice that will permanently be legible and easily readable throughout the life of the installation, shall be installed at the caravan electrical intake point. This notice shall give the nominal voltage and frequency for which the caravan's installation has been designed, and the rated current of the caravan installation.

608-07-03

All socket outlets shall have their nominal voltage clearly and indelibly marked.

608-08-03

16. Index of protection

Regulation 6 requires the correct selection of equipment for the environmental conditions that can foreseeably occur. The IP codes can be of assistance in determining the type of enclosure required to comply with Regulation 6. They give a means of specifying your requirements to manufacturers, or determining whether manufacturer's equipment is suitable for the conditions envisaged. This chapter explains the IP codes, and how the numbering system works.

Full information on the IP codes is given in BS 5490, the letters IP being followed by two numerals. On the continent, a third number is used to indicate the degree of protection against mechanical impact. (UTE C 20-010 French Standards.)

The following table, therefore, gives a brief outline of the third degree of protection, although it is not as yet part of BS 5490.

First characteristic numeral
Protection of persons against contact with live or moving parts inside the enclosure. Protection of equipment against the ingress of solid bodies.

0   No protection of equipment against the ingress of solid objects, and no protection against contact with live or moving parts.

1   No protection against deliberate access, but protection against solid objects exceeding 50 mm diameter.

2   Protection against contact with live and moving parts inside an enclosure by fingers and solid objects exceeding 12 mm. diameter and up to 80 mm. in length.

3   Protection against contact by tools and wires, or other solid objects exceeding 2.5 mm. diameter.

4   Protection against objects of a thickness greater than 1.0 mm.

5   Complete protection against contact with live or moving parts within the enclosure, and protection against the ingress of dust sufficient in quantity to interfere with the satisfactory operation of the enclosed equipment.

6   Complete protection against contact with live or moving parts, and against the ingress of dust.

Second characteristic numeral
Protection of equipment against the ingress of liquid.

0   No protection against the ingress of water.

1   Protection against drops of condensed water which will have no harmful effect on the enclosure.

2   Protection against drops of liquid which will have no harmful effect when the enclosure is tilted at any angle up to 15 ° from its normal position.

3   Protection against rain or spray at an angle up to 60 ° from the vertical position.

4   Protection against liquid splashed from any direction.

Second characteristic numeral continued

5   Protection against water projected from a nozzle against the enclosure from any direction.

6   Protection from water from heavy seas or water projected in powerful jets.

7   Protection against water entering the enclosure in harmful quantities, when the enclosure is immersed in water under defined conditions of time and pressure.

8   Protection against water entering the enclosure, when the enclosure is immersed for an indefinite period in water under specified pressure.

Third characteristic numeral

Protection of equipment against mechanical impact.

0   No protection

1   Protection against an impact energy of 0.225 joule, i.e. impact of 150 g from a height of 150 mm.

2   Protection against an impact energy of 0.375 joule, i.e. impact of 250 g from a height of 150 mm.

3   Protection against an impact energy of 0.500 joules, i.e. impact of 250 g from a height of 200 mm.

5   Protection against an impact energy of 2.00 joules, i.e. impact of 500 g from a height of 400 mm.

7   Protection against an impact energy of 6.00 joules, i.e. impact of 1.5 kg from a height of 400 mm

9   Protection against an impact energy of 20.00 joules, i.e. impact of 5 kg from a height of 400 mm.

Using the IP code

The letters IP can be followed by either two or three numerals, each numeral indicating the degree of protection either required or provided.

Where a degree of protection is not specified or provided, the numeral is replaced by an "X". This means that there may be some protection, but the degree of protection is not specified: e.g. IP2X indicates that the degree of protection against the ingress of liquid is not specified. The "X" is used instead of "0" since "0" would indicate that no protection was given.

Where a degree of protection is given, then protection is afforded against all the lower degrees: e.g. equipment specified as having IP44 protection would indicate that it was also protected against 3, 2, 1 and 0.

When the degree of protection against impact is agreed, and BS 5490 is amended, then the IP code will be expressed with three numerals e.g. IP 243 which indicates:

IP 2	First characteristic numeral
IP§ 4	Second characteristic numeral
IP§ § 3	Third characteristic numeral

17. Inspection and testing

Introduction
A complete chapter explaining Inspection and Testing is given in Part 2, and since this subject is a complete section of the regulations, the following is only a reminder of the requirements.

Requirement for testing
On completion of an extension, alteration or new installation, it has to be verified that, as far as possible, the requirements of the Wiring Regulations have been complied with by inspection and testing. Periodic inspection recommendations shall be given to the person ordering the work.
130-10

Applicable installations
Inspection and test have to be made to verify that the regulations have been complied with. The inspection and test can be carried out during the erection of the installation, or on completion of the installation, but shall be carried out before the installation is put into service.
The installations which must have inspection and testing carried out are as follows:
1. a new installation,
2. an alteration to an installation,
3. an extension or addition to an installation.

130-10, 711-01-01, 721-01-02

General requirements
Any tests carried out must not cause danger to persons, livestock or property, and must not damage equipment, even if the circuit tested is defective.
711-01-01

The information on which the installation design was based has to be made available to the person carrying out the inspection and testing, along with the drawings or schedules for the installation. The inspection and test have to verify that the requirements of the Regulations have been complied with; this includes the basic design requirements, such as:
the characteristics of the supply:
1. nominal voltage,
2. nature and frequency of current,
3. prospective short circuit current at the origin of the installation,
4. type and rating of protective device at the origin of the installation,
5. suitability of the requirements for the installation, including maximum demand,
6. the external phase earth loop impedance.

Other design requirements:
1. protection against direct and indirect contact,
2. protection against short circuit currents,
3. the requirements for overload protection,
4. correct size of earthing and protective conductors,
5. discrimination between protective devices.

711-01-01, 711-01-02, 311-01, 312-01, 313-01

INSPECTION AND TESTING

Drawings or schedules shall be given to the person carrying out the inspection and testing showing:
1. the position and type of isolators, switches, etc.,
2. the position and type of outlet points,
3. the position, type, and size of protective devices,
4. the size and type of conductors, together with the method of wiring,
5. any equipment likely to be damaged by a particular test.

Where symbols are used to show equipment etc. on drawings or in schedules, they are to comply with BS 3939.

Additionally, the person carrying out the inspection and test has to be given a statement advising the method used for protection against electric shock from indirect contact.
711-01-02, 514-09

Visual inspection

A visual inspection has to be carried out, to check that the materials are correctly erected, comply with British Standards, are the same as stated on the design drawings, and have not been damaged during erection. The inspection will include making certain that all connections are tight, conductors are correctly identified, and equipment is properly secured in place. It will also include checking protective conductors, bonding conductors, and supplementary bonding conductors, as well as checking that earthing conductors have been properly installed. Visual inspection includes checking that isolators, switches, and protective devices have been installed, and that such items have been correctly labelled where necessary.
712-01-03

Sequence of tests

The following tests have to be carried out in the sequence shown.
1. Continuity of protective conductors, including main and supplementary bonding conductors,
2. Continuity of final ring circuit conductors,
3. Insulation resistance,
4. Insulation applied on site,
5. Protection by electrical separation,
6. Protection by barriers or enclosures provided during erection,
7. Insulation of non-conducting floors and walls,
8. Polarity,
9. Earth fault loop impedance,
10. Earth electrode resistance,
11. Operation of RCDs.

If an installation fails any of the above tests, then any previous test already carried out, and affected by the fault indicated, shall be repeated after remedial work has been carried out.
713-01

Continuity of protective conductors

Every protective conductor has to be tested to make sure that it is electrically sound and properly connected. Where copper protective conductors are used, the test can be made with a d.c., ohmmeter. Where ferrous protective conductors are used, the test can be made with a 50 volt a.c., or d.c., supply.

Details of the tests required are given in Part 2 'Inspection and Testing.'

78 INSPECTION AND TESTING

The test comprises passing a current of up to 1.5 times the design current of the circuit under test through the protective conductor; the current need not, however, exceed 25 amps.
713-02

Continuity of ring circuit conductors
A test has to be made to make sure that a proper ring circuit has been installed. Additionally, the continuity of all the ring circuit conductors, including the protective conductor, has to be checked.
713-03

Insulation resistance
The insulation resistance between live conductors has to be measured before the installation is energised. Any electronic devices used in the installation shall be disconnected so that they are not damaged by the test voltage, and they should then be tested separately. The insulation resistance between each live conductor and earth has also to be measured, the PEN conductor in a TN-C system being considered as part of the earth.

The insulation resistance between live conductors, and live conductors and earth, shall be considered satisfactory if the main switchboard, and each distribution circuit when tested separately with all final circuits connected, but current-using equipment disconnected, has the insulation values given below:

Extra-low voltage circuits

Supplied from an isolating transformer to BS 3535: Test voltage 250 V d.c. Minimum insulation resistance 0.25 megohm.

Low-voltage up to 500 V

Test Voltage 500 V d.c. Minimum insulation resistance 0.5 megohm

Low-voltage between 500 V and 1000 V

Test voltage 1000 V d.c. Minimum insulation resistance 1 megohm.

Between SELV circuits and associated LV circuits

Test voltage 500 V d.c. Minimum insulation resistance 5 megohm.

713-04-01, 713-04-02, 713-04-03, 713-04-04

The insulation resistance of equipment with exposed conductive parts connected to protective conductors switched off in the previous test, shall be as prescribed in the British Standard for the equipment. The test shall be made between exposed conductive parts and live parts. Where there is no applicable British Standard, the insulation resistance shall be not less than 0.5 megohm.
713-04-06

Insulation applied on site
Where insulation is applied on site for protection against direct contact, it shall be verified that it will withstand the tests laid down in the British Standard for similar type tested equipment.
713-05-01

Where supplementary insulation is applied on site to protect against indirect contact, tests shall be made to verify that the degree of protection of the insulating enclosure is not less than IP2X, and that it will withstand an applied voltage test laid down in British Standards for similar types of equipment.
713-05-02

SELV

Where SELV provides protection against electric shock, the electrical separation of the separated circuit shall be inspected and tested.
The extra-low voltage circuits shall be checked to make sure that the nominal voltage does not exceed extra-low voltage, and that the source of supply is from one of the following:
1. safety isolating transformer manufactured and tested to BS 3535,
2. motor generator whose windings give the same degree of electrical separation as a BS 3535 transformer,
3. battery or motor driven generator,
4. electronic devices restricted to make sure that an internal fault cannot result in a voltage at the output terminals being greater than extra-low voltage, or if it so results in a greater voltage, it is immediately reduced on contact, to extra-low voltage.

The circuits and equipment shall also be checked that the comply with the SELV regulations.
713-06-01, 411-02-01, 411-02-02

Protection by electrical separation

Where separation of circuits is used for protection against electric shock, the circuits shall be checked to make sure that they comply with the regulations for protection by electrical separation.
713-06-02

Protection by barriers

Where protection against direct contact is made using barriers or enclosures, then they shall be inspected to make sure that the barriers and enclosures are firmly secured in place, and will maintain the degree of protection required. They must also be checked to make sure that they offer at least IP2X protection, and where there are horizontal top surfaces which are readily accessible, the minimum protection is increased to IP4X. The opening of the barrier or enclosure should only be by means of a tool, and should not be possible until the supply to live parts has been disconnected.
713-07, 412-03

Non-conducting floors and walls

Where a non-conducting location is used for protection against indirect contact, a check shall be made to make sure that the installation complies with the regulations, and with the specification for the installation, since this method of protection is not recognised by the regulations for general use.

Where the location has an insulating floor and walls, one or more of the following arrangements applies:
1. the distance between exposed conductive parts, and between exposed and extraneous conductive parts is not less than 2 m, or where out of arm's reach, 1.25 m,
2. obstacles, preferably insulating, are installed between exposed conductive parts and extraneous conductive parts. These are considered effective if they extend the distance between exposed and extraneous conductive parts to 2 m, or 1.25 m, if out of reach. If any obstacle is non-insulating, then it must not be in contact with earth, or with exposed and extraneous conductive parts,
3. extraneous conductive parts are insulated with material having adequate electrical and mechanical strength.

Where (3) is used for protection then the insulation resistance shall not be less than 0.5 MΩ when tested at 500 V d.c., and shall be capable of withstanding a test voltage of at least 2 kV r.m.s., a.c., and in normal use it shall not pass a leakage current exceeding 1 mA.
713-08-01, 713-08-02, 413-04-07

Polarity
It shall be verified that all protective devices, single pole switches, control switches, and the centre contact of Edison screw lampholders, are only connected in the phase conductor, and that socket outlets and similar accessories have been correctly connected.
713-09

Earth fault loop impedance
Where protection against indirect contact uses automatic disconnection involving the phase and cpc conductors, the relevant impedances shall be measured or calculated.

Where protection is provided only by limiting the impedance of the protective conductor, this impedance shall be measured.
713-10

Earth electrode resistance
Where protection involves the use of an earth electrode, the earth electrode resistance shall be measured.
713-11

Residual current devices
The effectiveness of residual current devices has to be verified by a test simulating a fault condition in the circuit. All loads normally supplied through the circuit breaker are disconnected, and a test is then made on the load side of the circuit breaker between the phase conductor and the cpc, so that a suitable residual current flows.
713-12

Alterations and additions to an installation
The rating and condition of the existing installation and its earthing arrangements shall be checked, to make sure that it is capable of carrying the additional load. The check shall be made before any alteration or addition, either temporary or permanent, is made to the existing installation.
130-09

The inspection and testing of any alteration or addition to an installation shall be the same as a new installation. Additionally, it shall be verified that every alteration or addition does not impair the safety of the existing installation.
721-01-01, 721-01-02

A Completion and inspection certificate has to be handed to the person ordering the work, after all defects found during the inspection and testing have been rectified.
Any defects found in related parts of the existing installation shall be reported in writing by the person or contractor responsible for the new work to the person ordering the work.
743-01

Periodic inspection and testing
The inspection and testing shall be carried out with careful scrutiny to make sure that the installation still complies with the regulations, and will not cause danger to persons or livestock, or damage to property. This may involve partial dismantling. The results shall be recorded in a report, and signed by the person carrying out the inspection. The report shall then be sent to the person ordering the inspection and test.
731-01-01, 731-01-02, 732-01-03

Following a periodic inspection and testing, a report on the inspection and test shall be handed to the person ordering the work to be done.
741-01-02, 744-01-01

The report shall contain a record of any dangerous conditions arising from non-compliance with the Wiring Regulations. Any limitations of the inspection and testing shall also be recorded in the report.
744-01-02

Frequency of testing
The frequency at which inspection and tests are carried out shall be determined by the type of installation, the way it is used and operated, external influences, and the frequency of maintenance.
  Danger shall not be caused to persons or livestock, or damage caused to property and equipment whilst the inspection and testing is carried out.
732-01-01, 732-01-02

Completion certificate
A completion and inspection certificate has to be handed to the person ordering the work, together with recommendations for periodic inspection and testing, after all defects found during the inspection and testing have been rectified. The certificates will be of the form shown in appendix 6. A competent person(s) shall sign the certificate, stating that the system has been designed, installed, inspected and tested in accordance with the Wiring Regulations.
741-01-01, 742-02-01, 742-02-02, 130-10

Any defects found in the installation shall be made good before the inspection certificate is issued
742-02-01

Testing notice
A notice in durable material which will remain easily legible throughout the life of the installation, shall be fixed in a prominent position as close to the main incoming distribution board as possible, and will contain the date of the last inspection, and the recommended date of the next inspection.

IMPORTANT
  This installation should be inspected and tested periodically, and a report on its
  condition obtained, as prescribed in the Regulations for Electrical Installations
  issued by the Institution of Electrical Engineers.
  Date of last inspection ....
  Recommended date of next inspection ....

This notice need not be installed for highway power supply cables where the installation is subject to a programmed Inspection and Testing procedure.
514-12-01, 611-04-04

18. Isolation

Object of isolation
Isolation is used to make live parts of an installation dead, so that work can then be carried out on those parts in safety.

What is an isolator
An isolator is an off load device; it is not designed to make or break load current, nor to make onto or break fault currents.

What is isolation
Cutting off the electrical supply from every source to an installation, a circuit, or an item of equipment.
One device can be used to perform the functions of:
1. isolation,
2. switching off for mechanical maintenance,
3. emergency switching,
4. functional switching.

Where a device is used for one or more of the above functions, it shall be verified that the device complies with the Regulations for each function it performs.
476-01-01, 530-01-03, 537-01

General requirements
Every installation shall be provided with a means of isolation and switching.
476-01-01

Non-automatic isolation and switching shall be provided to prevent danger associated with the electrical installation or electrically powered equipment and machines.
460-01-01

An isolator or a switch shall not be installed in a protective conductor. The exception to this is where an installation is supplied from more than one source of electrical energy, one of which requires earthing independently of the other sources, and where it is necessary to make sure only one means of earthing is used at anyone time. A switch is then allowed between the neutral point and the means of earthing, providing it is linked to disconnect and connect live conductors at the same time.
460-01-03, 460-01-04, 543-03-04

Where the Regulations require all the live conductors to be disconnected, it must not be possible to disconnect the neutral before the phase conductors are disconnected. It is recommended that the neutral be re-connected at the same time as the phase conductors, or before the phase conductors are connected.
530-01-01

No switch or circuit breaker shall be inserted in a neutral conductor in a TN or TT system, unless it is linked with the phase conductors.
530-01-02

Circuits to be isolated

To prevent or remove danger, isolators shall be installed to cut off all voltage from every installation, from every circuit, and from all equipment. The device used shall be suitably placed for ready operation.
130-06-01

Every circuit shall be provided with a means of isolation. Where the circuit is part of a TN-S or TN-C-S system, all phase conductors shall be isolated. Where the circuit is part of a TT or IT system, provision shall be made to isolate all live conductors.

One device may be used to isolate a group of circuits, e.g. an isolator feeding a distribution board.
461-01-01, 461-01-02, 461-01-03

Prevention of re-energisation

Precautions shall be taken to prevent any equipment being unintentionally or inadvertently energised.
461-01-04

Remote isolators

Where isolators are remote from the circuits or equipment they are intended to isolate, then both the following requirements shall be complied with.
1. The isolator shall be secured against inadvertent reclosure whilst being used to isolate the equipment or circuits it controls.
2. Where an isolator is secured against inadvertent reclosure by a lock or removable handle, the key or handle shall not be interchangeable with any others used for a similar purpose within the installation.

476-02-02

Motors

An efficient means of disconnection, which is readily accessible, easily operated, and so placed as to prevent danger, shall be provided for every motor.
130-06-02

Every motor circuit shall be provided with an isolator. The isolator shall disconnect the motor and its associated equipment, including any automatic circuit breaker.
476-02-03

Discharge lighting

Circuits or luminaires used for discharge lighting, where the open circuit voltage exceeds low voltage, shall use one or more of the following methods for isolation.
1. Self contained luminaires shall have, in addition to the switch normally used for controlling the circuit, an interlock which automatically disconnects the supply before access can be made to live parts.
2. In addition to the switch normally used to control the circuit, an effective local means of isolating the circuit from the supply shall be provided.

3. A lockable distribution board, lockable switch, or removable handle shall be provided. The locks and removable handle shall be unique within the same installation, and so placed or guarded that they can only be operated by skilled persons, and cannot inadvertently be returned to the ON position.
476-02-04

Where an autotransformer with an open circuit voltage not exceeding 1,500 volts r.m.s.,, is used on an a.c., 2 wire TN, or TT system to supply high voltage discharge lighting, provision shall be made to isolate both poles of the supply.
554-02, BS 559 1991

Isolation of discharge lighting circuits shall comply with the relevant IEE Regulations for isolation.
554-02, BS 559 1991

Isolation at the origin

Isolation shall be provided as near as possible to the origin of the installation, by a linked switch or circuit breaker capable of interrupting the supply on load.

This means that there should not be any other equipment upon which work may be required to be carried out, between the origin and the means of isolation provided. The means of isolation should comply with all of the relevant Regulations for isolation.

A main switch or circuit breaker at the origin of the installation shall switch the following conductors of the incoming supply:
1. both live conductors when the supply is single-phase a.c.,
2. all poles of a d.c., supply,
3. all phase conductors in a T.P., or T.P. & N., TN-S or TN-C-S system supply,
4. all live conductors in a T.P., or T.P.& N., TT, or IT system supply.

One switching device is allowed to isolate a group of circuits.
460-01-02, 461-01-01,
461-01-02, 461-01-02

Where an isolator is used in conjunction with a circuit breaker to enable maintenance of switchgear to be carried out, it shall either be interlocked with the circuit breaker, or be so placed and guarded that it can only be operated by skilled persons.
476-02-01

Where the supplier of electricity provides switchgear complying with the regulations at the origin of the installation, and agrees that it can be used to isolate the installation from the origin up to the main distribution at which isolation is provided, this agreement satisfies the requirement of the regulations for having a main switch at the origin.
476-01-01

Isolating switchgear

A switch or isolator shall not be placed in a protective conductor or PEN conductor. The exception to this is where an installation is supplied from more than one source of electrical energy, one of which requires earthing independently of the other sources, and where it is necessary to make sure only one means of earthing is used at anyone time.

A switch is then allowed between the neutral point and the means of earthing, providing it is linked so that it disconnects or reconnects live conductors at the same time as the earthing conductor.
460-01-03, 460-01-04

Where an installation is supplied from more than one source of electricity, each source shall be controlled by its own main switch. A permanent durable warning notice must, however, be installed, warning that all the switches need to be switched off if the whole installation is to be isolated.
460-01-02

A means of interrupting the supply on load shall be provided for every circuit and final circuit.
476-01-02

Identification
All isolators shall be durably marked to identify the circuits or installation they are controlling.
461-01-07

Where a single device is not capable of isolating the live parts of equipment or those contained within an enclosure, then either of the following precautions shall be taken:
1. a durable warning notice shall be placed so that any person gaining access to live parts will be warned to use the appropriate isolating devices, or
2. an interlocking arrangement is provided so that all the circuits concerned are isolated before access can be made..

461-01-05

Capacitors & inductors
Adequate means shall be provided to discharge capacitors or inductors to prevent danger. Similarly, capacitive circuits or inductive circuits shall be discharged to prevent danger.
461-01-06

Requirements for isolators
The isolating distance between contacts when in the open position shall not be less than that specified for disconnectors in BS EN 60 947-3. (Note: not yet specified, but can be taken to be at least 3 mm for 240 volts r.m.s. to earth installations).
537-02-01

Semiconductor devices shall not be used as isolators.
537-02-02

The isolated position shall only be indicated when the specified isolating distance has been achieved in each phase or pole.
The position of the isolator contacts or other means of isolation shall either be externally visible or clearly and reliably indicated
537-02-03

Isolators shall be selected and installed so that unintentional reclosure is prevented. Provision shall also be made to secure against the unauthorised and inadvertent operation of off-load devices used for isolation.
537-02-05, 537-02-06

ISOLATION

Isolation shall be achieved by the use of a single multipole device cutting off all the appropriate poles of the supply. All the appropriate poles of the supply can be isolated by single pole devices, providing they mounted adjacent to each other.
537-02-07

Where a link is installed in the neutral conductor in a T.P. & N. supply, it shall only be removable by the use of tools, or it shall be accessible only to skilled persons.
537-02-04

Accessibility

Isolators and switches shall be installed so that adequate and safe means of access and working space are provided to enable operation, maintenance, inspection and testing to be carried out. These facilities shall not be impaired by equipment being mounted in enclosures or compartments.
130-07, 513-01, 529-01-02

19. Mechanical maintenance

Switching off for mechanical maintenance
Switching off for mechanical maintenance is to stop mechanical movement of parts by electrical means, such as by a motor, in order to enable mechanical maintenance or cleaning of non-electrical parts of equipment, plant or machinery. It is also used for the replacement or cleaning of lamps. One device can be used to perform the functions of:
1. switching off for mechanical maintenance.
2. isolation.
3. emergency switching.
4. functional switching.

Where one device is used for one or more of the above functions, it shall be verified that the device complies with the regulations for each function it performs
476-01-01, 530-01-03, 537-01

Provisions for safety
Where mechanical maintenance may involve a risk of physical injury, provision shall be made for switching off for mechanical maintenance.
462-01-01

Precautions shall be taken in the selection and erection of devices, so that any equipment switched off is not unintentionally or inadvertently re-energised. (For instance, by mechanical shock or vibration)
537-03-03

Except where the switch is continuously under the control of the person carrying out the mechanical maintenance, provision shall be made to enable precautions to be taken to stop any equipment or machines being unintentionally or inadvertently started whilst maintenance is being carried out.
462-01-03

Identification
The equipment used shall be durably labelled and mounted in a position which is convenient for its intended use. i.e. local to, and within easy reach of the equipment it controls.
462-01-02

Position of devices
Devices shall be inserted in the main supply circuit, but if additional precautions which provide the same degree of safety as the main supply being interrupted are taken, the device can be inserted in the control circuit.
537-03-01

Requirements for devices
Devices used for switching off for mechanical maintenance, including control switches for such devices, shall have an externally visible contact gap, or clearly and reliably indicate the OPEN or OFF position only when the OFF or OPEN position has been achieved on each pole.

Devices used for switching off for mechanical maintenance shall be manually operated.
537-03-02

Switches used for switching off for mechanical maintenance shall be capable of switching off the full load current of that part of the circuit the switch controls.
537-03-04

Provision of devices
Every fixed motor shall be provided, for mechanical maintenance purposes, with a means of switching off: this device shall be readily accessible, easily operated, and so placed as to prevent danger.
130-06-02

Accessibility
Safe access and working space shall be provided for equipment that requires operation or maintenance to be carried out.
130-07

Access to equipment shall be arranged so as to give access to each connection. Additionall access shall be arranged so as to enable its operation, maintenance and inspection to be safely carried out. Access shall not be impaired by equipment mounted in enclosures or compartments.
513-01, 529-01-02

20. Overcurrent protection

General requirements
All electrical conductors shall be of sufficient size and current-carrying capacity.
130-02-03

Single pole fuses and circuit breakers shall only be placed in phase conductors. Additionally, no fuse or linked circuit breaker shall be installed in an earthed neutral. Where a linked circuit breaker is used, it must disconnect all the related conductors.
130-05-01, 130-05-02

Every installation and every circuit shall be protected against overcurrent by devices which will operate automatically, at values of current and time for the circuit, so that no danger is caused. The devices must have adequate breaking capacity, and, where necessary, making capacity. They shall be located so that there is no danger from overheating, arcing, or the scattering of hot particles, and they shall be able to restore the supply without danger.
130-03

Where switchgear or fusegear is supplied by the supply utility at the origin of the installation, and where they give permission for it to be used for isolation from the origin to the main installation switch, it will not be necessary to protect the tails from the origin to the incoming main switch of the installation with additional overcurrent devices.
476-01-01

Protection of live conductors shall be given by one or more devices which will interrupt the supply automatically in the event of an overload or short circuit current.
431-01-01

Where a common device is used for both overload and short circuit protection, it shall be capable of breaking (and, for circuit-breakers, making) any overcurrent at the point at which the device is installed.
432-02

Where the source is incapable of supplying a current exceeding the current-carrying capacity of the conductors, the conductors are considered to be protected against both overloads and short circuit currents, so overcurrent devices are not generally required.
436-01

Assessment of general characteristics
The prospective short circuit current at the origin of the installation shall be calculated, ascertained or determined.
313-01-01(iii)

The type and rating of the overcurrent protective device at the origin of the installation shall be ascertained.
313-01-01(iv)

Automatic reclosure
Where danger is likely to be caused by a protective device reclosing, the reclosure shall not be automatic.
451-01-06

Everything you need to know about Circuit Protection

In its 48 fully illustrated pages, the Crabtree Circuit Protecion Reference Manual provides complete technical information on the use of circuit breakers and residual current devices in domestic, commercial and industrial electrical installations. It also provides advice on protection against overcurrent and earth faults, electric shock and fire risk protection.

For your complimentary copy of the manual, contact us at the below address.

Crabtree Electrical Industries Limited
Head Office: Lincoln Works, Walsall, England WS1 2DN
Telephone: Walsall 0922 721202 Fax: Walsall 0922 721321
A Hanson Company

Crabtree
Electrical Excellence

Overcurrent detection
A device that will detect and disconnect an overcurrent in a phase conductor shall be provided. The device is not required to disconnect other live conductors, but precautions have to be taken where the disconnection of one phase conductor could cause danger, e.g. three-phase motor single phasing
473-03-01

There is no need to install overcurrent detection in neutral conductors that are part of a TN or TT system, if the neutral has the same cross-sectional area as the phase conductors.
473-03-03

When the neutral does not have the same cross-sectional area as the phase conductors in a TN or TT system, overcurrent detection shall be provided in the neutral, unless the neutral is protected against short-circuit currents by the protective device in the associated phase conductor, and the load is shared evenly between the phases of the circuit.
473-03-04

Where the requirements of Regulation 473-11 are not met, suitable overcurrent detection which will cause disconnection of the phase conductors, but not necessarily the neutral conductor, shall be provided in the neutral.
473-03-04

Discrimination
Where it is necessary to prevent danger, protective devices shall be sized so that the smaller rated device operates before with the larger rated device with which it is in series, in the event of a fault or overload. (This is known as 'discrimination'.)
533-01-06

Co-ordination of overload and short circuit protection
Co-ordination is required between overload devices and short-circuit protective devices, such that the energy let-through the short circuit protective device is within the withstand capacity of the overload device. Manufacturer's advice shall be obtained for motor starters.
435-01, 432-03

Energised fuse replacement
When fuses are likely to be removed or replaced whilst the circuit they protect is energised, they shall be of a type that can be removed or replaced without danger. (Live parts shrouded.)
533-01-03

Identification
Every overcurrent protective device shall have the nominal current rating of the circuit it protects either marked on it, or placed adjacent to it. The current rating to be indicated for semi-enclosed (rewirable) fuses shall be in accordance with the table given below[1].
533-01-01

To enable easy recognition of which protective device protects a circuit, the protective device shall be identified.
514-08

---

[1] It is preferable to check with the manufacturer of the fuse holder, otherwise excessive burning may occur when a fault occurs.

The size of fusewire used in semi-enclosed (rewirable) fuses shall be in accordance with the manufacturer's instructions. Where no instructions are available, the fusewire shall be selected from the following table. Cartridge type fuses are preferred.
533-01-04

Size of fuse wire for semi-enclosed fuses		
Size of fusewire in amperes	Nominal diameter of fusewire in mm	Approximate S. W. G.
3	0.15	38
5	0.20	36
10	0.35	29
15	0.50	25
20	0.60	23
25	0.75	22
30	0.85	21
45	1.25	18
60	1.53	17
80	1.80	15
100	2.00	14

Fuse replacement by unskilled persons

Where fuse links are likely to be replaced by unskilled persons, the fuses should preferably be of a type which cannot be replaced by a fuse having a higher fusing factor of the same nominal rating, or there is marked, either on the fuse or adjacent to it, the type of fuse link intended to be used.
533-01-02

Circuit breakers operated by unskilled persons

Where circuit breakers can be operated by unskilled persons, it should not be possible for the setting or calibration of their overcurrent releases to be modified without the use of a key or tool, which results in a visible indication of its setting.
533-01-05

Disconnection times

Where overcurrent protective devices are also used for protection against electric shock in TN and TT systems, they shall be selected so that their disconnection time is such that the permissible final temperature of the phase conductors and circuit protective conductors is not exceeded, when there is a fault of negligible impedance between a phase conductor and an exposed conductive part. The disconnection time must also comply with the shock protection regulations for the type of circuit and environmental conditions applicable, which, under normal conditions is 0.4 seconds for socket circuits, and 5 seconds for fixed equipment circuits, where the voltage to earth is 240 V.
531-01-01

Lampholders

The size of device protecting a lampholder and its wiring, which is not contained in earthed metal or insulating material having the ignitability characteristic 'P', as specified in BS 476 Part 5, or protected by its own protective device, shall be limited by the type of lampholder used as follows.

BS 5042, BS 6776 and BS 6702 type lampholders			Maximum amperes of protective device
BS 5042 Bayonet type	B15	SBC	6
	B22	BC	16
BS 6776 Edison Screw	E14	SES	6
	E27	ES	16
	E40	GES	16

553-03-01

21. Overcurrent protection - 'Fault currents'

Conductor protection
One or more protective devices that will operate automatically shall be used to protect phase and neutral conductors against prospective fault currents, and no fault current shall be allowed to persist indefinitely
431-01-01, 431-01-03

Withstand capacity
The characteristics of the fault current protective device have to be co-ordinated with the overload device, so that the energy let-through of the fault current protective device is within the withstand capacity of the overload device.
431-01-02, 435-01

Where a common device is used for both overload and fault current protection, it shall be capable of breaking (and for circuit breakers, making) any overload or fault current at the point the device is installed.
432-02

Devices installed to protect against fault current shall be capable of breaking (and for circuit breakers, making) the prospective short-circuit current available.
432-04

Limitation of mechanical and thermal effects
Fault current shall be interrupted by a protective device before such current can cause damage or danger due to thermal or mechanical effects in conductors or connections.

The nominal current rating of a protective device used for fault current protection may be greater than the current-carrying capacity of the conductor which the device is protecting.
434-01

Position of fault current protective devices
Fault current protective devices shall be placed at the point where a reduction in the conductor's current-carrying capacity occurs. Reduction in current-carrying capacity can be caused by:
1. reduction in conductor cross-sectional area,
2. change in ambient temperature,
3. contact with thermal insulation.,
4. change in the number of cables grouped together,
5. change in the method of installation,
6. change in the type of cable conductor.

473-02-01

A fault current protective device can be placed at any point on the load side of a reduction in current-carrying capacity of a conductor, providing:
1. the distance from the point of reduction in current-carrying capacity to the protective device does not exceed 3 metres, and
2. the conductors are so erected that the risk of short-circuit, the risk of fire, or the risk of danger to persons is reduced to a minimum, and

3. the protective device is not being used to give protection against indirect contact, if the earth fault loop impedance is too high.

473-02-02

A short-circuit protective device may be placed at a point other than where the current-carrying capacity of a conductor is reduced, provided that there is a protective device on the supply side of the change in current-carrying capacity, which will protect the conductors against short-circuit in accordance with Regulation 434-03-03.

473-02-03

Omission of fault current protection

Fault current protective devices need not be provided on the supply side of conductors connecting generators, transformers, rectifiers, or batteries with their control panels in which short-circuit protective devices are installed, provided the conductors are erected in such a manner that the risk of short-circuit, fire, or danger to persons is reduced to a minimum.

Certain measuring circuits, or circuits where disconnection would cause a greater danger than the short-circuit current need not be protected with short-circuit protective devices, provided that the conductors are erected in such a manner that the risk of short-circuit, fire, or danger to persons is reduced to a minimum.

Where a supplier provides fault current protective devices at the origin of the installation , and agrees that these may be used to protect the installation up to the main distribution point, there is no need to duplicate the protective devices at the origin of the installation. The first fault current protective device shall be installed at the main distribution point.

473-02-04

Prospective fault current

The prospective fault current shall be determined by measurement or calculation at every point of change in conductor current-carrying capacity. The calculations shall take into account maximum and minimum fault current conditions.

434-02, 533-03

Fault current protective devices

The breaking capacity of protective devices installed must not be less than the prospective fault current at the point the device is installed. (For three-phase circuits, the symmetrical short-circuit current is taken at the point at which the device is installed.)

434-03-01

Protective devices with a lower breaking capacity than the prospective fault current available at the point of installation can be installed, provided that:
1. there is a protective device, with the necessary breaking capacity, on the supply side of the lower rated protective device, and
2. the characteristics of both protective devices are co-ordinated, so that the energy let through the supply side device will not damage the lower rated device, or the cable conductors protected by both devices.

434-03-01

Fault current capacity of conductors

Where a protective device is used for both overload and fault current protection, and has a breaking capacity at least equal to the prospective fault current at the point it is installed, and the conductors

on the load side of the device have been sized in accordance with the overload regulations, it can be assumed that the conductors are protected against fault current. This assumption shall be checked where conductors are installed in parallel and it may not be valid for non-current limiting types of circuit breaker. In this case it shall be checked that the conductors are protected against fault current, by using Regulation 434-03-03
434-03-02

Where a protective device provides fault current protection, a check shall be made to make sure that the rise in temperature of the circuit conductors, caused by a fault at any point in the circuit, does not exceed the limit temperature for the conductors.

Where the protective device is also providing protection against indirect contact, and where the earth fault current is less than the short circuit current, a check shall be made to make sure that the temperature rise on the phase conductors under earth fault conditions does not exceed the limit temperature for the phase conductors.

The time t, in which a given fault current will raise the conductor temperature from its normal working temperature to its limit temperature, can be approximately determined from the following formula:

$$t = \frac{k^2 S^2}{I^2}$$

Values of k:
1. For pvc insulated copper conductors up to 300 mm^2 115.
2. For pvc insulated copper conductors over 300 mm^2 103.
3. For pvc insulated aluminium conductors up to 300 mm^2 76.
4. For pvc insulated aluminium conductors over 300 mm^2 68.
5. For micc copper conductors, pvc covered, or exposed to touch   115.
6. For impregnated paper insulated copper conductors   108.
7. For tin soldered joints in copper conductors   100.

Other values for k are given in Table K1A and K1B Part 4.

When the disconnection time t is 0.1 seconds or less, the energy let-through the protective device, as given by the manufacturer, must not exceed the thermal capacity of the conductor: i.e., $I^2 t$ shall be less than $k^2 S^2$. The value of $I^2 t$ (i.e., the energy let-through the protective device), given by the manufacturer, is not the $I^2 t$ of the equation in Regulation 434-03-03. Neither is it the calculated fault current multiplied by the disconnection time 't' obtained from the protective device characteristic.
434-03-03

Discrimination
Where necessary to prevent danger, protective devices which are in series, shall be sized so that the smaller rated device will operate before the larger rated device.
533-01-06

Conductors in parallel
Conductors installed in parallel shall be of the same type, construction, cross-sectional area, length and arrangement. There shall be no branch circuits throughout their length, and they shall be installed so as to carry equal currents. The cables shall be checked by calculation for being protected against fault currents, account being taken of what would occur if the fault only affected one of the conductors.
434-04

22. Overcurrents - 'Overloads'

Conductor protection
One or more protective devices that will operate automatically shall be used to protect phase and neutral conductors against possible overload currents.
431-01-01

Withstand capacity
The characteristics of the overload devices have to be co-ordinated with the short-circuit protective devices, so that the energy let-through of the short-circuit protective device is within the withstand capacity of the overload device.
431-01-02, 435-01

Overload protective devices may have a breaking capacity less than the prospective short-circuit current at the point at which they are installed. (But see previous paragraph.)
432-03

Where a common device is used for both overload and short-circuit protection, it shall be capable of breaking (and for circuit breakers, making) any overload or short-circuit current at the point at which the device is installed.
432-02

Limitation of temperature rise
Overload current shall be interrupted by a protective device before the overload current causes a temperature rise which could damage the conductors, insulation, joints, or terminations.
433-01

Duration of overload
Small overloads of long duration should not be allowed to occur, and circuits shall be designed so that such overloads are unlikely to occur. Large overloads of short duration are permissible, as in the case of starting current, or inrush current, when equipment is switched on.
433-01

Device size
The nominal rating of the protective device I_n must not be less than the full-load current (design current) of the circuit I_b.
433-02-01(i)

Conductor size
The current rating I_Z of the smallest conductor in the circuit must not be less than the nominal rating of the protective device I_n, (See BS 3036 derating for rewirable fuses.)
433-02-01(ii)

Device fusing current
The tripping (fusing) current I_2 for the protective device must not exceed 1.45 x I_Z for the smallest conductor in the circuit. (See BS 3036 derating for rewirable fuses.)
433-02-01(iii)

Devices that comply with requirements for fusing current
Providing the smallest conductor in the circuit has a current-carrying capacity not less than the nominal rating of the protective device, and its rating is not less than the full-load current (design current) of the circuit, the regulations for overload protection are satisfied, when general purpose type 'gG' BS88 Part 2 HRC fuses, BS 88 Part 6 HRC fuses, BS 1361 Cartridge fuses, BS 1362 cartridge fuses and BS 3871 part 1 or BS 4752 part 1 circuit breakers are used as the overload protective device.
433-02-02

Derating semi-enclosed fuses
Semi-enclosed fuse is often referred to as a BS 3036 fuse, or rewirable fuse. Due to the requirement that the current causing the effective operation of the protective device (I_2) must not exceed 1.45 times the current-carrying capacity of the smallest conductor in the circuit (I_Z), and due also to the fact that a semi-enclosed fuse has a fusing factor twice its normal rating, the minimum current-carrying capacity of the smallest conductor in the circuit, in order to guarantee overload protection shall be $\dfrac{I_n}{0.725}$ when used with semi-enclosed fuses, to comply with Regulation 433-02-01(iii).
433-02-01(iii), 433-02-03

Conductors in parallel
Providing conductors in parallel are of the same type, have the same cross-sectional area, are the same length, and are installed together following the same route, they can be protected by the same protective device. Socket outlet ring circuit cables are not considered to be cables in parallel.

The minimum required current-carrying capacity of each conductor will be:-

$$I_t = \frac{I_n}{\text{Number of conductors in parallel}}$$

433-03

Position of overload devices
Overload protective devices shall be placed at the point where a reduction in the conductor's current-carrying capacity occurs. Reduction in current-carrying capacity can be caused by:
1. reduction in conductor cross-sectional area,
2. change in ambient temperature,
3. contact with thermal insulation,
4. change in the number of cables grouped together,
5. change in the type of overload protective device.

473-01-01

Exception to normal overload position
Overload devices can be installed along the run of the conductors, providing there are no branch circuits or current-using equipment connected between the point the conductor size is reduced and the position of the overload device.
473-01-02

Omission of overload devices
No overload protective devices shall be placed in the secondary circuit of a current transformer.
473-01-03

An overload device is not required at the point where a reduction in the current-carrying capacity of a conductor occurs, if the protective device on the supply side of the reduction in conductor current-carrying capacity protects the conductor with the reduced current-carrying capacity.
473-01-04(i)

Overload protective devices are not required for conductors if the characteristics of the load are not likely to cause an overload in the conductors.
473-01-04(ii)

Overload devices are not required where a greater danger would be created with the unexpected opening of the circuit than would occur with the overload condition; in these circumstances overload alarms shall be installed.
473-01-04(iii)

Overload devices are not required at the origin of an installation if the supplier provides equipment at the origin, and agrees that his equipment can be used for overload protection between the origin and the main distribution point containing the next set of overload devices.
473-01-04(iv)

Discrimination
Where necessary to prevent danger, any intended discrimination in the operation between protective devices shall be achieved.
533-01-06

To avoid unintentional operation, the peak value of current may have to be taken instead of I_b when sizing conductors. For cyclic loads, the thermally equivalent constant load shall be worked out, and I_n, I_b, I_z, and I_2 shall be chosen accordingly. (Find r.m.s., value of peaks.)
533-02

Identification of overload devices
The intended nominal current of every protective device shall be provided on or near it.
The nominal current shall be related to the circuit being protected.
533-01-01

Overload replacement
Where non-skilled persons are likely to replace fuses, the fuses should preferably be of a type that cannot be replaced with one of a higher fuse rating. Additionally, it should have marked on it or adjacent to it, the type of fuse that shall be used, or be of a type that cannot inadvertently be replaced by fuse having a higher fusing factor than the one being replaced.
    Where a circuit breaker can be operated by non-skilled personnel, it shall not be possible to alter the setting or calibration of the breaker without using a key or tool. Any alteration must result in the display of a visible indication of its setting.
533-01-01, 533-01-02, 533-01-05

If a fuse is of the type that is likely to be removed or replaced whilst the supply is connected, that fuse shall be capable of being removed and replaced without danger.
533-01-03

Preferably, fuses shall be of the cartridge type. (BS 88, BS 1361, BS 1362.) The manufacturer's instructions shall be used for determining the size of fuse element for BS 3036 (rewirable fuses).
533-01-04

23. Protection against heat and fire

Location of equipment
The installation and location of fixed equipment shall make sure that the heat thereby produced does not cause a fire hazard to fixed materials installed adjacent to the equipment, or to materials which may foreseeably be placed in the same vicinity of the equipment. Any installation instructions provided by the manufacturer shall be conformed with.
422-01-01

Hazardous areas
Equipment shall be so constructed or protected, and special precautions shall be taken, when equipment is installed in an area susceptible to the risk of fire or explosion.
130-08-02

High temperature equipment
Where electrical equipment has a surface operating temperature that could cause a risk of fire, be harmful to adjacent materials, or produce harmful effects, one of the following methods of installation shall be utilised:
1. the equipment shall be mounted on a support, or placed in an enclosure which is capable of withstanding the temperature generated by the equipment without being set on fire, or creating a harmful effect. Any support used should limit the transmission of heat from the equipment by having a low thermal conductance, or
2. the equipment is screened by material which will withstand and not be affected by the heat generated by the equipment, or
3. the equipment is fixed so as to allow the safe dissipation of heat, and at such a distance from, other materials as to avoid fire and harmful effects.

422-01-02

Protection against arcing
Equipment that is likely to emit an arc or high temperature particles shall be installed by one of the following methods:
1. the equipment is totally enclosed in arc resistant material,
2. materials that could be affected are screened by arc-resistant screening material,
3. the equipment is so mounted that any emissions are safely extinguished at a safe distance from and before they reach materials which would be effected by the emissions.

422-01-03

Heat barriers in vertical ducts etc.
Where vertical channels, ducts or trunking contain cables, they shall be equipped with internal barriers to prevent the air at the top becoming excessively hot.
The distance between barriers shall be, either the distance between floors, or 5 metres, whichever distance is the less.
INF

Terminations

Live conductor terminations or joints made between live conductors shall be made in accordance to one of the following:
1. an electrical accessory or enclosure complying with British Standards,
2. an enclosure manufactured from material complying with the 'glow wire test requirements' of BS 6458 Section 2.1,
3. an enclosure using building materials which are considered to be incombustible when tested to BS 476 Part 4,
4. an enclosure which is partly or completely formed by the building structure, which has a BS 476 Part 5 ignitability characteristic 'P'.

422-01-04

Protection against burns

Table 42 specifies the maximum temperature which an accessible part of equipment that is within arm's reach must not exceed, unless a British Standard specifies the maximum temperature. Where any part of the fixed installation is likely, for short periods of time, to exceed the temperature limits specified in Table 42 under normal load conditions, it shall be guarded so that accidental contact with it is prevented.

Maximum temperature of equipment within arm's reach at full load

Part of equipment	Construction of accessible surface	Temperature allowed °C
Operation by hand or hand-held	Metallic Non-metallic	55 65
A touchable part, but not hand-held	Metallic Non-metallic	70 80
A part that does not need to be touched in normal use	Metallic Non-metallic	80 90

423-01

For safety reasons, equipment having electrically heated surfaces that can be touched, shall be provided with a switch.

476-01-02

Flammable liquid

Where electrical equipment, such as switchgear or transformers, in one location, contains more than 25 litres of a flammable liquid, such as oil, precautions shall be taken to prevent any burning liquid, smoke, and toxic gases spreading to other parts of the building.

422-01-05

> Note: One location does not mean each individual switch in a switchboard: it could comprise a complete switchboard in the corner of a factory, or a room containing several separate items of switchgear etc.

Prevention of overheating
Precautions shall be taken to prevent danger from overheating, arcing, or the scattering of hot particles during the operation of switchgear.
130-03

One or more of the following methods shall be used to protect the wiring system from the effects of heat arising from external sources, or from solar gain:
1. the installation shall be shielded,
2. the installation shall be installed out of range of the heat,
3. wiring materials that are suitable for the additional heat shall be selected,
4. the current-carrying capacity of the wiring is reduced,
5. the insulation is reinforced, or changed in the area affected by heat.

Cables or flexible cords installed within an accessory, appliance, or luminaire shall be suitable for the temperature in that equipment. Covering each cable core individually with glass sleeving is one way of satisfying this requirement.
522-02-01, 522-02-02

Reducing the risk of fire
The risk of the spread of fire shall be by selection of the appropriate wiring materials, and installing them so that the integrity of the building structure is not reduced. Where materials manufactured to a British Standard are used, and the standard does not specify testing for the propagation of flame, the wiring system shall be enclosed in non-combustible building material having the characteristic P, to BS 476 Part 5.
527-01

Fire barriers
Where cables, conduits, ducts, trunking, or other items of a wiring system, pass through walls, floors, roofs or ceilings of a building, any part of the hole that is left round the electrical material shall be made good to the same degree of fire resistance as that required for the element being passed through. Additionally, internal barriers that give the same degree of fire resistance shall be installed in conduits, trunking, ducting, busbars and busbar trunking, where the walls, floors, ceilings and roofs have a specified fire resistance. Where the wiring system is non-flame propagating, and has a maximum internal cross-sectional area of 710 mm^2, it need not be internally sealed.

If the fire resistance is for less than one hour, this regulation is satisfied if the sealing of the wiring system has been type tested in accordance with BS 476 Part 23.
527-02-01, 527-02-02

Any sealing arrangement used must comply with the following:
1. there shall be no incompatibility between the wiring system and the sealing material,
2. the wiring shall be permitted to expand and contract due to heat, without damage,
3. it shall be possible to remove the sealer without damage to the cables,
4. it shall withstand the same environmental conditions as the wiring system.
527-02-03

During erection or during an alteration, temporary sealing arrangements shall be made. Any sealing that is disturbed shall be reinstated as soon as possible. Each sealing arrangement shall be inspected during the installation to make sure that it conforms to the manufacturer's erection instructions. The inspection shall be recorded.
527-03-01, 527-03-02, 527-04

24. Protection against direct contact

Methods allowed
The following methods are illustrated in Figure 1.
1) Protection by insulation of live parts.
2) Protection by barriers or enclosures.
3) Protection by obstacles.
4) Protection by placing out of reach.

412-01.

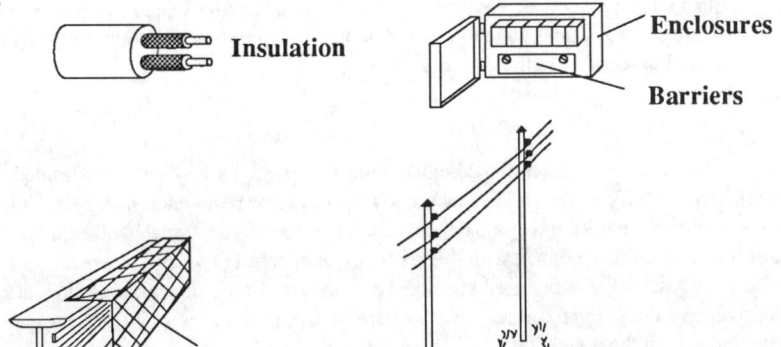

Figure 1 - Protection against direct contact

Method 1
Insulation shall be able to withstand any mechanical, electrical, thermal or chemical stresses to which it may be subjected; additionally, the insulation must only be removable by destruction. Insulation is to prevent contact with live parts when used in conjunction with protection against indirect contact.
412-02, 471-04

Method 2
Method two is intended to prevent or deter contact with live parts when used in conjunction with protection against indirect contact.

Barriers or enclosures shall be secured firmly in place, and must maintain the required degree of protection during normal service. Live parts shall be in an enclosure, any opening in which must not give access to objects larger than 12 mm diameter, and up to 80 mm long (IP2X). Where the opening is in a top surface of the enclosure, the opening must not exceed 1 mm in diameter (IP4X).
412-03-01, 412-03-02, 412-03-03, 471-05-01

Where openings in enclosures are required for replacement of parts, precautions shall be taken to prevent unintentional contact with live parts, and people shall be made aware that live parts accessible through the opening must not be touched.
412-03-01

With the exception of ceiling roses manufactured to BS 67, ceiling switches manufactured to BS 3676 and bayonet lampholders complying with BS 5490, or ES lampholder complying with BS 6776, one or more of the following requirements shall be satisfied where it is necessary to open an enclosure or remove a barrier:
1. it is possible to remove a barrier or open an enclosure only by a key or tool,
2. the supply is disconnected before access to an enclosure, or the removal of a barrier is possible, and the supply can only be restored after barriers are replaced or the equipment is re-closed,
3. contact with live parts is prevented by an additional barrier which can be removed only by a key or tool, and which prevents objects 12 mm diameter and up to 80 mm long coming into contact with live parts.

412-03-04

Method 3

This method does not stop intentional contact with live parts. Its application is limited to areas accessible only to skilled persons. Obstacles do not need to be removable with a key or tool, but shall be secured to prevent unintentional removal, and to stop unintentional bodily approach to live parts, or the unintentional contact with live parts, when operating live equipment.

Where only skilled or supervised instructed persons are allowed into an area, it is sufficient to provide against unintentional contact with live parts by using obstacles, or by placing out of reach in accordance with the regulations.

412-04-01, 412-04-02, 471-06, 471-13-01

Method 4

This method is intended to prevent unintentional contact with live parts. The application of Regulations 412-05-02 to 04 is limited to areas accessible only to skilled persons. Overhead lines between buildings shall be installed in accordance with the Electricity Supply Regulations.

412-05-01

Where bare live parts are out of arm's reach, but may still be accessible, they shall not be within 2.5 m of exposed or extraneous conductive parts, or bare live parts of other circuits.

412-05-02.

Where body movement is restricted in the horizontal plane by an obstacle which would allow access of objects larger than 12 mm diameter and 80 mm long, then arm's reach shall extend from that obstacle.

Where long or bulky objects are handled, the limit of arm's reach shall be increased by the length or size of the object being handled.

412-05-03, 412-05-04.

Protection by residual current device

A residual current device cannot be used as the sole means of protection against direct contact.

Where protection by insulation, by barriers or enclosures, by obstacles, or by placing out of reach is used and the RCD has a rated residual operating current of 30 mA and will operate in 40 milliseconds with a residual current of 150 mA, in accordance with British Standards, the RCD is recognised as reducing the risk of electric shock.

412-06-01, 412-06-02

25. Protection against indirect contact

General requirements
Where exposed conductive parts are earthed, the circuits shall be protected either by overcurrent protective devices, or by residual current devices, to prevent the persistence of dangerous earth leakage currents.

Where the prospective earth fault current is too low to give prompt disconnection of the circuit because the phase earth loop impedance is too large, then a residual current protective device shall be used for circuit protection.
130-04-03

Basic measure of protection
One of the basic measures of protection against indirect contact is equipotential bonding and automatic disconnection of the supply.
413-01

Basic rule for protection
The characteristics of protective devices have to be co-ordinated with the installation earthing arrangements and circuit impedances, so that the magnitude and duration of earth fault voltages appearing between simultaneously accessible exposed and extraneous conductive parts anywhere in the installation, is limited to a safe value.
413-02-04

Protection when the voltage to earth is 220 V to 277 V in a TN system
The basic rule for protection against indirect contact, inside the equipotential zone, can be considered satisfied, if the circuits comply with the following rules:
1. when a phase to earth fault occurs, every socket outlet and every circuit feeding portable equipment intended for manual movement, or hand-held Class I equipment, is disconnected within 0.4 secs, or
2. it is permissible to increase the disconnection time to 5 secs when the impedance of the protective conductor, from the end of the final circuit back to the earth bar where the equipotential bonding is carried out, is limited to a value of $50 \times Z_S \div U_o$ Ω, where Z_S is the earth loop impedance for a disconnection time of 5 secs. (Obtained from Tables ZS1 to ZS 7 Part 3.) The impedances of the protective conductor for different types of protective devices are given in Table PCZ 8 Part 3,
3. where 110 volt reduced voltage circuits that comply with the reduced voltage regulations, are used a phase to earth fault on a socket outlet, or on fixed equipment, is disconnected within 5 secs,
4. at every point of utilisation of fixed or stationary equipment, or distribution circuit, a phase earth fault is disconnected within 5 secs,
5. where fixed equipment is connected by means of a plug and socket, and where it can be guaranteed that the socket cannot be used for supplying hand-held equipment, a phase earth fault is disconnected within 5 seconds,

6. Where circuits requiring disconnection within 0.4 secs are connected to the same distribution board as circuits requiring disconnection within 5 secs, then one of the following protective measures is taken:
   1. from the main earth bar at which the main equipotential bonding conductors are connected, to the distribution board to which the mixed circuits are connected, the impedance of the cpc does not exceed $50 \times Z_S \div U_o \,\Omega$. (Where Z_S is the earth loop impedance for 5 secs disconnection time.) Values of maximum cpc impedances for different protective devices are given in Table PCZ 8 Part 3,
   2. equipotential bonding shall be carried out from the distribution board with the mixed circuits, to the same types of extraneous conductive parts as the main equipotential bonding.

413-02-08, 413-02-09, 413-02-12, 413-02-13

Where the voltage to earth is 240 V, then for fuses, the Z_S values in Table 41B1 can be used for 0.4 sec, and Table 41D for 5 sec disconnection times. Table 41B2 can be used for circuit breakers for 0.4 sec and 5 sec disconnection times. For both 0.4 sec and 5 sec disconnection times, Tables ZS1 to ZS7 in Part 3 can be used.

413-02-10, 413-02-11, 413-02-14

Protection at different voltages to earth in a TN system

The basic rule for protection against indirect contact, within the equipotential zone, can be considered satisfied, if the protective device for the circuit, and the phase earth loop impedance ensure that automatic disconnection of the supply occurs within a specified time, when a phase to earth fault occurs anywhere in the installation.

Item 1 in the paragraphs above is replaced with the following:

Regulation 413-02-04 can be considered to be satisfied if :

$$Z_S \leq \frac{U_o}{I_a}$$

where Z_S is the phase earth loop impedance, U_o is the voltage to earth, and I_a is the current causing the disconnection within the time specified in the following table (41A) of socket-outlet circuits, or circuits for portable equipment which can be manually moved, or is hand-held Class I equipment.

TABLE 41A
Maximum disconnection times allowed

Voltage to earth, U_o volts	Disconnection time 't' secs
120	0.8
220 to 277	0.4
277 to 400	0.2
Over 400	0.1

413-02-08, 413-02-09, 413-02-12, 413-02-13, 41A

Automatic disconnection when using RCDs in a TN system

Where the disconnection times for overcurrent devices cannot be met, where residual current devices are used to comply with the disconnection times, and where the installation is part of a TN system, the residual operating current $I_{\Delta n} \times Z_S$ must not exceed 50.

413-02-16

Where a circuit extends outside the earthed equipotential zone, and automatic disconnection of the circuit is provided by a residual current device, exposed conductive parts need not be connected to the cpcs of the TN system, providing they are connected to an earth electrode via the main earthing terminal, whose resistance is appropriate to the operating current of the RCD.
In this case, every circuit shall comply with $R_A \times I_{\Delta n} \leq 50$ V, where R_A is the resistance of the earth electrode added to the resistance of the cpcs connected to the exposed conductive part of the circuit.
413-02-17, 413-02-20, 413-02-18

Where several residual current devices are connected in series, the exposed conductive parts may be connected to separate earth electrodes for each RCD.
413-02-18

RCDs shall not be used for automatic disconnection with circuits incorporating a PEN conductor.
471-08-07

The risk of electric shock is reduced where earthed equipotential bonding and RCD protection is used. The operating current of the RCD shall be 30 mA, and it must disconnect a circuit within 40 msec when a residual current of 150 mA is applied, in accordance with BS 4293.
412-06-02

Z_S for different voltages between 220 and 277 V

Where the voltage to earth U_O is not 240 volts, the value of Z_S in Tables ZS1 to ZS 7 shall be modified as follows: Revised Design Z_S = (Table $Z_S \times U_O$) ÷ 240V.
INF

Equipotential zone

The values of Z_S given in Tables ZS1 to ZS 7 (41 B1, 41 B2, and 41D) are only applicable inside the equipotential zone created by the main equipotential bonding conductors. Furthermore, the limiting values of earth fault loop impedance and cpc impedance given by the formulae $Z_S \times I_a \leq U_o$ and $Z_{cpc} = 50 \times Z_S + U_o$, and $Z_S \times I_{\Delta n} \leq 50$ V, are only applicable inside the equipotential zone. Where the specified disconnection time cannot be met, 413-02-15 shall apply.
471-08-02, 413-02-15

Equipment outside equipotential zone

Where equipment is outside the equipotential zone, and that equipment can be touched by a person in direct contact with earth, and where the supply to the equipment is obtained from within the equipotential zone, disconnection shall occur within 0.4 seconds when the voltage to earth is 240 V, and an earth fault occurs. For other voltages to earth, the phase earth loop impedance shall ensure that the circuit is disconnected within the times specified in Table 41A.
471-08-03

Where a socket outlet rated up to 32 A is fitted in an installation, and where it can reasonably be expected that it may be used to supply portable equipment for use outdoors, or where a flexible cable, having a current-rating of 32 A or less, not connected through a socket outlet supplies portable equipment outdoors, it shall be protected by an RCD having a residual operating current not exceeding 30 mA and an operating time not exceeding 40 msec, when subjected to a residual current of 150 mA.
471-16-01, 412-06-02, 471-16-02

Compliance with Tables 41A to 41 D (TN system)

The intention of automatic disconnection is to prevent the magnitude and duration of dangerous voltages appearing between simultaneously accessible conductive parts, and includes all methods of earthing exposed conductive parts.

The limiting values of Z_s given in tables 41A to 41 D shall be applied within the equipotential zone where normal body resistance applies. In situations where a lower body resistance can be expected, lower values of Z_s shall be achieved, or an alternative method of protection shall be used.
471-08-03, 471-08-02

Normal body resistance

Normal body resistance can be expected where the skin is dry, or moist with perspiration, and involves contact with one hand and both feet. Low body resistance occurs where the hands or feet are wet, or a large portion of the body is in contact with conductive parts, or where the shock current path is not through the extremities of the body.
INF

Protection in TN systems

Protective conductors shall connect all exposed conductive parts of the installation to the installation main earthing terminal, and that terminal shall be connected to the means of earthing provided for the source of supply.
413-02-06

Protective devices allowed in TN system

Protective devices shall be either:
1. overcurrent or
2. residual current

RCDs must not be used with circuits using combined protective earthed neutral conductors i.e. PEN conductors. Where the system is TN-C-S the PEN conductor shall not be used on the load side of the RCD. Connection of the circuit protective conductor shall be made to the PEN conductor on the supply side of the RCD.
413-02-07, 471-08-07

Where the specified disconnection times cannot be achieved by the overcurrent protective devices, then either local supplementary equipotential bonding shall be used, or protection shall be given by an RCD, providing $Z_S \times I_{\Delta n} \leq 50$ V.
413-02-15, 413-02-16

Protection in TT systems

Exposed conductive parts protected by a single protective device are to be connected to a common earth electrode(s). Where several protective devices are in series, each may have its own earth electrode, and the exposed conductive parts of each circuit may be connected to that protective device's earth electrode. Protective conductors can be installed individually, in groups, or collectively.
413-02-18, INF

Protective devices allowed in TT systems

Protective devices shall be either residual current or overcurrent devices.
Residual current devices are preferred.
413-02-19

Bathrooms
Where equipment installed in a room containing a fixed bath or shower is not a SELV circuit, and where that equipment is simultaneously accessible with either other exposed conductive parts or extraneous conductive parts, the characteristics of the protective device for the circuit supplying the equipment shall be co-ordinated with the earthing arrangements, so that disconnection occurs within 0.4 seconds.
601-04-02

No electrical equipment shall be installed in the interior of a bath tub or shower basin, even if the supply is from a SELV source.
601-02

Protection by non-conducting location, or protection by means of earth-free local equipotential bonding shall not be used in a bath or shower room.
601-06

Reduced voltage systems
Reduced voltage systems using 110 volts r.m.s., between phases, shall use overcurrent devices in each phase conductor, or residual current devices to give protection by automatic disconnection against indirect contact, and all exposed conductive parts of the reduced voltage system shall be earthed. The value of Z_S at each point of utilisation shall ensure that disconnection occurs in 5 seconds. Where an RCD is used, $I_{\Delta n} \times Z_S$ must not exceed 50. Where circuit breakers or fuses are used, values of Z_S allowed are given in Table ZS 8 Part 3.
471-15-06

Agricultural installations
In agricultural buildings or in situations accessible to livestock, electrical equipment shall be of Class II construction, or suitably insulated.
605-11-01

Where automatic disconnection is used for protection against indirect contact, the disconnection time for final circuits supplying portable equipment intended for being moved during its use, or hand-held equipment, is reduced to 0.2 secs, when the voltage to earth is between 220 V and 277 V. Table ZS 9 Part 3 gives the maximum design value of Z_S for 0.2 disconnection time.
605-05-01

For distribution circuits, and final circuits for stationary equipment, the disconnection time is 5 secs, so Tables ZS1 to ZS 7 Part 3 are applicable.

Where circuits requiring disconnection in 0.2 secs and 5 secs are connected to the same distribution board, one of the following conditions shall be applied:
1. the impedance of the protective conductor from the main earth bar, to which the main equipotential bonding conductors are connected and the distribution board, shall not exceed $25 \times Z_S \div U_o$, where Z_S is the earth fault loop impedance for a disconnection time of 5 secs,
2. equipotential bonding shall be carried out at the distribution board to the same extraneous conductive parts as connected to the main equipotential bonding conductors.

Protection against indirect contact by limiting the impedance of the protective conductor is not allowed. Therefore, Table 41 C in the regulations, and PCZ 8 Part 3 cannot be used.

605-05-06, 605-05-05

The following formulas are changed so that the voltage is 25 V.
  $Z_S \times I_{\Delta n} \le 25$ V, Regulation 413-02-16
  $R_a \times I_a \le 25$ V, Regulation 413-02-20
  $R_b \times I_d \le 25$ V, Regulation 413-02-23
605-05-09, 605-06, 605-07-01

Caravans and motor caravans

Each mobile touring caravan shall be supplied from a BS 4343 socket outlet which is protected by a 30 mA RCD, with an operating time not exceeding 40 msec when subjected to a residual current of 150 mA.
608-13-05

Where protection by automatic disconnection of the supply is used for the internal installation of every caravan and motor caravan, protection shall be by a 30 mA double pole RCD with an operating time not exceeding 40 msec when subjected to a residual current of 150 mA.
608-03-02

Socket outlets shall incorporate an earthing contact, and a protective conductor shall be installed throughout each circuit of the caravan installation from the protective contact of the inlet.

Class II equipment can be used in the caravan, but protective conductors shall still be installed to each outlet point, (similar to house wiring).
608-03-02

Exceptions for protection against indirect contact

The following list gives those situations where bonding need not be carried out.
1. Overhead line insulator wall brackets, or any metal connected to them, providing they are out of arm's reach.
2. Inaccessible steel reinforcement in reinforced concrete poles.
3. Exposed conductive parts that cannot be gripped or contacted by a major surface of the human body, providing a protective conductor connection cannot be readily made, or be reliably maintained. This provision includes isolated metal bolts, rivet, nameplates up to 50 mm x 50 mm, and cable clips.
4. Fixing screws of non-metallic parts, providing there is no risk of them contacting live parts.
5. Short lengths of metal conduit or similar items which are inaccessible, and do not exceed 150 mm in length.
6. Metal enclosures used to mechanically protect equipment complying with the regulations for Class II (double insulated) equipment.
7. inaccessible, unearthed street lighting (furniture) supplied from an overhead line

471-13-04

Protective conductors

A bare conductor or single core cable can be used as a protective conductor, and where complying with Section 543, it need not be enclosed in conduit, trunking or ducting. Similarly the metal sheath, screen or armour of a cable can be used as a protective conductor, as can metallic conduit and trunking.
543-02-02, 521-07-03

Earthing conductors can be used for protective and functional purposes, but the protective qualities of the conductor must not be affected by the functional purpose.
546-01, 542-01-06

TN system disconnection times when U_0 between 220 V and 277 V

Equipotential Zone	Equipment type or location	Disconnection time allowed	Protective device allowed
Inside	Fixed	5 seconds	Overcurrent or R.C.D
	Sockets or portable equipment	0.4 seconds	
	Bathrooms	0.4 seconds	
	110V supplies	5 seconds	
Equipment outside fed from inside the equipotential zone	Fixed	0.4 seconds	Overcurrent or R.C.D
	110V supplies	5 seconds	
	Sockets for fixed equipment * see note	0.4 seconds	
	Sockets	0.4 seconds	R.C.D only

413-02-08, 413-02-09, 413-02-13,
471-08-03, 471-15-06, 471-16-01

* An overcurrent protective device can be used for socket outlets feeding fixed equipment, providing the socket outlet cannot be used for hand-held equipment.

Detrimental influences
The installation shall be checked to make sure that no detrimental influence can occur between the various protective measures in the same installation.
470-01-02

Supplementary bonding
When the overcurrent protective device cannot disconnect the circuit within the specified time in table 41A, then either:
 1. local supplementary bonding shall be installed, or
 2. protection shall be provided by a 30 mA RCD having an operating time not exceeding 40 msec, when subjected to a residual current of 150 mA.

413-02-15, 413-02-16

Maintainability
Where a protective measure has to be removed for maintenance, reinstatement of the protection shall be made without reducing the original degree of protection. Adequate access to all parts of the wiring system shall be possible for maintenance to be carried out safely.
529-01

26. Reduced voltage systems

Reduced voltage system
Where an extra-low voltage system cannot be used, and a SELV system is not needed, then a reduced voltage system can be used.
471-15-01

Maximum voltage
Reduced low-voltage circuits shall not exceed 110 V between phases, giving 63.5 V to earthed neutral for a three phase supply, and 55 V to the mid-point for a single phase supply.
471-15-02

Source of supply
The source of supply shall be:
1. double wound isolating transformer to BS 3535 Part 2, or
2. a motor generator whose windings provide the equivalent isolation of an isolating transformer, or
3. a diesel generator, or other independent source of supply.

471-15-03

The mid-point, or star point of the secondary winding of a transformer or generator shall be connected to earth.
471-15-04

Protection against direct contact
Protection against direct contact shall be the same as low voltage installations.
471-15-05

Protection against indirect contact
Overcurrent devices in each phase conductor, or residual current devices, shall be used to give protection against indirect contact. All exposed conductive parts of the reduced voltage system shall be earthed. The value of Z_S at each point of utilisation shall ensure disconnection occurs within 5 seconds. Where an RCD is used, $I_{\Delta n} \times Z_S$ must not exceed 50. Maximum values of Z_S for various types of protective device are given in Table ZS 8 Part 3.
471-15-06

For protective devices that are not given in Table ZS 8, the value of ZS is determined from the formula: $Z_S = U_o \div I_a$, where I_a is the current required to disconnect the protective device within 5 secs, and U_o is the voltage to earth, or to the neutral point.
471-15-06, 413-02-08

Plug and socket outlets
Plugs, socket outlets, and cable couplers shall have a protective conductor contact, and not be interchangeable with plugs, socket outlets, and cable couplers of different voltages or frequencies in the same installation.
471-15-07

27. Residual current devices

Fundamental requirement for safety
Where exposed conductive parts of electrical equipment may become charged with electricity due to the insulation of a conductor becoming defective, or due to a fault in the equipment, the exposed conductive parts shall be earthed by a method which will discharge the electrical energy without danger, or an equally effective measure that will prevent danger shall be used, i.e. Class II insulation.
130-04-01

Protective devices that will prevent the persistence of dangerous earth leakage currents shall be provided for every circuit.
130-04-02

Overcurrent devices shall be provided to stop the persistence of an earth fault current, and where an earth fault current is insufficient to cause the rapid operation of an overcurrent device, a residual current device shall be used to protect against dangerous earth fault currents.
130-04-03

Maximum phase earth loop impedance allowed
Where RCDs are used in TN systems to comply with the disconnection times required for the voltage to earth that is available, then $Z_S \times I_{\Delta n}$ must not exceed 50 V.

In a TT system, $R_A \times I_{\Delta n} \le 50 \text{ V}$, where R_A is the resistance of the protective conductors from the exposed conductive part to the earth rod, and includes the earth electrode resistance.
413-02-16, 413-02-20

PEN conductors
A PEN conductor shall not be used on the load side of an RCD. When the system is TN-C-S, circuit protective conductors shall be connected to the PEN conductor on the supply side of the RCD.
413-02-07

TT system installations
Residual current devices are preferred for protection against indirect contact. If a single RCD is used for protection, it shall be placed at the origin, unless the installation from the origin up to the RCD complies with the regulations for protection by Class II equipment.
413-02-19, 531-04

Socket outlets in TT system
Where automatic disconnection is used as protection against indirect contact in an installation forming part of a TT system, all socket outlets are to be protected by a residual current device, the product of $I_{\Delta n} \times R_A \le 50$.
471-08-06, 413-02-20

IT system installations
Where protection against indirect contact in an IT system uses RCDs, each final circuit shall be separately protected.
413-02-25

Electrical equipment outside the equipotential zone

Where fixed equipment is outside the equipotential zone, and that equipment can be touched by a person in direct contact with earth, and where the supply to the equipment is obtained from within the equipotential zone, disconnection shall occur within 0.4 seconds when an earth fault occurs, and where the voltage to earth is 220 to 277 V. For other voltages, the disconnection times shall be: 120 V - 0.8 secs: 400 V - 0.2 secs: over 400 V - 0.1 secs.
471-08-03

Where any socket outlets rated up to 32 A are fitted in an installation, and where it is possible that they may be used to supply portable equipment for use outdoors, or where a flexible cable, having a current-rating of 32 A or less, not connected through a socket outlet, supplies portable equipment outdoors, they shall be protected by a 30 mA RCD having a residual operating time not exceeding 40 msec, when subjected to a residual current of 150 mA, in order to comply with BS 4293.
471-08-04, 471-16-01, 471-16-02

Indirect contact

Where the disconnection times for protection against indirect contact cannot be achieved using overcurrent protective devices, protection shall be made by an RCD.

Where protection is provided by an RCD, $Z_S \times I_{\Delta n} \leq 50$ V.
413-02-15, 413-02-16

Direct contact

An RCD cannot be used as the only means of protection against direct contact, but it can be used to reduce the risk of direct contact, providing:
1. it has a residual operating current not greater than 30 mA, and has a residual operating time not exceeding 40 msec, when subjected to a residual current of 150 mA,
2. protection against direct contact is provided by insulation, barriers, enclosures, obstacles, or placing out of reach.

412-06-02

Reduced voltage systems

Automatic disconnection shall be used for protection against indirect contact by:
1. an overcurrent protective device in each phase conductor, or by
2. a residual current device, and

all exposed conductive parts shall be connected to earth. The Z_S at every point of utilisation (including socket outlets) shall ensure a disconnection time of 5 secs, where RCDs are used, then:
$Z_S \times I_{\Delta n} \leq 50$ V.
471-15-06

Caravan site installation.

Each caravan shall be supplied from a socket protected by a 30 mA RCD, which must not be the sole means of protection against direct contact. (i.e. protection against direct contact shall be by insulation, barriers or enclosures.) The RCD must disconnect within 40 msec, when subjected to a residual current of 150 mA. Not more than 6 sockets shall be protected by one RCD. Where the supply is PME, the protective conductor of each socket outlet shall be connected to an earth electrode, and the installation treated as a TT system.
608-13-05

Selection and erection of residual current devices
The RCD shall be capable of disconnecting all the phase conductors of the circuit at substantially the same time.
531-01-01
All phase and neutral conductors shall pass through the coil of the magnetic circuit. Protective conductors shall be outside the magnetic circuit, i.e. not pass through the coil of the magnetic circuit. Additionally, the RCD shall be capable of disconnecting all the phase conductors at the same time.
531-02-02, 531-02-01
The residual operating current $I_{\Delta n}$ shall not exceed $50\ V \div Z_s$
The loads and circuits protected by an RCD shall be so arranged, and the selection of the RCD shall be so made, that any leakage currents which may be expected to occur during normal use will not cause nuisance tripping of the RCD.
531-02-03, 531-02-04
Where separate auxiliary supplies are relied upon for the operation of an RCD and the RCD does not automatically trip if the auxiliary source of supply fails, one of the following conditions shall be complied with:
1. protection against indirect contact shall be maintained when the auxiliary supply fails, or
2. the device shall be regularly supervised, maintained and tested by skilled or instructed persons.

531-02-06
RCDs shall be located outside the magnetic fields of other equipment, unless it can be ascertained that the RCDs operation will not be impaired.
531-02-07
RCDs used with, but separately from, short circuit current protective devices, shall be able to withstand the short circuit current that would flow from a short circuit at the outgoing terminals of the RCD.
531-02-08
Where equipment in part of a TN system cannot comply with the disconnection times of Table 41A, then protection can be given by an RCD for that equipment. The exposed conductive parts for that part of the installation shall be connected to the TN earthing system cpc, or to an earth electrode. The earth electrode impedance shall be appropriate to the operating current of the RCD.
531-03

Discrimination
Where two or more residual current devices are installed in series, and are being used to protect against indirect contact, and where discrimination in their operation is required to prevent danger, then their characteristics shall be such that discrimination is achieved. In this case it becomes a TT system.
531-02-09
One way of achieving discrimination is to select devices with different tripping times.
INF

Protection not provided
Protection against indirect contact is not provided if there is no cpc connected to the circuit's exposed conductive parts, even if the RCD has an operating current of 30 mA.
531-02-05

28. SELV

Extra-low voltage
Extra-low voltage does not normally exceed 50 V r.m.s., a.c., or 120 V d.c., between conductors or between conductors and earth.
Definition.

Protection against direct and indirect contact
Using SELV is a means of providing protection against direct and indirect contact. SELV must not exceed 50 V r.m.s., a.c., or 120 V ripple-free d.c.
411-02-01

SELV sources
The extra-low voltage shall be obtained from one of the following sources.
1. A safety isolating transformer having no connection between the secondary winding and the core of the transformer, or the primary earthing circuit. The transformer must also comply with BS 3535, with no connection to earth on the secondary side.
2. A motor generator whose windings etc. provide the same degree of safety as an isolating transformer made to BS 3535.
3. A battery or other source which is independent of higher voltages.
4. Electronic devices manufactured and tested to British standards, the output terminal voltage of which cannot exceed extra-low voltage, even with an internal fault condition. A voltage higher than extra-low voltage is permissible at the output terminals with these electronic devices, providing that, where direct or indirect contact does occur, the voltage at the output terminals is immediately reduced to extra-low voltage. Compliance with this requirement can be verified by testing the output with a voltmeter having an internal resistance of at least 3000 Ω, the voltmeter indicating a voltage not exceeding extra-low voltage.

411-02-02

Where extra-low voltage is obtained from a higher voltage system, the device reducing the higher voltage must have the necessary electrical separation between the higher voltage and the extra-low voltage. (As between the primary and secondary windings of a BS 3535 transformer.)

Where extra-low voltage is provided from a higher voltage system through an autotransformer, potentiometer, semiconductor device etc., it is not a SELV system.
411-02-03

The regulations for Class II equipment, or for equivalent insulation shall be used for the selection and erection of a mobile source for SELV.
411-02-04

Segregation of live and exposed parts
Live parts of SELV circuits must not be connected to earth, protective conductors, or live parts of other circuits.
411-02-05

Exposed conductive parts of SELV circuits must not be connected to earth, to an exposed conductive part, or to the protective conductors or extraneous conductive parts of another system.
411-02-07

Where equipment has to be connected to extraneous conductive parts, any voltage that may appear on the extraneous conductive part must not exceed extra-low voltage.
411-02-07

Live parts of SELV equipment shall be electrically separate from higher voltages. The separation required is the equivalent of that obtained between the primary and secondary windings of a safety isolating transformer. Example: A SELV relay in a control panel using higher voltages. Additionally, a live part of a SELV system must not be connected to a live part, or to a protective conductor of another system, or to earth.
411-02-05

Segregation of conductors
SELV circuit conductors shall be kept physically separate from conductors of other circuits.
    Where conductors of SELV circuits cannot be kept separate from other circuits, then one of the following requirements shall be satisfied:
1. the conductors shall be insulated to the highest voltage present,
2. they shall be non-metallic sheathed cables, the sheath protecting them from the highest voltage present,
3. they shall be separated by an earthed metallic screen or sheath,
4. when grouped with other cables, or enclosed in a multicore cable, they shall either be insulated individually or collectively for the highest voltage present.

In (2) and (3) the basic conductor insulation beneath the sheath or screen need only be sufficient for the voltage of the SELV circuit.
411-02-06

Plugs and sockets
Plugs used for SELV circuits must not be capable of insertion into sockets of higher voltages, or of non SELV circuits.
    Sockets must not be capable of having plugs for other voltages used on the same premises inserted into them.
    The socket outlets shall not have a protective conductor connection.
411-02-10

Where a luminaire supporting coupler has a protective conductor contact, it shall not be installed in a SELV system.
411-02-11

Protection against touching live parts (direct contact)
Where the SELV exceeds 25 V a.c., or 60 V ripple-free d.c., one or more of the following has to be adopted for protection against direct contact:
1. protection by barriers or enclosures, giving protection against direct contact by the insertion of a solid body more than 12 mm in diameter, and up to 80 mm long, (IP2X).
2. the insulation shall be capable of withstanding 500 V d.c., (r.m.s., a.c.) for one minute.

411-02-09

If the nominal voltage does not exceed 25 V r.m.s., a.c., or 60 V ripple-free d.c., there is no need to insulate the cables, or to protect against direct contact, unless the normal body resistance is reduced, in which case the voltages of 25 V a.c., and 60 V d.c., must also be reduced. (Body resistance can be reduced by a large portion of the body being in contact with earth, or in wet conditions etc.)

411-02-09

Protection against indirect contact
If there is likely to be intentional or fortuitous contact between the exposed conductive parts of an extra-low voltage system and exposed conductive parts of any other system, that system shall be treated as a FELV system, and the FELV regulations shall be applied.

411-02-08

Voltage drop
The voltage at the terminals of fixed equipment must not be less than the minimum value specified in the British Standard relevant to the fixed equipment. Where no British Standard is available for the fixed equipment, the voltage drop within the installation must not exceed a value appropriate to the safe functioning of equipment.

Where the supply is given in accordance with the Electricity Supply Regulations 1988 & 1990, the previous requirements are satisfied if the voltage drop from the origin of the installation up to the terminals of the fixed equipment does not exceed 4% of the nominal voltage of the supply.

525-01

Other extra-low voltage systems
Where extra-low voltage is used, but not all the requirements for SELV are complied with, additional measures shall be taken to protect against both direct and indirect contact, as laid down in the regulations for FELV.

411-03

29. Sauna heaters

Type of appliance
Where electric hot air sauna heaters are used, they must comply with IEC Standard 335-2-53
603-01

Temperature zones
The temperature zones illustrated in Figure 1 shall be taken into account when assessing the general characteristics of the installation.
603-02

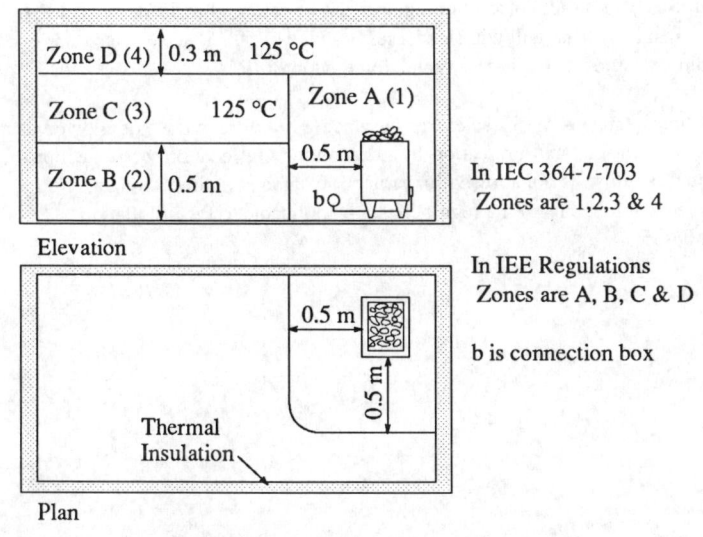

```
Zone D (4)  0.3 m   125 °C
Zone C (3)          125 °C    Zone A (1)
                              0.5 m
Zone B (2)  0.5 m             b
```
Elevation

In IEC 364-7-703
Zones are 1,2,3 & 4

In IEE Regulations
Zones are A, B, C & D

b is connection box

```
                      0.5 m
                      0.5 m
    Thermal
    Insulation
```
Plan

Figure 1 - Zones of ambient temperature

Electrical equipment
With the exception of luminaires, a thermostat and thermal cut-out, switchgear or accessories which are not built into the sauna heater shall be installed outside the sauna. Luminaires shall be so mounted in the sauna as to prevent overheating
603-08 , 603-09

The degree of protection for all equipment has to be IP24
603-06-01

Wiring
Only flexible cords having 150 °C insulation, and complying with BS 6141 shall be used, and shall be mechanically protected by double insulation, i.e. no metallic sheaths or metallic conduits are allowed for mechanical protection.
603-07-01

Installation
No equipment is allowed in Zone A other than the sauna heater and equipment directly associated with it. Equipment in zone B can have a standard temperature rating, there being no special requirements for the heat resistance of equipment. Any equipment installed in Zone C shall be suitable for an ambient temperature of 125 °C. In Zone D, only luminaires and the control devices for the sauna shall be installed. These, together with the cables, shall be suitable for an ambient temperature of 125 °C
603-06-02

Where SELV is used, protection against direct contact has to be provided by one of the following: insulation that will withstand a test voltage of 500 V a.c., r.m.s., for one minute, or barriers, or enclosures that will give protection to at least IP24.
603-03

Protection by means of obstacles, or by placing out of reach, is not allowed as a protection against direct contact, and protection by non-conducting location, or protection by earth free equipotential bonding, is not allowed for protection against indirect contact.

All other requirements of the regulations are applicable to the installation.
603-04 , 603-05

30. Selection and erection of cables

General
Proper materials shall be used, and all electrical conductors shall be of sufficient size and current-carrying capacity: to prevent danger, they shall be placed in such a manner as to be safeguarded, or insulated and where necessary, effectively protected. All joints and connections shall be properly constructed with regard to insulation, conductance, mechanical strength, and protection.
130-01, 130-02-03, 130-02-04, 130-02-05

Electrical equipment
This term is abbreviated to 'equipment' in the regulations, and includes wiring materials (such as cables), accessories, appliances, and luminaires.
Definitions

Selection of cables
Low voltage non-flexible cables shall be one of the following types:
1. non-armoured pvc-insulated to BS 6004, or BS 6346, or BS 6231,
2. armoured pvc-insulated to BS 6346,
3. split-concentric copper-conductor pvc insulated to BS 4553,
4. rubber insulated to BS 6007 or BS 6883,
5. impregnated-paper-insulated, lead sheathed to BS 6480,
6. armoured with thermosetting insulation to BS 5467,
7. mineral insulated to BS 6207,
8. consac cables to BS 5593.

Any of the listed cables sheathed with pvc, lead, h.o.f.r sheath, or an oil-resisting and flame retarding sheath may incorporate a catenary wire.
521-01-01

The phase conductors in a.c., circuits and live conductors in d.c., circuits shall have the following minimum sizes:
1. power and lighting circuits, 1.0 mm^2 copper or 16 mm^2 aluminium,
2. signalling and control circuit, 0.5 mm^2 copper, except for multicore cables having 7 or more cores, used for electronic equipment, when 0.1 mm^2 conductors are allowed,
3. bare conductors for power circuits, 10 mm^2 copper (16 mm^2 aluminium) and 4 mm^2 copper for control and signalling circuits,
4. insulated flexible connections and conductors, 0.5 mm^2 for any application except for electronic circuits outlined and complying with note 2,
5. insulated flexible connections and conductors for extra-low voltage circuits for special applications, 0.5 mm^2.

Connectors used for terminations for aluminium conductors, shall be suitable for this use.
524-01

Table 1
Cable selection

Cable type	Application	Comments
Fixed Wiring		
Single core pvc or rubber insulated unsheathed cable to BS 6004.	Used in conduit, trunking or ducting.	Not to be used in conduit underground.
Single core and flat pvc insulated and sheathed cable to BS 6004.	General indoor use. On exterior walls. Overhead between buildings. Underground in conduit.	Protect against severe mechanical stresses.
PVC insulated, armoured and pvc sheathed to BS 6346.	General indoor and outdoor applications.	Take precautions against damage when installed underground.
Mineral insulated cable to BS 6207.	General indoor and outdoor applications.	Requires pvc covering when exposed to weather or corrosion, or if installed in underground or in concrete ducts.
Flexible cords		
E.P. rubber insulated CSP (HOFR) sheathed to BS 6500 Ref: 3183TQ.	For use with heating appliances, such as immersion heaters.	Not to be used on portable domestic appliances.
V.R. insulated and braided unkinkable cord to BS 6500. Ref: 2213.	Indoors in domestic or commercial premises.	For portable appliances: only suitable for low mechanical stress.
PVC insulated, pvc sheathed, cords to BS 6500. Ref. 3183Y.	Indoor and outdoor use on domestic or commercial premises, including damp situations.	Not allowed outdoors for agricultural or industrial applications.

Where a conductor other than a cable is used for overhead lines at low voltage, it shall be selected from one of the following types:
1. hard-drawn copper or cadmium-copper conductors to BS 125,
2. hard-drawn aluminium and steel reinforced aluminium conductors to BS 215,
3. aluminium-alloy conductors to BS 3242,
4. conductors covered with pvc for overhead power lines to BS 6485 type 8.

521-01-03

Flexible cables used in an installation shall be selected from one of the following types:
1. insulated flexible cords to BS 6500,
2. rubber-insulated flexible cable to BS 6007,
3. PVC-insulated non-armoured flexible cable to BS 6004,
4. braided travelling cables for lifts to BS 6977,
5. rubber-insulated flexible trailing cables for quarries and miscellaneous mines to BS 6116.

521-01-01

Standard of cables

All cables shall comply with the current edition of the applicable British Standard. Where the British Standard takes account of a CENELEC Harmonisation Document (H.D., for short), then cables made to a foreign standard based on the same H.D., can be used, providing they are just as safe as if manufactured to the British Standard. This also applies to cables manufactured to an IEC Standard.

511-01-01, Preface.

If no standard exists, or if the cables are used in a way different from that which allowance is made in the standard, then the designer or the person responsible for specifying that the cables are used in that way, must certify that the degree of safety will not be less than that achieved by compliance with the regulations.

511-01-02

Current-carrying capacity

The current-carrying capacity of every cable shall be not less than the normal full-load current flowing through it, so that its operating temperature does not exceed the value given in the current rating tables for the type of conductor and insulation concerned.

Where a conductor operates at a temperature higher than 70 °C, it shall be verified that the equipment to which it is connected is suitable for that temperature.

523-01

Cables should not be installed in areas where thermal insulation is likely to be installed.

If cables have to be installed in such areas, then they must have their current-carrying capacities decreased. Where the cables are in contact with thermal insulation on one side only, the other side being in contact with a thermally conductive surface, the current-carrying capacities can be obtained from the tables in Part 3.

The derating factors for single cables totally enclosed in thermal insulation are given in the following table.

Table 2
Spacing of supports for cables in accessible positions

Overall diameter of cable in mm	Maximum spacing of cable supports in millimetres								
	Non-armoured pvc, or lead sheathed cables				Armoured multicore PVC or XLPE cable				Mineral insulated copper sheathed or aluminium sheathed cables	
	Generally		In caravans		Horizontal		Vertical			
	Horizontal	Vertical	Horizontal	Vertical	Copper	Aluminium	Copper	Aluminium	Horizontal	Vertical
Up to 9	250	400								
10 to 15	300	400	150	250	350		450		600	800
16 to 20	350	450	for all	for all	400	1200	550	550	900	1200
21 to 40	400	550	sizes	sizes	450	2000	600	600	1500	2000
41 to 60					700	3000	900	900		
over 60					1100	4000	1300	1300		

Note: The major axis is taken for flat cables. 'Horizontal' includes angles up to a maximum of 60° from the horizontal

SELECTION AND ERECTION OF CABLES

Length in mm of conductors totally enclosed in thermal insulation	Derating factor
50	0.89
100	0.81
200	0.68
400	0.55
500 or over	0.5

523-04

The current-carrying capacity of every cable shall take into account the maximum operating temperature likely to occur in normal service, including any heat transferred from accessories or appliances.

522-01-01

When determining the normal operating temperature of the cable, account shall be taken of the minimum ambient temperature, precautions being taken against mechanical damage occurring to the cable at low temperatures.

522-01-01, 522-08-01

The current-carrying capacity of a cable shall be reduced when grouped with other cables, installed in an ambient temperature higher than 30 °C, or in contact with thermal insulation.

App 4

Where a protective device is providing overload protection for a circuit, the current-carrying capacity of the circuit conductors must not be less than the nominal rating of the protective device. Additionally, where the fusing factor of the protective device exceeds 1.45 times its nominal rating, conductors having a higher current-carrying capacity shall be used, as determined by the formula:

$$\text{Calculated current-carrying capacity required } I_t = \frac{I_n \times \text{fusing factor}}{1.45}$$

433-02-01, 433-02-02, 433-02-03

The neutral conductor shall have a cross-sectional area not less than the phase conductor in a discharge lighting circuit. (fluorescent lighting etc.)

524-02-03

Mineral insulated cables shall not be used for discharge lighting circuits, unless surge arresters are used.

INF

Where the load of a three-phase and neutral circuit is not balanced due to the load, or where an imbalance occurs due to harmonic currents (3 rd harmonics from fluorescent fittings) or power factor, the neutral conductor shall be capable of carrying the current that is likely to flow through it. (Usually, it is safer to make the neutral the same size as the phase conductor.)

524-02-02

Cables connected in parallel shall be of the same type, size, length and disposition, and shall be arranged to carry equal currents.

523-02, 433-03

THE THREE MOST COMMON METHODS OF OVERHEAD WIRING BETWEEN BUILDINGS

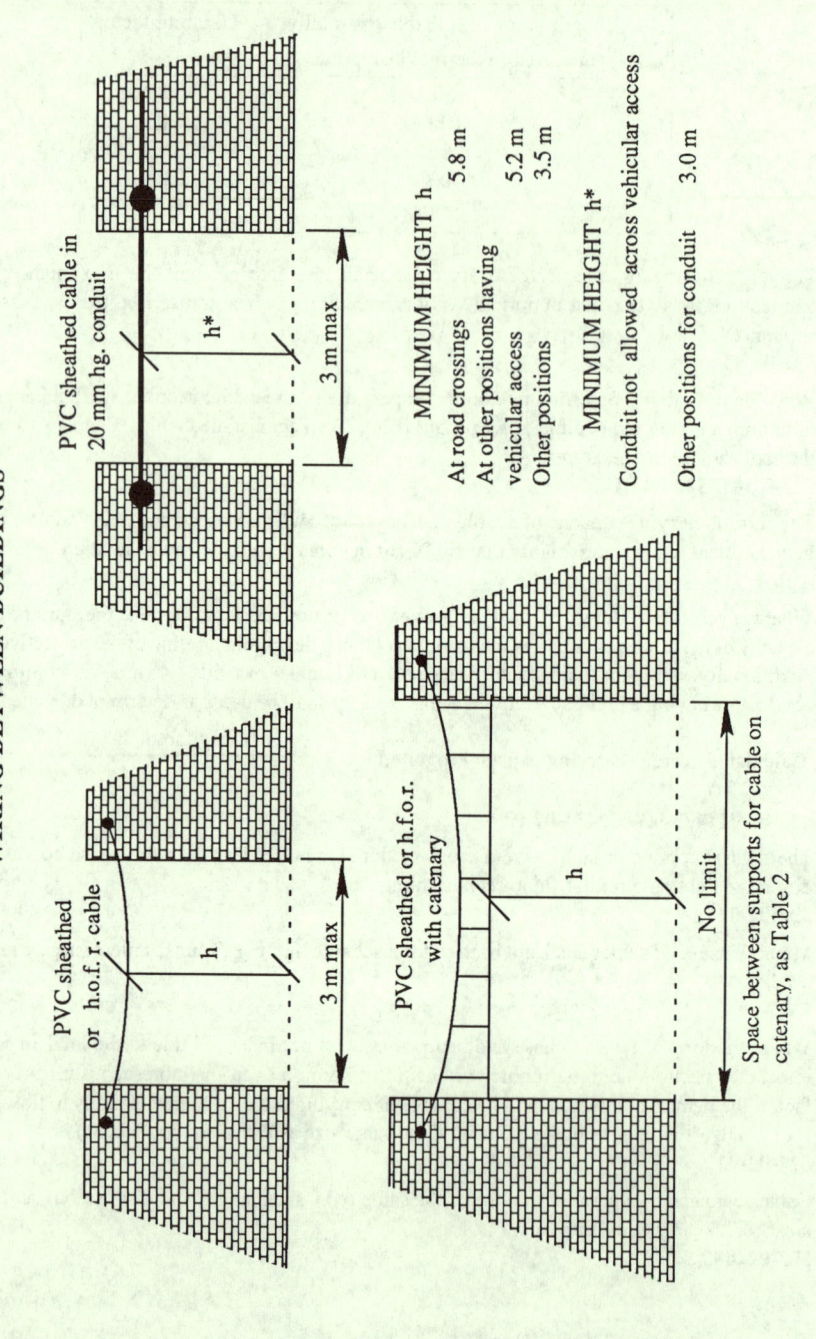

PVC sheathed cable in 20 mm hg. conduit

h*

3 m max

PVC sheathed or h.o.f.r. cable

h

3 m max

PVC sheathed or h.f.o.r. with catenary

h

No limit

Space between supports for cable on catenary, as Table 2

MINIMUM HEIGHT h

At road crossings 5.8 m
At other positions having vehicular access 5.2 m
Other positions 3.5 m

MINIMUM HEIGHT h*

Conduit not allowed across vehicular access

Other positions for conduit 3.0 m

Every cable shall be sized so that it can withstand the thermal and mechanical stresses imposed on it due to any current, or fault current it may have to carry.
521-03

Where cables in the same enclosure have different temperature ratings, they shall be sized and treated as though they all have the lowest temperature rating.
521-07-03

Rotating machines

Cables carrying the starting, accelerating, and load currents of a motor shall have a current carrying-capacity at least equal to the full load current rating of the motor. Where a motor is used intermittently, and for frequent starting and stopping, account shall be taken of the cumulative effect of the starting periods upon the temperature rise of the cables.
552-01-01

The rating of cables supplying rotors of slip-ring or commutator induction motors shall be suitable for the starting currents and load conditions. (Note: rotor currents often exceed the full load current rating of the motor.)
INF

Mechanical damage

All conductors and cables shall either be constructed to withstand, or protected against, any risk of mechanical damage to which they may be subjected in normal use.
522-06-01

Cables shall be protected against abrasion when passing through holes in metalwork (i.e., cables protected by bush or grommet).
522-06-01

When assessing the risk of mechanical damage to cables, allowance shall be made for the strains imposed during installation.
522-08-01

The number of cables drawn into an enclosure shall be such that no damage occurs to the cables during installation.
522-08-02

Conduit shall be completely erected before any cable is drawn in, unless a prefabricated system is being used.
522-08-02

Where cables are installed through timber joists in the ceiling or floor, they shall not be less than 50 mm from the top or bottom of the joist, unless mechanical protection that will stop the cable being penetrated by nails or screws etc. is provided. (150 mm long lengths of conduit need not be bonded 471-13-04). Alternatively, cables not protected by an earthed metallic sheath can be enclosed in an earthed steel conduit which is securely supported.
522-06-05

Cables and conductors used for fixed wiring shall be supported so that there is no mechanical strain, including the strain caused by the weight of the cable on them, or their terminations. (See table 2.)
522-08-01, 522-08-05

Table 3

Minimum internal bending radius for fixed wiring cables

Conductor Material	Conductor insulation	Cable finish	Overall diameter	Internal radius of bend given by diameter of cable x factor from below	
				For single core cables installed in conduit, trunking, or ducting	Other installation methods
Copper or aluminium	Rubber or p.v.c. Circular or stranded	Non-armoured	Less than 10 mm Between 11mm & 25mm Greater than 25mm	2 3 6	3 4 6
		Armoured	Any		6
Solid aluminium or shaped copper	P.V.C.	Armoured or non-armoured	Any		8
Copper or aluminium	Impregnated paper	Lead sheath	Any		12
Copper or aluminium	Mineral	All types micc cable sheaths	Any		6

Note: Flat cables have the factor applied to the major axis, i.e., width of cable, not thickness.

Supports for conductors shall be installed at intervals, if the conductor is not continuously supported, to stop the conductor being damaged by its own weight.
522-08-04

Where a wiring system is installed on a structure subject to vibration, the cable and its connections shall be suitable for the conditions encountered.
522-07

Cable buried directly in the ground shall be of circular construction, incorporate a metal sheath, or armour, or both, marked either by cable covers or marking tape. The cable shall be buried at a depth sufficient to avoid any damage by any ground disturbance likely to occur.
522-06-03

Cables installed in vertical conduits shall be supported at 5 metre intervals to avoid mechanical stress and compression of the cables. Any cable installed in a vertical trunking or duct shall be supported at 5 metre intervals.

Unarmoured cables installed in vertical runs, where they are inaccessible and not likely to be disturbed, shall be supported at 5 metre intervals if sheathed in rubber or pvc and at 2 metre intervals if lead covered.

Armoured or sheathed cables installed horizontally in inaccessible positions, where they are unlikely to be disturbed, need not be fixed providing the surface on which they are installed is smooth.
INF

Cable installation
Temperature
Where cables are installed in a heated part of a building, or in a heated floor, the maximum operating temperature of the heated floor, or that part of the building in which the cables are installed, shall be taken as the ambient temperature for the cables.
522-01-01

The insulation of conductors connected to busbars shall be suitable for the operating temperature of the busbar.
523-03

Where cables enter accessories, appliances, or luminaires, they shall be suitable for the temperatures likely to be encountered, or shall have each conductor sleeved with insulation suitable for the temperature encountered (e.g. glass sleeving), so that there is no risk of short circuit between conductors or an earth fault.
522-02-02

Internal barriers shall be provided in every vertical channel, duct, ducting or trunking, at the ceiling of each floor, or at 5 metre intervals, whichever is the less, to limit the temperature rise.
INF

Fire barriers
Where cables pass through a wall or floor that is a fire barrier, the hole round the cable shall be sealed to the same degree of fire resistance as the wall or floor.
Where conduits, ducts, channels or trunking pass through a fire barrier, the interior of the conduit, duct, channel or trunking, shall also be sealed with suitable fire resistant material. Steel conduit, trunking or ducting having an internal area of 710 mm^2 need not be internally sealed.

Where a fire barrier is installed in a trunking, there is no need to fix an additional barrier at that point to stop the air becoming excessively hot.
527-02-01, 527-02-02

Eddy currents
Steel wire armoured, or steel tape armoured, single core cable shall not be used on a.c., supplies.
521-02-01

Metallic sheaths and/or non-magnetic armour of single core cable shall be bonded together at both ends of their run. Alternatively where the conductors exceed 50 mm^2, and have a an insulating sheath, they can be bonded at one point, and insulated at the unbonded ends; the length of such cables shall be limited so that when carrying full load current, the voltage from metallic sheath or armour to earth shall not exceed 25 V, and shall not cause corrosion or danger and damage when the cables are carrying fault currents.
523-05

All phase and neutral conductor circuits installed in ferrous enclosures on an a.c., supply shall be contained in the same enclosure, (conduit, trunking etc.). Conductors entering ferrous enclosures shall not be separated by ferrous material, or provision shall be made to prevent eddy currents, (i.e. cut a slot between individual holes and then braze with brass rod).
521-02

Bending radii
The internal radius of every bend in a non-flexible cable shall ensure that conductors and cables are not damaged.
522-08-03

Lift shafts
Only cables that form part of a lift installation are allowed to be installed in a lift or hoist shaft.
528-02-06

Cables installed in walls, floors and ceilings
Where cables are installed through timber joists in the ceiling or floor, they shall not be less than 50 mm from the top or bottom of the joist, unless mechanical protection that will stop the cable being penetrated by nails or screws etc., is provided (150 mm long lengths of conduit need not be bonded, 471-13-04). Alternatively, cables not protected by an earthed metallic sheath can be enclosed in an earthed steel conduit which is securely supported.
522-06-05

Where a cable is to be installed in a wall or partition at a depth of less than 50 mm, it shall be placed in an area within 150 mm from the top of the wall or partition, or within 150 mm of an angle formed by two walls or partitions, as illustrated by the shaded area in figure 1.
522-06-06

Where the cable feeds a point or accessory on the wall outside the above zones, the cable shall be installed in straight vertical or horizontal runs to the point or accessory, as illustrated by 'A' and 'C' in figure 1.
522-06-06

SELECTION AND ERECTION OF CABLES 131

Where the above requirements are not practicable, the cable shall be enclosed in earthed metallic conduit, trunking, ducting, or shall incorporate an earthed metallic covering, all complying with the regulations for protective conductors, as illustrated by 'B' in figure 1.
522-06-07

Note: For house wiring, check the NHBC Rules in which there are slight differences from 522-06-06.

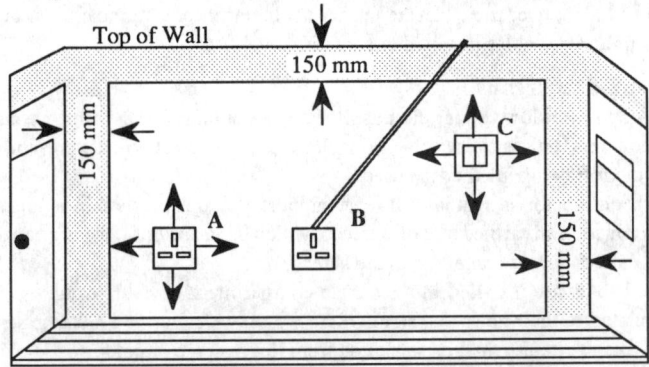

Figure 1 - Installation Zones

Other than protective conductors, non-sheathed cables shall be enclosed in conduit, trunking or ducting.
521-07-03

Cables shall be sheathed and/or armoured when installed in underground ducts, conduit, or pipes.
522-06-01

Cables shall be sheathed and/or armoured, or in conduit or other enclosure, giving suitable protection when installed on walls outdoors.
522-06-01

Cables buried directly in the ground shall have a metallic sheath and/or armour, be marked with cable covers or marking tape, and be buried at sufficient depth to avoid damage to the cable.
522-06-03

Sufficient protection shall be given to a wiring system buried in a floor, so that it is not damaged.
522-06-04

Cables installed overhead between buildings must comply with the Electricity Supply Regulations 1988 (Modified 1990).
412-05-01

Low voltage cables installed between buildings shall be placed beyond risk of mechanical damage.
522-06-01, 522-08-05

In areas inaccessible to vehicular traffic, cables may be installed in heavy gauge conduit (minimum size 20 mm without joint in span), or in other enclosure, giving suitable protection (See page 126). Cables and conductors shall be supported throughout their length, so that they and their terminations are not subjected to undue mechanical strain. This includes the weight of the cable or conductor.
522-08-05

Vermin and sunlight
Where cables are liable to attack by vermin, they shall be of a suitable type, or suitably protected. Cables exposed to direct sunlight shall be resistant to damage by ultra violet light, or shielded from the solar radiation.
522-10 , 522-11

Note: Insulated and sheathed pvc cables are alright since the inner cores are protected against direct sunlight, as are single core cables installed in conduit.

Category 1,2 & 3 circuits
Low voltage and extra-low voltage circuits shall not be contained in the same wiring system as high voltage conductors, unless every cable is insulated to the highest voltage present, and unless one or more of the following methods is used:
1. each conductor is insulated to the highest voltage present in a multicore cable, or contained in an earthed metallic screen, which has a current-carrying capacity equal to the largest conductor contained in the screen, or
2. The cables are installed in a separate compartment of cable trunking or ducting, and insulated to their normal system voltage, or have an earthed metallic screen.

Extra-low voltage circuits shall be separated from low voltage circuits.
528-01-01, 528-01-02

Where Category 1, 2 and 3 circuits are used in the same installation, precautions shall be taken to prevent electrical contact between the cables of the various categories.
528-01-03

Category 1 circuit- a circuit operating at low voltage, and supplied directly from a mains supply system.
Category 2 circuit- telecommunication, radio, telephone intruder alarm, bell and call, or data transmission circuits supplied at extra-low voltage.
Category 3 circuit- fire alarm or emergency lighting circuit.

Definitions
Category 3 circuits shall be segregated from Category 1 circuits. Category 3 fire alarm circuits shall be segregated from category 3 emergency lighting circuits. These requirements to be in accordance with BS 5266, BS 5839 and BS 6701 where appropriate.
528-01-04

Unless Category 2 circuits are insulated to the highest voltage present, or segregated from Category 1 circuits by a continuous partition, they shall not be installed with Category 1 circuits. If the partition is metallic the partition shall be earthed.
528-01-05

Unless a Category 3 circuit is contained in a mineral insulated cable which has a current rating for exposed to touch, Category 3 circuits shall be segregated from all other circuits by a continuous partition, when installed in a trunking, duct, ducting or a channel. The partitions shall remain continuous at outlet points, and shall be earthed, if metal.
528-01-06

Where multicore cables, flexible cables, or flexible cords contain both Category 1 and Category 2 conductors, the Category 2 conductors shall either be:
1. individually insulated to the highest voltage present, or
2. grouped and enclosed in an insulating sheath, or

3. grouped and enclosed in an earthed metallic screen having an equivalent current-carrying capacity to the cores of the Category 1 circuits.

528-01-08

Note: Care is needed if Category 1 and Category 2 circuits are installed in the same cable, conduit or trunking etc., since this could lead to a breach of Regulation Numbers 331-01-01 and 512-05-01.

INF

In no circumstance shall Category 1 and Category 3 circuits be contained in a common multicore cable, flexible cable, or flexible cord.

528-01-04

Joints and terminations

Where Categories 1 and 2 circuits are terminated at a common box, accessory, switchplate or box, the conductors and terminals of each category shall be effectively segregated by a partition. The partition shall be earthed if it is made from metal.

528-01-07

Joints shall be accessible for inspection unless they are compound filled, encapsulated, or made by welding, brazing, or compression lugs, and contained in suitable enclosures. Apart from the exceptions listed above, and the cold tail connection made between a cold tail and a heating element, joints shall be accessible for inspection.

526-04

Every cable termination or joint shall be:
1. electrically and mechanically sound,
2. protected against mechanical damage and vibration,
3. shall not cause harmful mechanical damage to the conductor, or place a strain on the fixings of the connection.

Mechanical clamps and compression lugs must retain all the wires of the conductor.

526-01

Terminations shall be of the correct size and type for the conductor to be used, and suitably insulated for the nominal voltage present. They must take into consideration the temperature attained by the conductor connections and its effect on conductor insulation. Where connections or joints are soldered, they must take into consideration mechanical stress, creep, and temperature rise under fault conditions.

526-02-01

Note: Where soldered connections are used, then the 'k' factor used in fault calculations shall be based on the soldered joint, if this is lower than the 'k' factor of the conductor.

INF

All terminations and joints in conductors shall be made in the appropriate enclosure manufactured to British Standards. Any enclosure used shall provide adequate mechanical protection against external influences.

526-03-02, 526-03-01

Contact with non-electrical services

Metal conduit, trunking, ducting and the metal sheaths or armour of cables operating at low voltage, or bare protective conductor which might come into contact with other metalwork, shall either be

134 SELECTION AND ERECTION OF CABLES

effectively bonded to the other metalwork, or effectively segregated from it, irrespective of any equipotential bonding carried out at the origin of the installation.
528-02-05

Protection shall be provided against indirect contact, and against the hazards that are likely to arise when a wiring system is located in close proximity to non-electrical services.
528-02-01

Unless a wiring system is suitably protected, and derated, where appropriate, it shall not be installed in the vicinity of a service which produces heat, smoke, fumes, or condensation likely to be detrimental to the wiring.
528-02-02, 528-02-03

Where non-electrical services and electrical services are installed close to each other, it shall be so arranged that work being carried out on either service will not cause damage to the other.
528-02-04

Moisture and corrosion

All cables likely to be exposed to weather, corrosive atmospheres, or other adverse conditions, are to be constructed or protected to prevent danger.
130-08-01

Wiring systems shall not be exposed to rain, water, or steam unless designed to withstand these items. When exposed to the weather, or in damp situations, all materials used shall be corrosion-resistant, enclosures shall be damp-proof, and every cable and joint shall be protected against the effects of moisture.
522-03-01, 528-02-03

Drainage points shall be provided in a wiring system where moisture may collect due to condensation. Drainage points shall be suitably located to allow the harmless escape of the condensation.
522-03-02

Protection against mechanical damage shall be provided if the wiring is subjected to waves.
522-03-03

Materials shall not be placed in contact with each other, where they are liable to initiate electrolytic action. For example, precautions shall be taken to maintain electrical continuity when aluminium conductors are connected to brass or to other metals with a high copper content.
522-05-02, 522-05-03

Wiring systems shall be suitably protected where they are subjected to corrosion and deterioration from corrosive or polluting substances.
522-05-01

Aluminium conductors shall not be used for connections to earth electrodes, or to water pipes subject to condensation.
542-03-02, 547-01

Hazardous areas

All cables in surroundings susceptible to risk of fire or explosion shall be constructed or protected to prevent danger.
130-08-02

SELECTION AND ERECTION OF CABLES

Where cables are installed in potentially explosive atmospheres, the additional precautions detailed in BS 5345 shall be taken.
110-01-01(xi)

Voltage drop

The voltage drop in the installation shall be such that the voltage at the terminals of fixed equipment is not less than that value required by the British Standard for such equipment.

Where no British Standard exists for the equipment, then the voltage drop within the installation must not exceed a value appropriate to the safe functioning of the equipment.

Where the supply to the premises is provided in accordance with the Electricity Supply Regulations 1988 (amended 1990), the above requirements are deemed to be complied with if the voltage dropped between the origin of the installation and the fixed equipment does not exceed 4% of the nominal supply voltage. With a 415/240 V supply, this is 9.6 V single-phase, and 16.6 V three-phase.
525-01-01, 525-01-02

When calculating voltage drop, starting currents or inrush currents can be ignored.
525-01-02

Cable colours

The colour combination green/yellow shall be used only for protective conductors. The single colour green or yellow alone or any bi-colour other than the combination green/yellow shall not be used in flexible cables or cords.
514-03, 514-07-02

Protective conductors shall be coloured green and yellow. The phase conductors of multi-phase a.c., circuits shall be red, yellow and blue, with the neutral coloured black. The phase conductor of single-phase a.c., circuits shall be red and the neutral black.

For d.c., supplies, the positive of a two wire circuit will be red and the negative black. The positive of a d.c., three wire circuit will be red, the middle wire black and the negative blue. The functional earth for telecommunications is cream.
Table 51A

Every single core and multicore non-flexible cable conductor shall be identified at its terminations, and preferably throughout its length, red, yellow or blue indicating the phase conductors and black the neutral conductor. Paper-insulated cables, pvc insulated multicore auxiliary cables, or thermosetting insulated multicore cables can have cores numbered in accordance with British Standards, providing that 1, 2 & 3 are used for phase conductors, and 0 for the neutral conductor. Micc cables shall be identified at their terminations by tapes, sleeves or discs of the appropriate colour.
514-06-01, Table 51A

Red yellow and blue may be used as phase colours for cables up to final circuit distribution boards, but single-phase circuits from a final distribution board shall be coloured red and black.
Table 51A

Bare conductors and switchboard busbars shall, where necessary, be identified by the application of tape, sleeves or disks in accordance with Table 51 A, as specified above.
514-06-03, 514-06-04

SELECTION AND ERECTION OF CABLES

Flexible cords
Every core of a flexible cable or cord shall be identified throughout its length, Brown for phase, Blue for neutral, Green/yellow for the protective conductor. (See cable colours.)
514-07-01

Excessive tensile and torsional stresses to conductors and connections shall be avoided in a flexible wiring system installation.
522-08-06

Flexible cords shall not be used for fixed wiring unless the installation complies with the requirements of the regulations, such as complying with all the requirements for a fixed wiring installation, as well as with the additional requirements for flexible conductors.
521-06

Portable equipment shall be connected with flexible cable or cords of a length that will avoid mechanical damage. Cookers rated over 3 kW are not considered portable.
INF

Where a flexible cord supports, or helps to support a luminaire, the maximum weight that can be placed on the cord is as follows:

Conductor size mm^2	Maximum weight
0.5	2 kg
0.75	3 kg.
1.0	5 kg.
1.5	5 kg.
2.5	5 kg.
4.0	5 kg.

Table 4H3A

31. Selection and erection of equipment

General
To prevent danger, proper materials shall be used; in addition, equipment shall be properly constructed, installed and protected, and shall be capable of being maintained, inspected and tested. Equipment must also be suitable for the maximum power it is required to handle.
130-01, 130-02-01, 130-02-02

Standard of equipment
All equipment shall comply with the current edition of the applicable British Standard or CENELEC H.D..
511-01-01

Where a British Standard takes into account a CENELEC Harmonization Document, then equipment manufactured to a foreign standard based on the same Harmonization Document can be used, providing it is no less safe than if manufactured in accordance with the relevant British Standard.
The same comments apply to equipment manufactured to an IEC Standard.
511-01-01, Preface

Location of equipment
Equipment shall be located so that any heat generated by the equipment will be dissipated, and not present a fire hazard to the adjacent building materials, or to materials likely to be placed in the same location.
422-01-01

Where electrical equipment has a surface operating temperature that could cause a risk of fire could be harmful to adjacent materials, or could produce harmful effects, one of the following methods of installation shall be utilised:
1. the equipment shall be mounted on a support, or placed in an enclosure which is capable of withstanding the temperature generated by the equipment, without being set on fire, or creating a harmful effect. Any support used shall limit the transmission of heat from the equipment by having a low thermal conductance, or
2. the equipment is screened by material which will withstand and not be affected by the heat generated by the equipment, or
3. the equipment is fixed so as to allow the safe dissipation of heat, and at such a distance from other materials as to avoid fire and harmful effects.

422-01-02

Provision shall be made for portable appliances and luminaires to be fed from an accessible and convenient adjacent socket.
553-01-07

Where equipment in one location contains more than 25 litres of flammable liquid, precautions shall be taken to contain the liquid, and to stop burning liquid or the gases of combustion spreading to other parts of the building.
422-01-05

Service from the most comprehensive electrical supply and distribution system in the United Kingdom

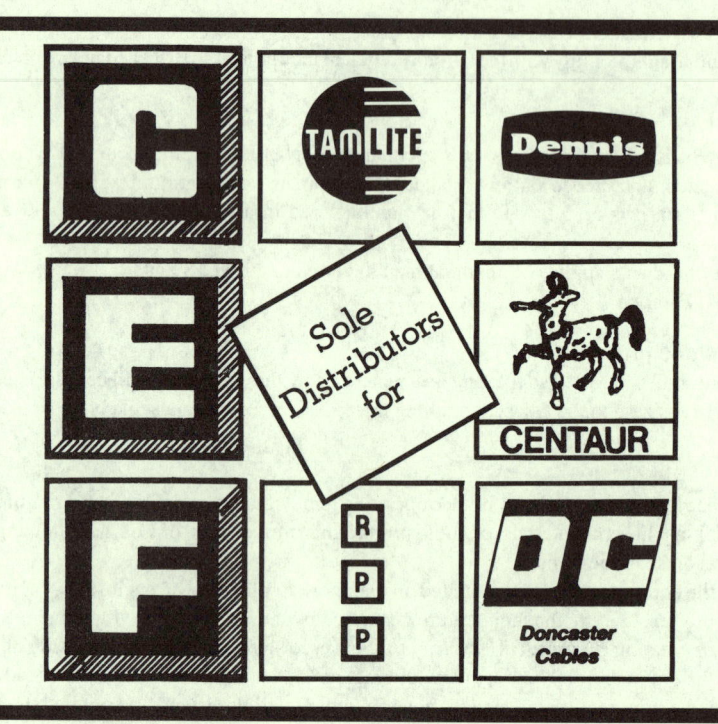

TAMLITE

Dennis

CEF

Sole Distributors for

CENTAUR

RPP

DC Doncaster Cables

switch on to

city electrical factors ltd

Over 280 branches nationwide

Enclosures for cables shall be suitable for the extremes of ambient temperature to which they will be exposed. Where a non-metallic or plastics box is used for suspension of a luminaire, it shall be suitable for the suspended weight and the expected temperature when the luminaire is working.
522-01-01, 522-01-02, 422-01-01

A means of switching off shall be provided for equipment having electrically heated surfaces that can be touched.
476-01-02

Suitability of supply
All equipment shall be suitable for the nominal voltage of the supply, the design current, and the current likely to flow under short-circuit, overload, or earth fault conditions.
512-01, 512-02

Equipment designed to work at a given frequency shall match the frequency of the supply, (e.g., frequency of contactor coils), and the power demanded by the circuit. (e.g., h.p. of motor suitable for the power output required).
512-03, 512-04

Compatibility and external influences
Equipment installed shall not cause harmful effects on other equipment, or on the supply, and shall be suitable for the environmental conditions likely to be encountered.

Where the equipment is not suitable for the environmental conditions, it shall be provided during erection, with suitable protection, which must not interfere with the equipment's operation.

The degree of protection that is given shall take into account the simultaneous occurrence of different external influences.
512-05, 512-06-01,
512-06-02, 331-01

Examples of the characteristics likely to have a harmful effect on other electrical equipment or services include:
1. transient overvoltages,
2. rapidly fluctuating loads,
3. starting currents,
4. harmonic currents from fluorescent lighting loads and thyristor drives,
5. surge currents of transformers,
6. mutual inductance,
7. d.c. feedback,
8. high frequency oscillations,
9. earth leakage currents,
10. additional connections to earth for equipment in order to avoid interference with its operation.

INF

Where equipment that is installed is likely to affect the supply, the supplier of electricity shall be consulted
331-01

Accessibility and maintainability

With the exception of joints that are allowed to be inaccessible, all equipment shall be accessible with sufficient working space to enable safe operation, maintenance, inspection, and testing to be carried out.
130-07, 513-01,529-01-02

The quality and frequency of maintenance shall be assessed, so that the reliability of equipment chosen is appropriate to its intended life: so that the protective measures for safety remain effective during the intended life of the installation and so that any periodic maintenance or inspection and testing can be carried out safely.
341-01

Voltage drop

The voltage drop in the installation shall be such that the voltage at the terminals of fixed equipment is not less than that value required by the British Standard for such equipment.
Where no British Standard exists for the equipment, then the voltage drop within the installation must not exceed a value appropriate to the safe functioning of the equipment.

Where the supply to the premises is provided in accordance with the Electricity Supply Regulations 1988 (amended 1990), the above requirements are deemed to be complied with if the voltage dropped between the origin of the installation and the fixed equipment does not exceed 4% of the nominal supply voltage. With a 415/240 V supply, this is 9.6 V single-phase, and 16.6 V three-phase.
525-01-01, 525-01-02

When calculating voltage drop, starting currents or inrush currents can be ignored.
525-01-02

Devices for protection against undervoltage shall comply with the following regulations.
535-01

Where a reduction in voltage, or loss and subsequent restoration of the supply could cause danger, suitable precautions shall be taken, such as provision of No-Volt or undervoltage releases.
451-01-01

Every motor shall be provided with a means of preventing the automatic restarting of the motor if there is a drop in voltage, or if the supply fails, and if the unexpected automatic restarting would cause danger. This does not apply to those instances where there is a brief interruption of the supply, and where failure of the motor to restart would cause a greater danger. In this case other safety precautions would be required.
552-01-03

Where damage to equipment or to the installation may be caused by a drop in voltage, and where the drop in voltage will not cause danger, then one of the following arrangements shall be chosen:
1. precautions against any foreseen damage shall be taken, or
2. it shall be confirmed with those responsible for the operation and maintenance of the installation that the foreseen damage is an acceptable risk.

451-01-02

Providing equipment can withstand a brief reduction or loss of voltage without danger, then a suitable time delay may be incorporated in the operation of an undervoltage protective device controlling the equipment.

The instantaneous disconnection by control or protective devices must not be affected by any delay in the opening or closing of contactors, and the characteristics of the undervoltage protective device shall be compatible with the requirements for the starting and use of the associated equipment.
451-01-03, 451-01-04, 451-01-05

The reclosure of a protective device shall not be automatic if it is likely to cause danger.
451-01-06

Moisture and corrosion

Wiring systems shall not be exposed to rain, water, or steam, unless suitably designed to withstand these. When exposed to the weather, or in damp situations, all materials used shall be corrosion-resistant, enclosures shall be damp-proof, and every cable and joint shall be protected against the effects of moisture.
522-03-01, 522-05-01

Precautions are to be taken to maintain electrical continuity when aluminium conductors are connected to brass or other metals with a high copper content, or to any other dissimilar metals that are in contact with each other.
522-05-03

Where metalwork is installed in a corrosive atmosphere, it shall be protected to withstand corrosion. Non-metallic materials shall not be placed in contact with materials that will cause chemical deterioration of the wiring.
522-05-01

Provision shall be made for the harmless escape of any water which may accumulate in a wiring system due to condensation. Where the equipment is subject to waves, mechanical protection shall be provided.
522-03-01, 522-03-02

Contact with non-electrical services

Metal conduit, trunking, ducting, and the metal sheaths or armour of cables operating at low voltage, or bare protective conductor which might come into contact with other metalwork, shall either be effectively bonded to the other metalwork, or effectively segregated from it, irrespective of any equipotential bonding carried out at the origin of the installation.
528-02-05

Protection shall be given against indirect contact, and against the hazards that are likely to arise when a wiring system is located in close proximity to non-electrical services.
528-02-01

Unless a wiring system is suitably protected, and derated where appropriate, it shall not be installed in the vicinity of a service which produces heat, smoke, fumes, or condensation likely to be detrimental to the wiring.
528-02-02, 528-02-03

Where non-electrical services and electrical services are installed close to each other, it shall be so arranged that work being carried out on either service will not cause damage to the other.
528-02-04

Careful selection is needed to avoid any harmful influence between the electrical installation and other services; additionally, where there is likely to be any detrimental influence on the equipment of different services, those services shall be effectively segregated from the electrical installation.
515-01

Joints and terminations
Where Categories 1 and 2 circuits are terminated at a common box, accessory, switchplate or box, the conductors and terminals of each category shall be effectively segregated by a partition. The partition shall be earthed if it is made from metal.
528-01-07

Joints shall be accessible for inspection unless they are compound filled, encapsulated, or made by welding, brazing or compression lugs, and contained in suitable enclosure. Apart from the exceptions listed above and the cold tail connection made between a cold tail and a heating element, joints shall be accessible for inspection.
526-04

Every cable termination or joint shall be:
1. electrically and mechanically sound,
2. protected against mechanical damage and vibration,
3. shall not cause harmful mechanical damage to the conductor, or place a strain on the of the connection.

Mechanical clamps and compression lugs must retain all the wires of the conductor.
526-01

Terminations shall be of the correct size and type for the conductor to be used, and suitably insulated for the nominal voltage present. They must take into consideration the temperature attained by the conductor connections, and its affect on conductor insulation. Where connections or joints are soldered, they must take into consideration mechanical stress, creep, and temperature rise under fault conditions.
526-02-01

Note: Where soldered connections are used, then the 'k' factor used in fault calculations shall be based on the soldered joint, if this is lower than the 'k' factor of the conductor.
INF

All terminations and joints in conductors shall be made in the appropriate enclosure manufactured to British Standards. Any enclosure used shall provide adequate mechanical protection against external influences.
526-03-02, 526-03-01

Agricultural installations
In agricultural buildings and areas accessible to livestock, electrical equipment shall be of class II construction, or protected by a suitable insulating material, and suitable for all the environmental conditions likely to occur.
605-11

Where cables are liable to attack by vermin, they shall either be of a suitable type, or suitably protected.
522-10

Outside installations

Where cables are installed in positions where they may be exposed to direct sunlight, they shall be of a type that is resistant to ultra-violet light.
522-11

Note: Insulated and sheathed pvc cables are alright since the inner cores are protected against direct sunlight, as are single core cables installed in conduit.

Dusty atmospheres

Enclosures for conductors, joints, and terminations shall be protected against the ingress of foreign bodies, protection being given in accordance with the IP codes.

Where the dust conditions are onerous, additional precautions shall be taken to stop accumulated dust reducing the heat dissipated from the wiring system.
522-04-01, 522-04-02

Motors

Where motors are subject to frequent starting, the effect of the temperature rise on equipment and conductors caused by the starting current shall be taken into account.
552-01-01

The circuits supplying the rotors of slip-ring motors, or the rotors of commutator induction motors, shall be suitable for the starting and load conditions.
INF

Motors exceeding 0.37 kW shall be provided with overloads.
552-01-02

Every motor shall be provided with a means of preventing the automatic restarting of the motor if there is a drop in voltage, or if the supply fails, and if the unexpected automatic restarting would cause danger. This does not apply to those instances where there is a brief interruption of the supply, and where failure of the motor to restart would cause a greater danger. In this case other safety precautions would be required.
552-01-03

Every motor or electrically powered equipment shall be provided with a means of switching off which shall be accessible, easily operated, and so placed so as to prevent danger. (i.e., not having to lean across moving parts to operate the switch.).
130-06-02, 476-01-02

Transformers

The common terminal of an autotransformer connected to a circuit having a neutral conductor shall be connected to the neutral conductor.
551-01-01

A linked switch shall be provided for disconnecting a step-up transformer from all live conductors of the supply. Additionally, a step up transformer shall not be used in an IT system.
551-01-03, 551-01-02

Switchboards and switchgear

The current-carrying capacity and the temperature limits of busbars, busbar connections, and bare conductors which form part of a switchboard, shall comply with BS 5486.
INF

The nominal current rating of low voltage equipment such as switchgear, protective devices, and accessories is normally based on the conductors connected to the equipment not exceeding a working temperature of 70 °C. Where the operating temperature of conductors exceeds 70 °C, the current rating of equipment may need to be reduced by an amount specified by the manufacturer of the equipment.
512-02

Discharge lighting

Every switch used to control discharge lighting shall be designed in accordance with BS 3676. Alternatively, its nominal rating shall be twice the total steady current of the circuit. Where used for both discharge lighting and tungsten lighting, its nominal rating shall be twice the total steady current of the discharge lighting plus the total current of the tungsten lighting.
553-04-01, INF

Final circuits for discharge lighting shall be capable of carrying the total steady current, harmonic currents, and any fault current likely to flow..
512-02

Socket outlets

With the exception of administration buildings such as site offices and toilets, every socket outlet used on a construction site shall comply with BS 4343.
604-12-02

Socket outlets mounted on a vertical wall shall be at a height sufficient to minimise the risk of mechanical damage to the socket outlet, plug, or cable, during insertion and withdrawal of the plug.
553-01-06

With the exception of 12 V SELV circuits and shaver sockets complying with BS 3535, there shall be no socket outlets or provision for connecting portable equipment in a room containing a fixed bath or shower.
601-10-02

Where socket outlets rated up to 32 A are fitted in an installation, and where it is possible that they may be used to supply portable equipment for use outdoors, it shall be protected by a 30 mA protective device which will disconnect a 150 mA fault within 40 msec.
471-16-01

Ceiling roses

Ceiling roses shall not be used in circuits exceeding 250 volts, nor shall they have more than one outgoing flexible cord, unless designed for multiple pendants.
553-04-02, 553-04-03

Immersion heaters

Every water heater shall be provided with a thermostat.
554-04

It is recommended that where immersion heaters are fitted to storage vessels in excess of 15 litres capacity, they shall be supplied from their own separate circuit.
INF

Instantaneous water heaters

The metal water pipe through which the water supply to the heater is provided shall be solidly and mechanically connected to all metal parts (other than live parts) of the heater or boiler. There must also be an effective electrical connection between the metal water pipe and the main earthing terminal, independent of the circuit protective conductor.
554-05-02

The supply to the heater shall be through a double pole linked switch. The switch shall be either separate from, and within easy reach of the heater, or incorporated in the heater. The wiring from the heater to the switch shall be direct, without using a plug and socket outlet.

Where the heater is installed in a room containing a fixed bath or shower, the switch shall be out of reach of a person using the bath or shower: an insulated cord operated pull switch would be acceptable.
554-05-03

The installer shall check or confirm that no single pole switch, fuse, or non-linked circuit breaker is installed in the heater circuit between the origin of the installation and the heater, before the heater is connected.
554-05-04

Cables

All cables and conductors shall either be suitably constructed, or mechanically protected to withstand any mechanical damage to which they may be subjected in normal conditions of service
522-06-01

Lampholders and luminaires

The maximum voltage allowed for a filament lamp lampholder is 250 V.

BC and SBC lampholders shall comply with, and have the temperature rating T2 specified in BS 5042.

The outer contact of ES lampholders shall be connected to the neutral in TN and TT systems, as will lampholders mounted on a track system.
553-03-02, 553-03-03, 553-03-04

Luminaire supporting couplers shall not be used for the support or connection of other equipment.

The weight of a lighting fitting suspended from a pendant luminaire shall not damage the flexible cable or its connections.

Where an extra-low voltage luminaire has no facilities for connecting a protective conductor, it shall only be used as part of a SELV system
553-04-04, 554-01

Building structures

The cable support and protection system employed where a building is likely to have structural movement shall allow for the movement of conductors, so that they are not damaged.

Where the building structutre is flexible or unstable, a flexible wiring system shall be installed.
522-12

32. Socket outlet circuits

General
Every installation shall be divided into circuits to avoid danger in the event of a fault, and to facilitate safe operation, inspection, testing, and maintenance.
314-01-01

The installation shall be broken down into circuits, such that the number of final circuits installed, and the number of points connected to each final circuit will ensure compliance with the regulations for current-carrying capacity of conductors, overcurrent protection, isolation, and switching.
314-01-03

Where more than one final circuit is installed, each circuit shall be connected to a separate way in the distribution board, and shall be electrically separate from every other final circuit, e.g., neutrals must not be borrowed from other circuits. This is to prevent an isolated circuit from being indirectly energised.
314-01-04

BS 1363 13 A socket outlets
General requirements
A ring or radial circuit, with or without spurs, can feed an unlimited number of socket outlets, or feed permanently connected equipment, providing any immersion heater fitted to a storage vessel having a capacity exceeding 15 litres, or permanently connected heating appliances used as part of a space heating installation, are supplied from their own separate circuits.

The area covered by a circuit shall be determined by the known or estimated load, which must not exceed the nominal rating of the circuit protective device, and the floor area covered shall not exceed the area given in Table 1.

The number of socket outlets shall be such that every portable luminaire and appliance can be fed from an adjacent and conveniently accessible socket outlet.
INF, 553-01-07

Table 1 already takes into account diversity within the circuit, so no further diversity shall be applied. Where more than one final ring circuit is installed, the load shall be distributed among them. The cable sizes given in Table 1 may need to be increased if the ambient temperature is likely to exceed 30 °C, or if more than two circuits are bunched or grouped together.

Determination of cable size
When more than two ring circuits are grouped together, or when the ambient temperature exceeds 30 °C, or when the conductor is totally enclosed in thermal insulation, the cable size required has to be calculated. The cable size is determined by calculating the current-carrying capacity required for the conductors from the following formula:

$$I_t = \frac{I_n \times 0.67}{G \times A \times T}$$

SOCKET OUTLET CIRCUITS

Only the correction factor applicable is placed in the formula. No derating factor is used if the cables are in contact on one side only with thermal insulation, since current-carrying capacity tables are provided for this condition. No further diversity is allowed, since this has already been taken into account in multiplying I_n by 0.67.

Spurs off the ring
The total number of fused spurs allowed is unlimited, but the following minimum cable sizes shall be used if a fused spur feeds socket outlets.
  1.5 mm^2 for pvc insulated copper conductor cables.
  1 mm^2 for micc copper conductor cable.

Fused spurs are connected to the ring circuit through a fused connection unit, the fuse of which must not exceed 13 A, or the rating of the cable taken from the fused spur.

The number of non-fused spurs must not exceed the total number of socket outlets and stationary equipment connected directly in the ring circuit.

A non-fused spur can only feed one single or one twin socket outlet, or one permanently connected item of equipment.

The supply for a non-fused spur can be taken from the origin of the circuit in the distribution board, or from a junction box, or from socket outlet connected directly to the ring circuit, the conductors used being the same size as the ring circuit.

Permanently connected equipment shall be controlled by a switch or switches, separate from the equipment, in an accessible position, and protected by a fuse not exceeding 13 A, or a circuit breaker not exceeding 16 A.
INF

BS 4343 16 A socket outlets
The requirements can be summarised in a list:
  1. the circuit shall be a radial circuit,
  2. the protective device for the circuit shall be 20 A, and the maximum demand for the circuit, having allowed for diversity, must not exceed 20 A,
  3. the number of socket outlets is unlimited,
  4. the circuit conductor size is determined in the same way as for other circuits, the conductor size being obtained from the tables in Appendix 9, after derating factors for grouping, ambient temperature, contact with thermal insulation, and the type of protective device used, if applicable, have been taken into account,
  5. the current-carrying capacity of the cable shall be not less than the nominal rating of the overcurrent protective device protecting the circuit,
  6. the current rating of socket outlets used shall be 16 A, and of a type appropriate to the number of phases, circuit voltage, and earthing arrangement applicable.
INF

BS 196 socket outlets
The requirements are listed as follows:
  1. the circuit can be a radial circuit or a ring circuit with spurs,
  2. the number of socket outlets is unlimited,
  3. the protective device shall have a nominal rating not exceeding 32 A, and the maximum demand for the circuit after diversity has been taken into account must not exceed the nominal rating of the protective device, or 32 amps.

4. no diversity is to be allowed for permanently connected equipment. Such equipment shall be assumed to be working continuously,
5. the total current taken by a fused spur must not exceed 16 A,
6. the circuit conductor size is determined in the same way as for other circuits, the conductor size being obtained from the tables in Part 3, after derating factors for grouping, ambient temperature, contact with thermal insulation, and the type of protective device used, if applicable, have been taken into account,
7. the current-carrying capacity of the cable shall be such that it is not less than 0.67 times the rating of the overcurrent protective device for ring circuits, or not less than the nominal rating of the overcurrent protective device protecting the circuit for radial circuits,
8. the current-carrying capacity of the conductor for a fused spur is determined by the total load on the spur, which load, in any event, must not exceed 16 A,
9. non-fused spurs are not allowed, and a fused spur is to be connected to a circuit through a fused connection unit,
10. the rating of the fuse in the fused spur unit must not exceed 16 A or the current rating of the cable used for the spur,
11. permanently connected equipment shall be locally protected by fuse or circuit breaker whose rating does not exceed 16 A, and controlled by a switch or switches, separate from the equipment, in an accessible position,
12. only 2-pole-and-earth contact plugs, with single pole fusing on the live pole, are to be used in a circuit where one pole of the supply is earthed,
13. only 2-pole-and-earth contact plugs with double pole fusing are to be used if the circuit has neither pole earthed: e.g. the supply from a double wound transformer having the mid point of its secondary winding earthed.

See paragraphs on SELV, FELV, and reduced voltage systems.

Shock protection when U_0 is 240 V

Where the socket outlet is to be used for hand held equipment indoors, and where the supply voltage to the socket outlet circuit is 240 V, the regulation for protection against indirect contact can be satisfied if the circuit disconnection by the protective device is achieved in 0.4 seconds when an earth fault occurs. Where the socket outlet is used for hand held equipment outdoors, the circuit shall be protected by a 30 mA RCD.
413-02-08, 413-02-09

Where a socket outlet feeds an item of fixed equipment indoors, and where it can be guaranteed that the socket outlet will not be used for hand-held equipment, then the disconnection time can be 5 seconds. Where the socket outlet is installed outdoors, the disconnection time shall be 0.4 seconds.
413-02-09, 413-02-13

When the supply voltage is 110 volts between phases, and 55 volts to the earthed midpoint for single-phase supplies, or 63.5 volts to the earthed neutral point for three-phase supplies, protection against indirect contact is provided if the circuit protective device disconnects the supply in 5 seconds when an earth fault occurs.
471-15-06

Where the voltage to earth is 240V r.m.s., a.c., and protection is given by an overcurrent protective device, the disconnection times of 0.4 or 5 seconds will be achieved if the phase earth loop

SOCKET OUTLET CIRCUITS

impedance at every outlet on the circuit complies with the appropriate Tables ZS1 to ZS 7 in Part 3, or Table 41 B1/B2 and 41D in the regulations.
413-02-10

Except when the voltage is 240 V, where the voltage to earth is in the range 220 V to 277 V, the impedance values of Tables 41B1/B2, and 41D, and ZS1 to ZS 6 shall be adjusted by multiplying the Z_s value from the table by the actual voltage to earth, and dividing the result by 240V.
INF

The phase earth loop impedance values given in Tables ZS 1 to ZS 7 are applicable for conventionally normal body resistance. Where the body resistance is expected to be reduced, the values of Z_S shall be reduced appropriately.
471-08-01

Where RCDs are used to comply with the disconnection times of 0.4 or 5 seconds, then the residual operating current in amps multiplied by Z_s in ohms must not exceed 50 V.
413-02-16

Bathrooms

Unless the circuit is derived from a 12 V SELV safety source or a shaver socket detailed below, socket outlets shall not be installed in a room containing a fixed bath or shower, and there shall be no provision for connecting portable equipment.

Where a shower is installed in a room that is not a bathroom, sockets shall be at least 2.5 metres from the shower cubicle.

Providing a shaver socket has been manufactured to BS 3535: Part 1, it can be installed in a bathroom for the use of electric shavers. In this case, the protective conductor of the final circuit feeding the shaver socket shall be connected to the earth terminal of the shaver socket.
601-09-01, 601-10-01, 601-10-02, 601-10-03

Providing socket outlets are part of a SELV circuit which is derived from a safety source with a nominal voltage not exceeding 12 V, they can be installed in a room containing a fixed bath or shower.

The safety source for the SELV circuit shall be out of reach of a person using the bath or shower, the socket outlets must not have any accessible metal parts, and shall be insulated or protected with barriers or enclosures against direct contact.
601-03-01, 601-10-01

Equipment not allowed

No electrical equipment shall be installed in the interior of a bath tub or shower basin.
601-02-01

Equipment to be used outside

Unless the circuit is a SELV circuit or a reduced low-voltage circuit, where socket outlets rated up to 32 A are fitted in an installation, and where it is possible that they may be used to supply portable equipment for use outdoors, they shall be protected by a 30 mA RCD, having an operating time not exceeding 40 msec, when subjected to a residual current of 150 mA.
471-16-01, 471-08-04, 412-06-02

SOCKET OUTLET CIRCUITS

SELV systems
The plugs used on SELV circuits shall be incapable of insertion into any socket outlet of another voltage in the same premises. The socket outlets must also prevent plugs of other voltages from being inserted into them, and shall not have a protective conductor connection.
411-02-10

FELV systems
The socket outlets used in the same premises for functional extra-low voltage shall not accept plugs used for other voltage systems.
471-14-06

Reduced voltage systems
A protective conductor contact shall be provided in plugs, sockets, and cable couplers used on a reduced voltage system. Additionally, they shall not be interchangeable with plugs, sockets or cable couplers of other voltages used in the same installation.
471-15-07

Protective conductors
Except for socket outlets connected to a SELV circuit, a protective conductor shall be provided at every socket outlet.
471-08-08

The earthing terminal of each socket outlet shall be connected by a separate protective conductor to an earthing terminal in the associated box or enclosure when the protective conductor is formed by conduit, trunking, ducting, or the metal sheath and/or armour of a cable.
543-102-07

Switching
Plugs and socket outlets shall not be selected for emergency switching.
537-04-02

Other than for emergency switching, plugs and socket outlets up to 16 amps can be used for switching; if the socket outlet exceeds 16 amps, it can be used to switch a.c., circuits, providing it has the appropriate breaking capacity for the use intended.
537-05-01, 537-05-02

Caravan site installations.
The electrical supply point to a caravan pitch shall be located not more than 20 m from any point which it is serving on the pitch.
608-13-01

The caravan supply point shall contain an IPX4 BS 4343 16 A socket outlet, with its key position at 6h. The socket shall be placed at a height between 0.8 m and 1.5 m from the ground. There shall be not less than one socket outlet for each caravan pitch. Socket oultets with a higher current rating shall be provided where the demand of the caravan pitch exceeds 16 A.
608-13-02, 608-13-03

Each socket outlet shall be of the two pole and earthing contact type, and shall be protected individually, or in groups of not more than six sockets, by a 30 mA RCD complying with BS 4293.

Each socket shall be protected individually by an overcurrent device. The protective conductor must not be bonded to the PME earth terminal.

Where the supply is PME, each RCD shall be connected to an earth electrode, and the installation shall be considered to be a TT installation; and shall comply with the regulations for TT systems for protection against indirect contact.
608-13-02, 608-13-03,

608-13-04, 608-13-05

Socket outlets shall incorporate an earthing contact when they are not supplied by an individual winding on an isolation transformer.
608-08-02

Every plug and socket outlet used to connect a caravan to a caravan site installation, shall be of the two pole and earthing contact type to BS 4343, with the key position at 6h.
608-07-01

TT installations
Where automatic disconnection is used for protection against indirect contact, every socket outlet shall be protected by a 30 mA RCD, which will trip in 40 msec at 150 mA residual current.
471-08-06

Requirements for plug and socket outlets
With the exception of SELV sockets, every plug and socket outlet shall be of the non-reversible type, with provision for connecting a protective conductor. It must not be possible for any pin of a plug to make contact with live parts, whilst any other pin is completely exposed, nor shall it be possible for the pin of a plug to be engaged with a live contact in any socket outlet which is not of the same type as the plug top. A plug top which contains a fuse shall be non reversible.
553-01-01, 553-01-02

The position of sockets mounted on a vertical wall or surface shall minimise the risk of damage to the socket outlet, plug, or the flexible cable, whilst the plug is being inserted or withdrawn.
553-01-06

With the exception of toilets and offices, every socket outlet and cable coupler used on a construction site shall comply with BS 4343.
604-12-02

Plug and socket outlets shall conform to the appropriate British Standard.
553-01-03

Plug and socket outlets shall not be used for instantaneous water heaters.
554-05-03

In domestic installations, plug and socket outlets shall be shuttered, and should preferably comply with BS 1363.
553-01-04

In single-phase a.c. , or two wire d.c circuits, a plug and socket may be used, even though it does not comply with BS specifications 1363, 546, 196 or 4343, providing the nominal operating voltage does not exceed 250 V, and the socket is used for the following purposes:
  1. special circuits used to distinguish their function, or circuits having special characteristics

in which the socket is used to prevent danger, or
2. shaver sockets incorporated into shaver supply units or lighting fittings complying with BS 3535 or, of a type complying with BS 4573, when used in a room other than a bath or shower room, or
3. electric clock sockets designed specifically for that purpose, and incorporating a cartridge fuse not exceeding 3 A in the plug top, which complies with BS 646 or BS 1362.

553-01-05

A conveniently accessible socket outlet shall be provided where portable equipment is likely to be used, taking into consideration that the length of flex on portable equipment is usually only 1.5 m long.

553-01-07

Cable couplers

With the exception of Class II equipment, or SELV equipment, cable couplers shall be non-reversible, and shall be provided with the a connection for a protective conductor. The cable coupler shall also be arranged so that the shrouded pins of the coupler are connected to the supply, the exposed pins being connected to the equipment to be used.

553-02

High earth leakage current equipment

Where socket outlets are used for equipment having a high leakage current, the socket outlets and the protective conductors installed to them shall comply with the Data Transmission regulations given in an earlier chapter.

TABLE 1

Cable sizes and maximum floor area allowed for BS 1363 socket outlet circuits

Type of circuit	Overcurrent device allowed		Minimum conductor size in mm^2		Maximum floor area in
	Rating	Type	PVC	MICC	square metres
Ring	30/32 A	Any	2.5	1.5	100
Radial	30/32 A	Cartridge fuse or circuit breaker	4	2.5	50
Radial	20 A	Any	2.5	1.5	20

33. Switching

General
A main linked switch or circuit breaker shall be provided for every installation. This circuit breaker, which shall be installed as close to the origin as possible, shall be capable of interrupting the supply on load. The same device must also be capable of being used for the isolation of the supply, by complying with the regulations for isolation.

A main switch or circuit breaker at the origin of the installation shall switch the following conductors of the incoming supply:
1. both live conductors when the supply is single-phase a.c.,
2. all poles of a d.c., supply,
3. all phase conductors in a T.P., or T.P. & N., TN-S, or TN-C-S system supply,
4. all live conductors in a T.P., or T.P.& N., TT, or IT system supply.

460-01-02, 461-01-01, 461-01-03

Where the electricity supplier provides at the origin of the installation, switchgear that complies with the regulations and gives permission for it to be used to isolate the electrical installation from the origin up to the main switch of the installation, the requirements of 460-01-02 above are complied with.
476-01-01

One switching device is allowed to isolate a group of circuits.
461-01-01, 476-01-02

Where more than one source of supply feeds an installation, each source shall be provided with a main switch, together with a warning notice advising that all switches have to be operated to switch off the installation. Alternatively, an interlock system can be used instead of the warning notice.
460-01-02

A switch shall be provided for every circuit and final circuit, and shall be capable of interrupting the circuit on-load, and under any foreseeable fault conditions.
314-01-02, 476-01-02

Where it is necessary for safety reasons, switches shall be provided to enable circuits, or parts of the installation, to be switched independently from each other.
476-01-02, 314-01-02

Where the regulations require all the live conductors to be disconnected, it must not be possible to disconnect the neutral before the phase conductors are disconnected. It is recommended that the neutral be re-connected at the same time as the phase conductors, or before the phase conductors are connected.
530-01-01

An isolator or switch shall be provided to cut off the supply from every installation, every item of equipment or circuit, as may be necessary to prevent danger. The device provided shall be suitably placed for ready operation.
130-06-01

Single pole switches
Single pole switches shall only be inserted in the phase conductor.
130-05-01, 530-01-02

Appliances
A means of interrupting the supply on load shall be provided for every appliance or luminaire not connected to the supply by a plug and socket outlet complying with Regulation 537-05-02. The means of interrupting the supply shall be in a readily accessible position, and so placed that the operator is not put in danger.
476-03-04

The means of interrupting the supply to an appliance or luminaire may be mounted on, or incorporated in, the appliance or luminaire.
476-03-04

Heating and cooking appliances
A means of interrupting the supply on load shall be provided for every appliance which may give rise to a hazard in normal use. The device provided shall be so placed in an accessible position, that it can be operated without putting the operator in danger.
476-03-04

One means of interrupting the supply may be used where two or more appliances are installed in the same room .
476-03-04

Bathrooms
Switches and control devices shall be mounted so that they are inaccessible to a person using a fixed bath or shower.

The only switches or controls that are allowed to be accessible in a room with a fixed bath or shower are: shaver sockets manufactured in accordance with BS 3535, the switches or controls of instantaneous water heaters manufactured to BS 3456, insulating cords of cord-operated switches which comply with BS 3676, switches which form part of a circuit derived from a SELV safety source with a nominal voltage not exceeding 12 V r.m.s., a.c., or d.c., and switches or controls operated by mechanical actuators with insulating linkages.
601-08

The safety source for the SELV circuit shall be out of reach of a person using the bath or shower; the switches must not have any accessible metal parts, and shall be insulated or protected with barriers or enclosures against direct contact.
 601-03,

No electrical equipment shall be installed in the interior of a bath tub or shower basin.

Installations carried out on the surface shall not use metal conduit, metal trunking, a cable with an exposed metallic sheath , or an exposed earthing or bonding conductor.
601-02, 601-07

Motors
Every motor shall be provided with an efficient means of switching off; this shall be readily accessible, easily operated, and so placed as to prevent danger.
130-06-02, 476-02-03

Every motor shall be provided with a means of switching off for mechanical maintenance.
462-01-01

Transformers
A linked switch that disconnects all the live conductors shall be provided for disconnecting a step-up transformer from the supply.
551-01-03

Discharge lighting
Every switch which is controlling only discharge lighting, must have a nominal current rating of at least twice the steady current of the circuit, unless it has been manufactured to the appropriate section of BS 3676, and is marked accordingly.

Where a switch controls a circuit containing filament lamps as well as discharge lighting, it shall have a nominal current rating of the current for the filament lamps, plus twice the steady current of the discharge lighting.
INF

Plugs and socket outlets
A plug and socket may be used as a switching device providing its rating does not exceed 16 amps. A socket having a rating exceeding 16 amps may be used for switching a.c. circuits, providing it has the required breaking capacity for the use intended.
Plugs and sockets should not be selected for emergency switching.
537-05, 537-04-02

Fireman's switch
A fireman's emergency switch shall be provided in the low voltage circuit for interior and exterior discharge lighting which exceeds low voltage. (See emergency switching).
476-03-05

Switches prohibited
A switch must not be installed in a protective conductor. The only exception to this regulation is where an installation receives its supply from more than one source. (See Isolation - General requirements.)
543-03-04, 460-01-03

A switch shall not be installed in the outer conductor of a concentric cable.
546-02-06

Semiconductor devices
Semiconductor devices can be used for control and functional switching, providing they are suitable for the nominal voltage of the installation, the design current (full load current), the frequency of the supply, and the power requirements of the equipment. The device must not cause harmful effects on other equipment, nor impair the supply; it shall be suitable for any external influences likely to be encountered.
512-06

Accessibility
Every switch which requires operation or attention during normal use shall be so positioned as to give adequate working space, and safe access to the equipment for operation, inspection and maintenance.
130-07, 513-01, 529-01-01

34. Trunking

Material
Trunking shall be manufactured to BS 4678 The material of insulated trunking shall have the ignitability characteristic "P" as specified in BS 476 Part 5.
521-05

Corrosion & exposed to weather
Precautions are to be taken in damp situations against corrosion that can be caused by metal trunking coming into contact with building materials containing chemicals, of which magnesium chloride, lime, and acidic woods are example.
522-05-01

Metal trunking and associated fixings exposed to the weather, or installed in damp situations, shall be made from corrosion resistant material, and shall not be in contact with metals likely to cause electrolytic action.
522-05-01, 522-05-02

Entries into trunking shall be so placed or protected that the ingress of water is prevented. (Trunking exposed to the weather shall be of the weatherproof type; it is important to make certain that entries are not made in the top face of the trunking, and that any entry is made waterproof, by, for example, flanged coupling and lead washer if the entry is for conduit.
522-03-01

Heat barriers
Vertical trunking shall be equipped with internal barriers to prevent the air at the top of the trunking becoming excessively hot. As a recommendation, the distance between barriers shall be either the distance between floors, or 5 metres whichever distance is least.
422-01

Fire barriers
Where trunking passes through walls or floors, the hole round the trunking shall be made good to the same degree of fire resistance as that of the wall or floor. Where the wall or floor has a designated fire resistance, it shall be internally sealed to maintain the fire resistance of the floor or wall. The sealing material shall be compatible with the wiring system, permit thermal movement of the cable without reducing the quality of the seal, be removable without damaging the cables, and have the same degree of resistance to environmental conditions as the wiring system.
527-02

Trunking supports
Trunking shall be so supported, and of such a type, that it is suitable for the risk of the mechanical damage which it is likely to incur during normal use. Alternatively, it shall be protected against such mechanical damage.
522-08-04, 522-08-05

Spacing of supports for trunking

The following table does not apply to lighting suspension trunking, or trunking using special strengthening couplers.

Trunking size mm × mm	Maximum distance between supports in metres			
	Metal		Insulating	
	Horizontal	Vertical	Horizontal	Vertical
16 × 16	0.75	1.00	0.50	0.50
25 × 16	0.75	1.00	0.50	0.50
38 × 16	0.75	1.00	0.50	0.50
38 × 25	1.25	1.50	0.50	0.50
38 × 38	1.25	1.50	0.50	0.50
50 × 38	1.75	2.00	1.25	1.25
50 × 50	1.75	2.00	1.25	1.25
75 × 50	3.00	3.00	1.50	2.00
75 × 75	3.00	3.00	1.75	2.00
Larger sizes	3.00	3.00	1.75	2.00

Trunking as a protective conductor

The cross-sectional area of trunking used as a protective conductor shall be not less than that given in the formula of Regulation 543-2, or by Table 54G
543-01-01
Where the trunking is common to several circuits, its cross-sectional area shall be determined by either of the following criteria:
 1. using the formula of Regulation 543-01-03, using most unfavourable values of fault current and disconnection time for each of the circuits it contains, or
 2. using the conductor with the largest cross-sectional area in Table 54G.

543-01-02
Protective conductors shall be protected against mechanical and chemical deterioration, and against electrodynamic effects.
543-03-01
When trunking is used as a cpc, the joints do not need to be accessible. (In such cases bonding across joints to be certain of permanent and reliable continuity is advisable.)
543-03-03

Busbar trunking

When used as a protective conductor, the electrical continuity of enclosed metallic busbar trunking systems shall be protected against mechanical, chemical, and electrochemical deterioration, and shall have a cross-sectional area not less than that given either by the formula or by Table 54G, and the enclosure shall allow connection of other protective conductors at every tap-off point.
543-02-04

Trunking shall not be used as a PEN conductor.
543-02-10

Category 1, 2 and 3 circuits.
When enclosed in the same trunking, Category 1 circuits shall be partitioned from Category 2 circuits, unless the Category 2 circuits are insulated to the highest Category 1 voltage present.
528-01-05

Category 3 circuits, when enclosed in the same trunking as other Category circuits, shall be separated by continuous partitions, unless the Category 3 circuits are wired in micc cable.

Partitions shall also be installed at any common outlet to the trunking system, which outlet accommodates Category 3 circuits with Category 1 or Category 2 circuits.
528-01-06, 528-01-07

Where controls or outlets for Category 1 and Category 2 circuits are mounted in or on common enclosures, the cables and connections of the two Categories shall be partitioned by rigidly fixed screens or barriers.
528-01-07

Termination of trunking.
At the termination of trunking, non-sheathed cables or sheathed cables that have had the outer sheath removed, shall be enclosed in fire resistant material, unless they are terminated in a box, accessory, or luminaire complying with British Standards.
The building structure may form part of the enclosure referred to above.
526-03-02, 526-03-03

Cables in trunking
Any type of cable installed in a vertical trunking shall be supported at regular intervals, usually at 5 metre intervals.
INF

The internal radius of every trunking bend shall comply with the minimum bending radius for cables given in Table 3 - 'Selection of Cables'.
522-08-03

Socket outlets
The earthing terminal of each socket outlet shall be connected by a separate protective conductor to an earthing terminal in the trunking, when the socket outlet is mounted in the trunking.
543-02-07

Wiring capacity of trunking
See Table CCT 1 in Part 3.

PART TWO

EXPLANATIONS AND WORKED EXAMPLES OF THE REGULATIONS

SPACE FOR NOTES AND AMENDMENTS:

Contents and index

1. **Introduction** 166
   - Electricity at Work Regulations 167
   - Health and Safety at Work Etc. Act 166
   - Legality of the regulations 166
   - Object of the regulations 166
   - Scope of the regulations 166
   - Voltage ranges 168
2. **Assessment of the installation** 169
   - Compatibility 170
   - Environmental conditions 170
   - Maintainability 170
   - Maximum demand 169
   - Purpose and structure of the installation 170
   - Supply characteristics 169
3. **Selecting protective device** 171
   - Advantages and disadvantages of devices 171
   - Overload discrimination 173
   - Short-circuit discrimination 174
4. **Sizing conductors** 177
   - Ambient temperature 185
   - Application of grouping factor 'G' 181
   - Application of the thermal insulation factor 'T' 187
   - Cables in parallel 179
   - Circuits and cables for which derating is not required 185
   - Confusion between factors 178
   - Correction for mixing cables with different operating temperatures 186
   - Derating factors 180
   - Determination of final conductor temperature 186
   - Fault current 188
   - Fusing factor 177
   - General rules 187
   - General when circuits not subject to simultaneous overload 183
   - Grouping factor 181
   - Lightly loaded cables 184
   - Motor overloads 178
   - Omission of overload devices 180
   - Overload selection 177
   - Overloads along the run of a conductor 179
   - Positioning overload devices 179
   - Protective device not a semi-enclosed fuse 182
     - Circuits not subject to simultaneous over 182
     - Circuits subject to simultaneous overload 182
   - Relationship between I_n, I_b, I_z and I_2 177
   - Resistance of conductor at another temperature 188
   - Semi-enclosed fuse factor 'S' 187
   - Semi-enclosed fuse factor 178
   - Socket outlet circuits 184

161

4. Sizing conductors continued
- Thermal insulation --- 186
- When overload protection is not required --- 184
- When protective device is a semi-enclosed fuse --- 183
  - Circuits not subject to simultaneous overload --- 183
  - Circuits subject to simultaneous overload --- 183

5. Examples of sizing conductors --- **189**
- Example 1 - Selecting section board fuse rating --- 189
- Example 2 - Grouping of cables in conduit with HRC fuse --- 190
- Example 3 - Grouping with a BS 3036 fuse --- 190
- Example 4 - Ambient temperature correction with mcb --- 191
- Example 5 - Ambient temperature correction with BS 3036 fuse --- 191
- Example 6 - Thermal insulation; cable in contact on one side only --- 192
- Example 7 - Thermal insulation; cable totally enclosed --- 192
- Example 8 - Cables not subject to overload --- 192
- Example 9 - Lightly loaded conductors --- 193
- Example 10 - Problem involving all the factors --- 193
- Comments --- 195

6. Voltage drop --- **196**
- An explanation of the regulations --- 196
- Basis of voltage drop tables --- 196
- Calculating voltage drop --- 198
- Combing the factors --- 198
- Correcting for load power factor --- 197
- Correction for conductor operating temperature --- 197
- Extra-low voltage --- 196
- Motor circuits --- 196
- Worked examples --- 199
  - Example 1 - Single-phase voltage drop --- 199
  - Example 2 - Three-phase voltage drop --- 199
  - Calculating circuit length --- 200
  - Example 3 - Motor circuit --- 200
  - Correction factor examples --- 201
  - Example 4 - Power factor in motor circuit --- 201
  - Example 5 - Correcting for both power factor and conductor temperature --- 201
- Summary --- 202

7. Protection against electric shock --- **203**
- Arm's reach --- 204
- Barriers or enclosures --- 204
- Insulation --- 204
- Obstacles --- 204
- Protection against direct contact --- 203
- Protection against indirect contact --- 205
  - Automatic disconnection --- 208
  - Calculations --- 210
  - Earthed equipotential bonding and automatic disconnection --- 205
  - Equipotential bonding --- 205
  - Metal framed windows --- 207
  - Socket outlet circuits --- 211
  - Supplementary bonding --- 206
  - Temperature effect on Z_S --- 210
- Residual current devices (RCD) --- 205

7. Protection against electric shock continued
- **Worked examples** ---211
  - Example 1 - Protection by limiting impedance of ring circuit cpc ---211
  - Example 2 - Protection by limiting impedance of radial socket circuit cpc ---211
  - Example 3 - Protection by limiting phase earth loop impedance ---212
  - Example 4 - Different way of working out example 3 ---212
  - Example 5 - Protection by limiting phase earth loop impedance ---212
  - General ---213

8. Circuit protective conductors ---214
- Armoured cables ---221
- Calculating the size of circuit protective conductor ---215
- Calculations for larger installations ---217
- Conduit and trunking ---221
- Determination of 'k' ---216
- Determination of Z_S in large installation ---219
- Determination of 't' ---216
- Impedance tables ---217
- Options available ---214
- Phase earth loop ---214
- Protection of conductors ---219
- Protective conductor sizes in twin & cpc cable ---219
- Relationship between earth fault and short-circuit ---215
- Resultant cable size 'S' ---217
- Sizing cpcs by using the PCZ tables from Part 3 ---219
- Small conductors ---215
- Table 54G ---217
- Two values of Z_S ---220
- What can be used as a protective conductor ---218

9. Circuit protective conductors examples of calculations ---222
- Example 1 - Conduit as a circuit protective conductor by calculation ---222
- Example 2 - Conduit as a circuit protective conductor using Table 54G ---223
- Example 3 - Armoured cable using Table M54G from Part 4 ---223
- Example 4 - Twin and cpc cable by calculation ---223
- Example 5 - Twin and cpc cable using PCZ Table Part 3 ---224
- General ---224

10. Fault current regulations ---225
- Application of the fault current regulations ---229
- Breaking capacity lower than short-circuit current ---228
- Cable size independent of protective device rating ---225
- Calculating disconnection time- ---226
- Calculations redundant ---227
- Determination of prospective short-circuit current at relevant point ---225
- Disconnection time 0.1 seconds or less ---228
- Illustration of where fault current regulations apply ---230
- Omission of protective devices ---227
- Protective device allowed at different position than the relevant point ---226
- What the formula means ---228

11 Maximum fault current lines ---231
- Dangers using fault current lines ---232
- Producing fault current lines ---231

12.. Short-circuit current calculation theory — 233
- Calculation sheet — 238
- Checking conductor withstand capacity — 237
- Checking withstand capacity of switchgear etc. — 237
- Co-ordination between protective devices — 239
- Conductor temperature to be used in calculations — 236
- Determination of maximum short-circuit current — 234
- Determination of minimum short-circuit current — 235
- Information required before starting — 233
- Reason for minimum short-circuit current being used for conductors — 235
- To determine the impedance to short-circuit current — 233
- Two calculations required — 233

13. Short-circuit current calculations - examples — 241
- Example 1 - Co-ordination between protective devices — 241
- Example 2 - Three-phase distribution — 242
- Comment on example 2 — 244
- Example 3 - Maximum short-circuit current — 245
- Example 4 - Socket outlet ring circuit — 245
- Energy let-through a rewirable fuse — 247

14. Isolation and switching — 248
- Fundamental requirements for safety — 250
- Isolation — 248
- Isolation by non-electrical personnel — 248
- Object and purpose — 248
- Requirements for isolators — 249
- Types of switching — 248
- Warning notices required — 249
- What can be used as an isolator — 250
- Where required — 249

15. Mechanical maintenance — 251
- General requirements — 251
- Labelling — 251
- Type of device — 252
- What it means — 251
- Where required — 251

16. Emergency switching — 253
- Colour of emergency switches — 254
- Colour of fireman's switch — 254
- Definition — 253
- Emergency stopping — 254
- Exterior and interior lighting — 254
- Fireman's switch — 254
- Operation of emergency switches — 253
- Where required — 253

17. Other means of switching — 255
- Circuits and appliances — 255
- Cooking appliances — 255
- Difference between switching and interrupting the supply — 255
- Functional switching — 255

18. Swimming pools --- 256
- Additional requirements --- 259
- Classification of Zones --- 256
  - Zone A (0) --- 256
  - Zone B (1) --- 258
  - Zone C (2) --- 258
  - Zone dimensions --- 257
- Electric heating --- 259
- Equipment outside Zones A, B & C (Zone D) --- 259
- Local supplementary bonding --- 259
- Relaxation of IEC Standard for Zone B --- 258
- Size of supplementary bonds --- 260

19. Inspection and testing --- 261
- Alterations and additions --- 276
- Checking the design --- 262
- Completion Certificates --- 275
- Earth electrode test --- 270
- Earth fault loop impedance --- 272
- Ferrous protective conductors --- 266
- Information to be provided --- 261
- Inspection certificate samples --- 277/278
- Inspection certificates --- 276
- Inspection: visual and physical --- 262
- Insulation resistance --- 268
- Non-ferrous protective conductors --- 265
- Periodic inspection and testing --- 276
- Periods of inspection --- 276
- Polarity --- 272
- Protective conductors --- 265
- RCD testing --- 275
- Residual current device --- 274
  - principle of operation --- 274
- Ring circuit continuity --- 266
  - Test method 1 --- 266/269
  - Test method 2 --- 268/269
- Sequence of tests --- 265
- Test Certificate --- 277
- Testing --- 265
- Test schedule --- 278
- When to start inspection --- 261

1. Introduction

Scope
The IEE Wiring Regulations cover all general commercial, industrial, agricultural and horticultural installations operating at a voltage up to 1000 volts; they also cover construction sites, street lighting, caravans, and their sites.

The general rules of the Wiring Regulations apply to hazardous areas, but do not include the extra precautions that have to be observed with such installations as outlined in BS 5345.

The Wiring Regulations do not cover installations for electric railways, motor cars, ships, mines and quarries, aircraft, off-shore installations, or the generation, transmission and distribution of energy to the public.

Object
The main object of the IEE Wiring Regulations is to provide protection of persons, property and livestock from hazards associated with the electrical installation. This is achieved by giving the requirements for protection against fire, shock, burns and injury from mechanical movement of electrically actuated equipment, where such equipment is controlled by electrical devices intended to prevent such accidents.

They are not intended to instruct untrained persons, or to provide for every conceivable circumstance that may arise. The regulations contain the Fundamental Requirements for Safety and give the methods and practices that will in general satisfy the Fundamental Requirements for Safety.

Compliance with the Fundamental Requirements for Safety will generally satisfy the statutory requirements applicable to electrical installations.

Legality
The IEE Wiring Regulations are not mandatory but are accepted as satisfying the requirements of the Electrical Supply Regulations 1988, as amended, and the Electricity at Work Regulations 1989 for general installations up to 1000 V. If a contract specifies that the work will be carried out in accordance with the IEE Wiring Regulations, then this would be legally binding, and the Wiring Regulations then become a legal requirement.

Health and Safety at Work etc. Act.
Electrical installations must be carried out to a suitable standard to satisfy the Health and Safety at Work etc. Act. Your attention is drawn to Section 6.(3) 'It shall be the duty of any person who erects or installs any article for use at work in any premises where that article is to be used by persons at work, to ensure, so far as is reasonably practicable, that nothing about the way in which it is erected or installed makes it unsafe, or a risk to health, when it is being used, cleaned or manufactured by a person at work.'

Installations carried out to the IEE Wiring Regulations should satisfy the Health and Safety at Work etc. Act as to being of a suitable standard.

Electricity at Work Regulations 1989

There is insufficient space available to give a full explanation of these regulations in this book. The following outline gives an indication of their requirements. A fuller explanation of these regulations is given in the 'Handbook on the Electricity at Work Regulations' by the author.

The Electricity (Factories Act) Special Regulations 1908 and 1944 were revoked by the Electricity at Work Regulations 1989 on the 1st April 1990.

The 1989 regulations were made under the Health and Safety at Work etc. Act 1974 and reinforce the requirements of the Health and Safety at work Act with respect to the use of electricity at work. They are also applicable to mines and quarries.

The 1989 regulations apply to employers, the self employed and employees, all of whom will be responsible where matters are within their control, these duties being in addition to those imposed by the Health and Safety at Work etc. Act.

Under the 1989 regulations the responsibilities can be summarised as follows:
1. to ensure at all times that the electrical system is of such construction as to make it safe,
2. the electrical system is maintained so that it remains safe,
3. that no person is in danger whilst using, operating, maintaining, or working near such a system,
4. that any equipment provided for the purpose of protecting persons at work, on or near electrical equipment, shall be suitable for that purpose, shall be maintained and shall be properly used,
5. that the system has sufficient strength and capability for any short-circuit currents, overload currents and voltage surges etc., that can foreseeably occur,
6. that where necessary, protection is provided to disconnect any excess current before danger can occur,
7. that every joint and connection in a system is mechanically and electrically suitable so as to prevent danger,
8. that there is adequate working space, access and lighting for working on or near such a system,
9. that provision is made for cutting off the supply or isolating equipment as may be necessary to prevent danger,
10. the equipment is suitable for any environmental conditions to which it may reasonably and forseeably be exposed,
11. that persons are competent for the duties they have to perform,
12. that no person shall be allowed to work on or near any live conductor where a danger could arise, unless it is unreasonable in all the circumstances for the equipment or conductor to be dead, and it is reasonable in all the circumstances for that person to be at work on or near it while it is live, and suitable precautions are taken to prevent injury,
13. when equipment and conductors are made dead, suitable precautions must be taken to ensure that they do not become electrically charged.

The regulations apply to all voltages from the lowest to the highest available. They apply to all electrical systems whenever manufactured, purchased, or taken into use, even if the manufacture pre-dates the regulations. Existing equipment can still be used, even if it has been made to a standard that has since been modified; its replacement will only become necessary when it becomes unsafe, or falls due for replacement.

The regulations are applicable to the design of an installation as well as to the operation and maintenance of it. They also apply where any person is working near an electrical system, and as such, they are applicable to managers, mechanical and civil engineers and to any other person who has personnel working with or near electricity, who are under his control.

By definition, in the Electricity at Work Regulations, portable electrical tools, portable electrical equipment and test instruments become part of the system. So that any person using such equipment is covered by the regulations, as is any person in charge of other personnel using such equipment. In practice almost everyone is working near or using electricity at work, the regulations therefore cover such persons as typists, computer operators, photocopier operators and labourers digging trenches on a building site, to name but a few.

General

The IEE Wiring Regulations only take into account established equipment, materials and methods of installation, but do not preclude the use of new materials, inventions or designs, providing the degree of safety is not less than that required by the regulations.

Any departures from the regulations should be noted on the Completion Certificate, when the contract is finished.

Licensing or other authorities exercise control over certain premises, and whereas the IEE Wiring Regulations are still applicable to those premises, the additional requirements of the appropriate authority must be determined.

Voltage ranges

The voltage ranges covered by the IEE Wiring Regulations are:

Extra-low voltage Not exceeding 50 V a.c., or 120 V d.c., between conductors or between any conductor and Earth.

Low voltage From 50 V a.c., to 1000 V a.c., between conductors or from 50 V a.c., to 600 V a.c., between any conductor and Earth.
From 120 V d.c., to 1500 V d.c., between conductors or from 120 V d.c., to 900 V d.c., between any conductor and earth.

2. Assessment of the installation

Before a start can be made on the design or carrying out an installation, an assessment has to be made of the installation, which involves the characteristics which are applicable to that installation.

Maximum demand
The first requirement is to work out the total electrical load that will be connected to the installation, and then to decide what diversities may be applicable. This is best worked out from the knowledge one has of the installation, or from experience. For instance: there may be two groups of motors that cannot possibly run at the same time, such as the re-circulation pumps and storm water pumps in a sewage works. In this instance, only the set of motors that has the largest demand needs to be included in the maximum demand for the installation.

Where no information is available, tables can be used to make an assessment of the diversity that will be applicable to that type of installation.

When assessing the maximum demand, starting currents or in-rush currents of equipment can be ignored.

Supply characteristics
The next step is to determine how the maximum demand is going to be supplied. Most installations will obtain their electrical supply from the local supply company. The first approach is to check with the supply company that they can give a supply and the type of supply they will provide. This will include determining the number and types of live conductors, i.e. single-phase or three-phase four wire, and the type of earthing arrangement, i.e., the type of system of which the installation will be part. This is important since the type of protection required is determined by the type of system. For instance, if the supply is TT then the installation will require its own earth, and to satisfy the requirements for protection against indirect contact, an RCD will be required at the mains position, and all socket circuits in the installation must be supplied through an RCD, such that $Z_S I_{\Delta n} \leq 50$.

The supply company must also be asked for the supply characteristics to enable the protection within the installation to be designed. These characteristics are still required even if the supply is from a private source such as a generator. The characteristics required can be listed as follows:

1. nominal voltage,
2. nature of current and frequency - (a.c., or d.c., 50 Hz or 60 Hz etc.),
3. the single-phase and three-phase prospective short circuit current at the origin of the installation,
4. type and rating of the overcurrent protective device at the origin of the installation,
5. the earth fault loop impedance external to the origin of the installation.

If the supply is to be three-phase and neutral then two values of prospective short-circuit current are required: the three-phase symmetrical short circuit current to give the maximum fault current that will flow, and the phase to neutral short-circuit current to give the minimum that will flow. The importance of this latter item is covered in the Chapter on Fault Currents. This information is required to enable the correct type of protective devices to be installed within the installation.

Purpose and structure of the installation

Having determined the maximum demand and the characteristics of the supply, there are two items that have still to be considered.

The first is the purpose for which the installation is intended. This will include such items as: the type of building structure: whether the building will be used for handicapped people: are there any hazardous areas: are corrosive substances being used? The purpose of the installation is required to enable the correct selection of equipment and wiring materials to be made.

The next stage is to determine the distribution arrangement: in other words, the structure of the installation. This includes the earthing arrangement, i.e. the type of system. It also involves dividing the installation into circuits, so that danger is avoided in the event of a fault, and to ensure that the operation, maintenance, inspection and testing of the installation can be carried out without danger.

Separate circuits have to be provided for those parts of the installation that need to be separately controlled to avoid danger. This is to ensure that other circuits remain energised when there is a fault on one circuit. This means making certain there is discrimination between protective devices. Consideration has to be given as to what would happen when a protective device operates. For instance, what would happen if all the lights went out due to a fault in a power circuit. Could there be an accident due to the lights going out?

Each final circuit has to be connected to a separate way in a distribution board, so that it is electrically separate from every other final circuit. This means that where the final circuits are single phase and are taken from a three-phase and neutral distribution board, each circuit must have its own neutral conductor.

Environmental conditions

The environmental conditions applicable to the installation have to be taken into account when selecting equipment and wiring materials. Consideration must be given to the ambient temperature, mechanical stresses, corrosive substances and any other conditions which can foreseeably be seen to affect the installation now and in the future.

Compatibility

Consideration has to be given to what effects the installation is likely to have on other equipment and services. For instance, installing mains cables in close proximity to telemetry cables can affect the signals carried by the telemetry cables. Starting currents of large motors can cause a dip in voltage which will affect electronic control circuits. The supplier of the electricity will also need to be informed of any equipment which is likely to affect their supplies, thus affecting other customers connected to the same service.

Maintainability

When selecting the protective devices for the installation, consideration has to be given to the frequency and quality of maintenance that the installation will receive throughout its intended life.

This is required so that the protective measures taken to ensure safety and the reliability of the equipment selected remain effective throughout the intended life of the installation. In addition, consideration has to be given to any periodic inspection, testing, maintenance of and repairs to the installation that may be necessary during its intended life.

3. Selecting protective devices

The next stage in the installation design is to select the type and size of protective devices for each final circuit and distribution circuit. There are several types of protective device available, each type having its own particular characteristics. The various types available are listed as follows:
    cartridge fuses manufactured to BS 88. Known as HRC or HBC fuses,
    cartridge fuses manufactured to BS 1361. Known as cartridge fuses,
    semi-enclosed fuses manufactured to BS 3036. Known as rewirable fuses,
    miniature circuit breakers manufactured to BS 3871 Part 1. Known as mcbs,
    moulded case circuit breakers manufactured to BS 4752. Known as mccbs.
Each type of device has advantages and disadvantages over the other types, and a brief outline of these is given later. Whichever type of device is selected, it is better to use that type of device throughout the installation, since discrimination has to be achieved between successive devices that are in series. For instance, the operation of a protective device in a final circuit from a distribution board, should not cause operation of the protective device in the circuit feeding that distribution board. If it did, all the other circuits in the distribution board would be made dead, causing loss of those services.

When selecting the current rating of the protective device for a fixed load circuit, such as, a lighting circuit or a heating circuit, the current rating of the protective device must not be less that the total load connected to that circuit.

When selecting a protective device for a circuit that has an initial in-rush or starting current, such as a motor or transformer, the protective device must take into account the starting current or in-rush current, otherwise it will operate. In many such cases the protective device is not providing overload protection, but fault current protection. A table for selecting HRC fuses for motors is given in Part 4.

The selection of protective devices has also to take into account fault currents, which is covered in more detail in a later Chapter.

Advantages and disadvantages of devices
The BS 88 fuse which is generally called an HRC fuse is made in two types. Type 1 rated at 240 volt, which will interrupt a fault current up to 40,000 amps, and is generally used in domestic installations. Type II fuse rated at 415 volt for industrial installations which will interrupt a fault current up to 80,000 amps. This breaking capacity is available with all sizes of the fuse from 2 amps to 1250 amps, making it more than adequate for the normal industrial installation, since it is only on very large installations that the fault current will exceed 80,000 amps.

This is one of the advantages the BS 88 fuse has over other devices. A further advantage is that when it has operated and is replaced, the circuit is protected by a brand new device. The disadvantage is that it cannot be controlled: the fuse element is either whole or broken, and has to be replaced manually. Additionally, it cannot be operated by remote control or give in itself, indication that it has operated.

The BS 1361 fuse, generally referred to as a cartridge fuse is made in two types: Type 1 rated at 240 volts for domestic installations, and having a breaking capacity of 16,500 amps; Type II rated at 415 volts for industrial installations, having a breaking capacity of 33,000 amps. The advantages

HAWKER SIDDELEY

CAM_{LOC} DISTRIBUTION FUSEBOARD COMPLETELY RE-DESIGNED FROM THE GROUND UP

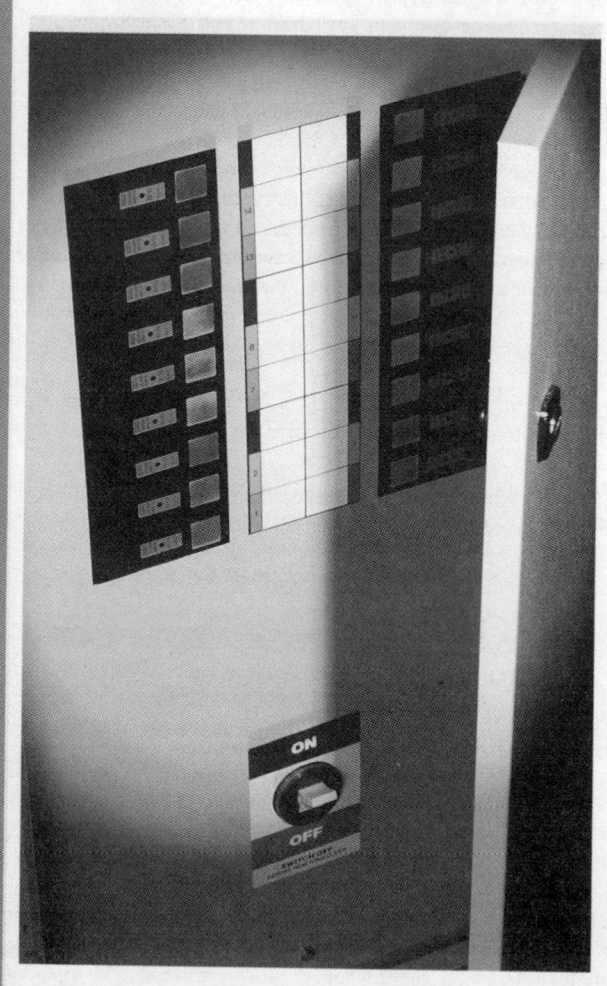

CAM_{LOC} is an advanced design of Distribution Fuseboard. The optimum features of the high performance HRC Fuse and the layout of m.c.b. Distribution Boards, have been combined to create a Fuseboard incorporating all the advantages of these tried and tested systems.

- Three distinct ratings of Cam action release Fuse Holders, with common external dimensions offering a mixed rating capability of 16A, 32A and 63A.

- Integral On-Load Isolator rated at either 63A or 125A.

- Fuse Carrier Ganging facility available for 3 phase supplies.

- 50kA through fault short circuit rating. Tested in accordance with BS 5486 Part 11 1989.

- Shrouded Busbar system fully rated at 200A.

- Reduced installation time due to unique configuration and ease of cabling.

HAWKER SIDDELEY FUSEGEAR

For more information and a copy of the CAM_{LOC} brochure contact the Marketing Department at Hawker Fusegear Limited, Burton-on-the-Wolds, Leics. Phone: 0509 880737

PANELMASTER

ELECTRICAL PROTECTION TECHNOLOGY

A GREAT COMPANY IN GREAT COMPANY

and disadvantages are the same as for the BS 88 fuse. In this case, however, there is a further disadvantage in that its breaking capacity is only 33,000 amps for the industrial fuse, and the maximum fuse size available is 100 amps. Type II fuses are generally used by the electricity supply company.

The BS 3036 fuse which is usually called the rewirable fuse, or semi-enclosed fuse, is only rated at 240 volts and should only be used in domestic installations. Its breaking capacity is limited, in the case of Type S1 fuses to 1,000 amps, Type S2 to 2,000 amps and Type S3 to 4,000 amps. It suffers the same disadvantages as the previous fuses, with the additional disadvantages of lower breaking capacity, and of only interrupting the fault current when it passes through zero. The maximum size of fuse available is 100 amps.

The BS 3871 type 1 miniature circuit breaker, referred to as an mcb, has a range of breaking capacities, from M1 - 1000 amps to M9 - 9,000 amps. There are, however, now on the market mcbs that go up to 16,000 amps. These are used in domestic and industrial installations. They have a low breaking capacity compared to the BS 88 fuse. They have the advantage that the circuit can be re-energised quickly by pressing the switch on the circuit breaker. They do have a further disadvantage, in that, each time it operates due to a fault current at its rated value, it is weakened. The time will come, when it is no longer capable of interrupting the fault current for which it is rated. Unfortunately, there is no indication on the breaker that this point has been reached.

The BS 4752 moulded case circuit breaker, referred to as an mccb, is manufactured with a range of breaking capacities. These breakers are available in larger current ratings and breaking capacities. Mccbs are often used in conjunction with the BS 3871 mcbs for lower currents. They have the disadvantage that most mccbs have only a short-circuit performance category of P1; that is, they are only designed to break the fault current twice. The actual test at the mccbs rated breaking capacity is, to open and after a time delay ,to close onto the fault and immediately to break it again. So having operated twice at their rated braking capacity, they should be replaced.

The only reason that such mccbs do not fail in practice, is that they do not have to exceed their rated breaking capacity. They have the advantage that they are easily switched back on again after a fault, and can be operated by remote control. They can also be made to give remote indication showing whether the breaker is closed or open. Time delays can be fitted, so that discrimination with smaller rated devices is achieved. Additionally ,the rating of overload protection in the breaker can be changed.

Overload discrimination

Having selected the protective devices for the size and type of load, the protective device size in the section board has to be worked out. Item (A) in Figure 1

Figure 1 - Distribution diagram

The current rating of the protective device in 'A' will be determined by the total connected load on the final distribution board, reduced by any diversity that is allowed. Where the final distribution board contains equipment that has a peak current at start-up such as the starting current of motors, the number of items that may start at the same time has to be determined, together with the length of time the starting currents will persist.

These starting currents are then added to the total current load on the final distribution board, to give the total current demand at start-up. It is now necessary to check the protective device characteristic for the total current demand and the time the starting currents will persist. This is to ensure that the protective device will not operate on start-up, as shown in Figure 2.

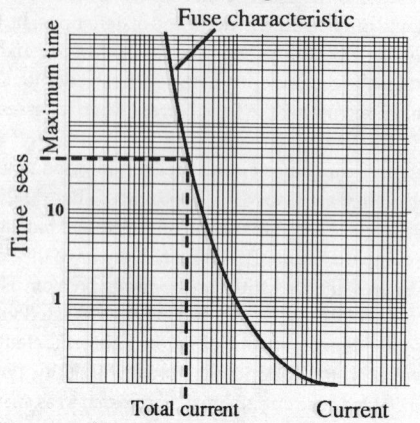

Figure 2 - Selecting protective device

Where the total current intersects the characteristic, read off the time on the time axis, as shown in Figure 2. This must be greater than the time taken for the motors to start, otherwise the protective device will trip.

Where starting currents are not involved, there will be no problem with discrimination between overload devices of the same type. Overload devices should be selected so that the current causing disconnection does not exceed 1.45 times the nominal rating of the protective device. It therefore follows that $2A \times 1.45$ will always be less than a $4A \times 1.45$, and no problem with discrimination will occur.

Short-circuit discrimination

Although protective devices may discriminate with each other under overload conditions, they may not do so under fault current conditions, and so a check has to be made to ensure that they discriminate under the latter conditions.

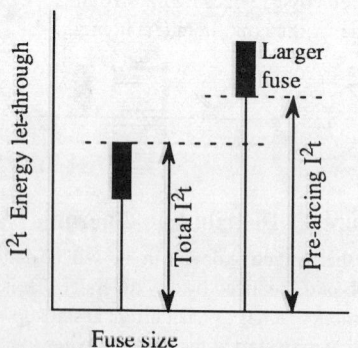

Figure 3 - HRC fuse discrimination

As far as fuses are concerned, this means checking that the total energy (I^2t) let-through the smaller device does not exceed the pre-arcing energy (I^2t) of the larger device, for discrimination to be effective. See Figure 3.

The total I^2t let-through the smaller device should not be too close to the pre-arcing value of the larger device. If it is, the larger device is likely to fail if there is a second fault on the circuit, since it will have been weakened by the first fault.

Where circuit breakers are concerned, it is a case of considering their characteristics, to ensure that the fault current flowing through the smaller device is less than the current needed for the instantaneous trip of the larger device as shown in Figure 4.

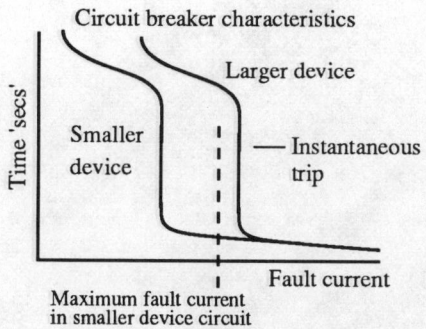

Figure 4 - Discrimination with mcbs

BICC Pyrotenax

Pyrotenax for Life...

...Safety,
...Preservation,
...Support Systems

Pyro Twist...

...a New Twist to a Proven Product

FIREPROOF Pyrotenax® CABLES

BICC Pyrotenax Limited
P.O. Box 20, Prescot
Merseyside L34 5GB, England.
Tel: 051 430 4000 Fax: 051 430 4004
Telex: 628811 Station Code BPTXP

A BICC Cables company.

BICCGroup

4. Sizing conductors

Overload selection

The majority of circuits in the installation will require protection against fault currents, but only those circuits where an overload can occur will require overload protection.

An overload is caused by a circuit carrying more current than that for which it was designed, as in the case of a 3kW load being connected to a circuit designed for 1kW, or a motor being overloaded.

An overload occurs in a circuit which is a healthy circuit, whereas a short-circuit is a fault condition, which occurs between live conductors, (which includes the neutral), in conductor connections, or in the equipment. In general, live conductors must be protected by one or more devices for the automatic interruption of the supply, in the event of an overload, or fault current.

The devices installed must break any overload current flowing in the circuit, before such a current causes a temperature rise detrimental to the conductors, their insulation, joints, and terminations, or any material surrounding or in contact with the conductors.

Relationship between I_n, I_b, I_z, and I_2

The nominal current rating of the protective device is designated by I_n, the design current (full-load current of the circuit) by I_b and the current-carrying capacity of the conductors by I_z. The calculated value of current-carrying capacity required for the conductor is given by I_t.

The nominal current rating of the protective device, or the setting of the overload device, must be not less than the design (full-load) current of the circuit, i.e.

$$I_n \text{ must be greater than or equal to } I_b$$

The nominal current rating or setting of the overload device must not be greater than the current-carrying capacity of the smallest conductor in the circuit, i.e.

$$I_z \text{ must be greater than or equal to } I_n$$

There is a further requirement, that the current causing the effective operation of the protective device, referred to as I_2, must not exceed 1.45 times the current-carrying capacity of the smallest conductor in the circuit, i.e.

$$I_2 \text{ must not exceed } 1.45\, I_z$$

Fusing factor

The factor 1.45 referred to in the regulations is the fusing factor of the device obtained from the equation:

$$\text{Fusing factor} = \frac{I_2}{I_n}$$

which is:

$$\text{Fusing factor} = \frac{\text{Current causing operation of the protective device}}{\text{Nominal rating of protective device}}$$

Devices that have a fusing or tripping factor in excess of the 1.45 factor do not comply with the above requirements, and compensation has to be made for the increase in temperature rise of the conductors, caused by the higher overload current which will flow through the circuit, before it is interrupted by the protective device.

Semi-enclosed fuse factor

One device which does not comply with the above requirement is the BS 3036 fuse, more commonly known as the 'rewirable fuse', or more correctly, 'semi-enclosed fuse', because it has a fusing factor of 2.

The regulations can be used to determine the degree of compensation that must be applied to the cables. This compensation allows for the higher overload current that will flow before the device interrupts the circuit, i.e.

$$I_2 = 1.45 I_z \quad \text{———(1)}$$

From the previous equations: Fusing factor for BS 3036 fuse = 2

$$\text{Therefore } 2 = \frac{I_2}{I_n}$$

Substituting for I_2 in equation (1) as follows:

$$2 I_n = 1.45 I_z \quad \text{i.e. } I_n = 0.725 I_z$$

This is more conveniently expressed in the form:

$$I_z \text{ minimum} = \frac{I_n}{0.725}$$

This shows that the cable size has to be increased when using semi-enclosed fuses as overload devices.

H.R.C. (high rupturing capacity) fuses referred to in the regulations as BS 88 fuses, cartridge fuses referred to as BS 1361 fuses, mccbs and mcbs, all comply with the requirement that the fusing current does not exceed 1.45 I_z, providing the current-carrying capacity of any conductor in the circuit is not less than the nominal rating of the protective device.

The fusing factor for the H.R.C. fuse is often referred to as having a fusing factor of 1.5, but this is in an open test rig, and when installed in an enclosure, its fusing factor can be considered to be 1.45.

Confusion between factors

Confusion often arises between the two factors of 1.45 and 0.725. The factor 1.45, is the amount by which the nominal rating of the device has to be multiplied, to give the overload current that must flow, in order to disconnect the circuit.

Technically speaking, the 0.725 factor is the amount by which current-carrying capacity of the smallest conductor in the circuit has to be be multiplied, to give the nominal rating of the protective device, when its fusing factor is 2.

Since the application of the 0.725 factor in this way is impractical, the nominal rating of the device is divided by 0.725 to give I_t. It is commonly referred to as the derating factor for semi-enclosed fuses (S). (S is a symbol used in a formula given later.)

Motor overloads

Fuses and circuit breakers are not the only means of protecting a circuit against overload: the most common alternative is the overload in a motor starter. The rules are still the same, i.e.

I_n greater than or equal to I_b
I_z greater than or equal to I_n
I_2 less than or equal to 1.45 I_z

The last item does not usually pose a problem, but since most starter overloads have a variable setting, the maximum setting must be used in the calculations, since overloads are always subject to adjustment, particularly at a later date.

Cables in parallel

Conductors are allowed to be installed in parallel from the same protective device, providing the conductors are of the same type, have the same cross sectional area, are the same length, follow the same route, and have the same disposition, and there are no branch circuits throughout their length. The sum of the current-carrying capacities of all of the cables in parallel can be equal to the nominal rating of the protective device. This can be expressed as follows: (See Figure 1.)

$$I_t = \frac{I_n}{\text{Number of cables in parallel}}$$

The above cannot be applied to ring circuits since they are not parallel circuits.

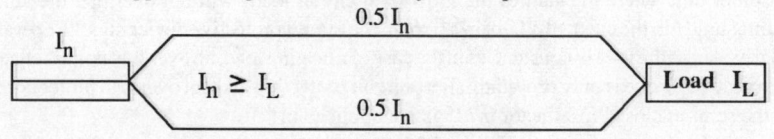

Figure 1 - Protection of conductors in parallel

Positioning overload devices

Overload protection devices should be placed at the point where a reduction occurs in the current-carrying capacity of the conductors. When considering where to install overload protective devices, the circuit must be looked at to determine the factors that will cause a reduction in current-carrying capacity. For instance, if there is a change in the cross-sectional area of the conductor, or a change in the method of installation, overload devices may be required at the point of change. Similarly, if there is a change in the type of cable or conductor, overload protection may be required.

The environmental conditions can affect the current-carrying capacity of a conductor: for example, a change in the ambient temperature along the route the conductor takes, or contact with thermal insulation. Grouping with other cables will also affect the current-carrying capacity of the conductor.

Overloads along the run of a conductor

Under certain circumstances overload devices can be installed along the run of a conductor, as shown in Figure 2.

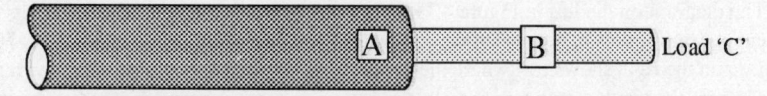

Figure 2 - Overload device installed along the run of a conductor

The overload device can be installed at point 'B', providing there are no branch circuits or outlets for the connection of current using equipment between the points 'A' and 'B'.
A typical example of this arrangement is a motor circuit where the starter with its overloads is installed adjacent to the motor.

Omission of overload devices

Under certain circumstances overload protection devices may be omitted as shown in Figure 3.

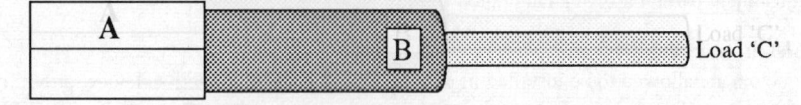

```
┌─────────┬──────────────────────────┬──────────────
│    A    │            B             │         Load 'C'
└─────────┴──────────────────────────┴──────────────
```

Figure 3 - Overload devices can be omitted in certain circumstances

Overload devices need not be provided at point 'B', if the protective device at point 'A', protects the conductors between 'B' and the load 'C'.

An overload protective device is not required if the characteristics of the load are not likely to cause an overload in the conductors. For example, an overload device is not required in the tails to a consumer unit where the sum of the individual circuit loads will not overload the tails to the consumer unit, (in the event of a fault in a circuit, the circuit protective device should operate first). Similarly, where the load is constant, as in the case of a heating load, an overload is unlikely to occur. The protective device is only providing short-circuit protection and not overload protection; in this case, the semi-enclosed fuse factor 0.725 is not applicable.

No overload protective device is required where the unexpected opening of the circuit would give rise to a greater danger than the overload condition. In these situations audible alarms warning that the overload has occurred should be installed.

Finally, overload devices are not required in the secondary circuit of current transformers. Open-circuiting or the sudden opening of the circuit, can cause a dangerously peaking high voltage which could damage the current transformer.

Derating factors

The object of derating factors is to ensure that the temperature of the conductor does not rise to a temperature which is greater than the temperature the conductor insulation can withstand, when carrying its full rated current, and takes into consideration the temperature rise of the conductor during an overload.

The current-carrying capacity of conductors given in the current rating tables is based on the heat generated by the current flowing through the conductor being dissipated from the conductor, so that the conductor operating temperature is as stated in the current rating tables.

When conductors are grouped with other conductors carrying current, or are installed in an ambient temperature higher than 30 °C, or are in contact with thermal insulation, the rate of flow of heat from the conductor is reduced.

The diagram on the left in Figure 4, shows that when the conductor is carrying its full rated current, the rate of flow of heat out of the conductor leaves the conductor temperature at 70 °C. The diagram on the right shows that where the conductor is surrounded by insulation, the rate of flow of heat from the conductor is reduced; this reduction will raise the conductor temperature above 70 °C. The unit of heat is the Joule, which is Watt seconds, but Watt Seconds are equal to I^2Rt. Since time is constant and the current I, is the current needed for the load, it can be seen that the only item that can be changed is the resistance of the conductor. The resistance of the conductor is reduced the larger the conductor is made. Derating factors should not be ignored, because there will be cases where their application indicates the need to install a larger conductor.

SIZING CONDUCTORS

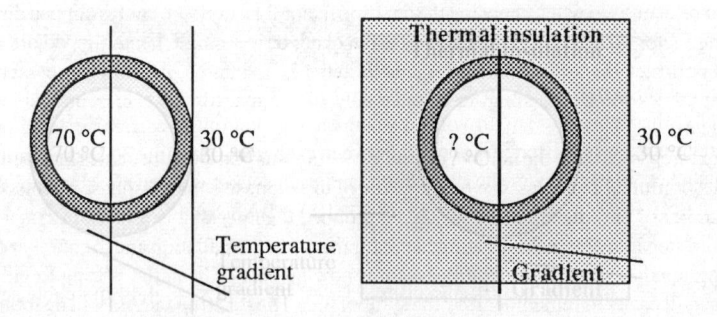

Figure 4 - Rate of flow of heat from a conductor

Circuits to equipment may be grouped together, be subjected to an ambient temperature higher than 30 °C, or be in contact with thermal insulation. Any of these conditions will necessitate applying the appropriate derating factor, so that the conductor's operating temperature does not exceed the value given in the current rating tables.

There are four derating factors to be taken into account: grouping, ambient temperature, contact with thermal insulation, and type of protective device used. These derating factors are often referred to as C_g, C_a, C_i and C_4 and can be easily remembered by the word G.A.T.S., detailed as follows:
G = Grouping Factor = C_g.
A = Ambient temperature = C_a.
T = Thermal insulation = C_i.
S = Semi-enclosed fuse = C_4, or F_4.

The current ratings given in the tables of the regulations are based on the free flow of the heat generated in the conductor to the surrounding atmosphere. The first three factors listed above, can interfere with the natural flow of heat from the conductor, thus raising the conductor's temperature above its designed working temperature. The last factor is concerned with the type of protective device used, and the current which causes the effective disconnection of the circuit.

The first three derating factors are designed to increase the conductor size so that with the restricted flow of heat, the conductor temperature does not exceed the designed working temperature, when each derating factor is considered in turn.

Grouping factor
Not all cable types need derating for grouping, the exception being non-sheathed mineral insulated cables which are not exposed to touch. With the exception of non-sheathed micc cables in CR 16, derating factors for grouping should be applied to all the CR tables in Part 3 of this Handbook.

Application of grouping factor G
The first step is to determine the method of installation of the cables:
1. enclosed in conduit, trunking, or bunched and clipped direct,
2. in a single layer clipped direct to, or lying on a non-metallic surface,
3. multicore cables in a single layer on a perforated metal cable tray,
4. single core cables in a single layer on a perforated metal cable tray,
5. multicore cables in a single layer on a ladder support system,
6. micc cables installed on cable tray.

SIZING CONDUCTORS

Care has to be exercised when choosing the derating factor for multicore cables clipped direct, since the derating factor is different if the cables are bunched, or in a single layer clipped direct, or just lying on a non-metallic surface. With regards to items 2, 3, 4 and 5, it is also necessary to know whether the cables will be touching, or whether they will be spaced with at least one cable diameter between them.

The next step is to count the number of single core circuits or multicore cables grouped together. The circuits or multicore cables can be a mixture of three-phase or single-phase circuits or cables; it is the number of circuits or cables which are grouped together that is important.

The third step is to obtain the derating factor for the type of installation and the number of circuits or cables grouped together. The derating factor is now given as 'C_g in the Wiring Regulations; it is however, still easier to remember it by the symbol 'G'. The derating factors will be found in table GF 1 in Part 4 of the Handbook.

The fourth step is to look at the type of protective device for the circuit, since the formula used to determine the calculated current-carrying capacity required for the conductor I_t, is dependent upon the type of protective device.

Protective device not a semi-enclosed fuse

When a BS 88, BS 1361, or a circuit breaker to BS 3871 Part 1 or BS 4752 Part 1 is used as the protective device, there are two methods of determining the tabulated current-carrying capacity required for the cable. This depends upon whether more than one circuit can be overloaded at any one time, or whether the grouped circuits are not subjected to simultaneous overload.

Circuits subject to simultaneous overload

In this instance, more than one circuit can be overloaded at the same time. The first method uses formula 1. The current-carrying capacity is determined by dividing the nominal rating of the fuse I_n by the derating factor G.

$$\text{Formula 1} \quad I_t \text{ to be not less than } \frac{I_n}{G}$$

Having calculated I_t, the cable size is chosen from the single circuit column of the appropriate current rating table; the cable chosen must have an actual tabulated (I_{tab}) current-carrying capacity equal to, or greater than, I_t calculated from the equation. Single phase circuits are chosen from the single-phase column and three-phase circuits from the three phase column.

Circuits not subject to simultaneous overload

Although any circuit in the group can carry overload current, providing it can be guaranteed that only one of the circuits will be carrying overload current at any one time, the following calculation can be used to obtain the calculated current-carrying capacity required for the conductors. This formula must not be used for socket circuits, since overloads cannot be guaranteed to be confined to one circuit.

Two calculations have to be made, and the one giving the highest calculated current-carrying capacity I_t for the conductors, is the one that is used to size the cables.

$$\text{Formula 2} \quad I_t \text{ to be not less than } \frac{I_b}{G}$$

Note the difference between formulae 1 and 2: I_n is replaced with I_b.

SIZING CONDUCTORS

$$\text{Formula 3} \quad I_t \text{ to be not less than} \quad \sqrt{I_n^2 + 0.48 I_b^2 \left(\frac{1-G^2}{G^2}\right)}$$

Where G is the derating factor and I_b is the design or full load current for the circuit.

Whichever value of I_t is the larger, from either formula 2 or 3, this is the value to use for sizing the cable.

If derating factors for ambient temperature or thermal insulation are applicable, the value of I_t obtained from any of the above formulae is divided by the applicable derating factor, to find the revised current-carrying capacity I_t required for the cables.

When protective device is a semi-enclosed fuse (BS 3036)

Sizing the cables is carried out in exactly in the same way as for the other types of protective device and again two methods are available.

Circuits subject to simultaneous overload

This time formula 4 includes the derating factor 0.725.

$$\text{Formula 4} \quad I_t \text{ to be not less than} \quad \frac{I_n}{0.725 \times G}$$

Again having worked out I_t the cable size is chosen from the single circuit column for either single-phase or three-phase in the appropriate table.

Circuits not subject to simultaneous overload

Again the highest value of I_t obtained from either formula 5 or 6 is used to size the cables.

$$\text{Formula 5} \quad I_t \text{ to be not less than} \quad \frac{I_b}{G}$$

$$\text{Formula 6} \quad I_t \text{ to be not less than} \quad \sqrt{1.9 I_n^2 + 0.48 I_b^2 \left(\frac{1-G^2}{G^2}\right)}$$

As before, formula 4 is used when more than one circuit can be overloaded at the same time; formulae 5 and 6 are used when it can be guaranteed that not more than one circuit will be carrying overload current at any one time; socket outlet circuits are not covered by this directive. The largest value of I_t obtained from either formula 5 or formula 6 is the one used to determine the cable size.

If ambient temperature or thermal insulation derating factors are applicable, then I_t is divided by the applicable derating factor, to determine the tabulated value of the current-carrying capacity required for the cable.

General (Circuits not subject to simultaneous overload)

The second method of sizing cables (Formulae 3 and 6) for grouping will not be available for socket outlet circuits or motor circuits, since it will be difficult to guarantee that not more than one circuit will carry overload current at any one time. It can be used for circuits where the load is constant, such as heating circuits and lighting circuits.

When overload protection is not required

When the circuit conductors are either not required to carry overload, or because of the characteristics of the load, they are not likely to carry overload current, then I_n is replaced by I_b in formula 1.

$$\text{Formula 7} = \text{(i.e Formula 1 becomes)} \frac{I_b}{G}$$

Formula 7 is also used where a circuit is protected by a semi-enclosed fuse, since the 0.725 factor is only required when overload protection is being provided.

This means that the protective device is providing fault current protection and not overload protection. An example of the application of this formula is the heating circuit, such as, an immersion heater, where its load is fixed; it is therefore unlikely to carry overload current. This means that a 15A semi-enclosed (rewirable) fuse does not have to be divided by 0.725 to determine the I_t for the cable when feeding a 3 kW immersion heater.

Socket outlet circuits

The problem with socket outlet circuits, is that the designer of the installation, has no knowledge of what the customer may plug into the socket outlets in the future. The designer cannot therefore assume that the socket outlet circuits will not be overloaded. Neither can he assume that simultaneous overloads will not occur.

The correct way to derate for socket outlet ring circuits when more than two circuits are grouped together is to use the following formula.

$$I_t = \frac{\text{Size of protective device} \times 0.67}{G} = \frac{I_n \times 0.67}{G}$$

No further diversity can be allowed, since this has already been taken into account by reducing the protective device size by 0.67.

As far as radial socket outlet circuits are concerned, the protective device size is divided by the derating factor to obtain the current-carrying capacity I_t required for the cable, i.e.

$$I_t = \frac{I_n}{G}$$

Lightly loaded cables

One of the most useful relaxations to the derating for grouping, is when circuits are lightly loaded. The relaxation works on the principle that the restriction of heat loss from the fully loaded cable by the lightly loaded cable is counteracted by the conductor in the lightly loaded cable acting as a heat sink, since its temperature is only just above ambient temperature. This relaxation will be found most useful in trunking runs which carry large numbers of control circuits, each of which carry only a few amps.

If it is known that a cable will not carry more than 30% of its grouped rating, that cable can be excluded from the number of circuits or cables grouped together when working out the number of circuits or cables that may properly be grouped together. This means they are not included, when counting the remaining cables or circuits that are grouped together.

In these circumstances it is necessary to count all the circuits or cables grouped together and work out the tabulated current-carrying capacity required I_t for the lightly loaded circuit or cable. This is done by dividing I_n by G and then selecting the cable size required from the tables.

Having worked out the cable size for the lightly loaded circuit, the following calculation can be made to see whether it can be excluded from the number of circuits grouped together, when determining the derating factor for the remaining circuits.

$$\text{Grouped current-carrying capacity \%} = \frac{100 \times I_b}{I_{tab} \times G}$$

If the full load current (design current I_b) of the lightly loaded conductor does not exceed 30% of the cables grouped current-carrying capacity, it need not be counted when working out the number of cables grouped together for the rest of the group, for determining the 'new' grouping factor.

Each circuit can be considered in turn so that the derating factor gradually becomes less onerous as each circuit is excluded from the group.

The cable size worked out using the grouping factor for the lightly loaded cable is the size of cable that has to be used for the lightly loaded circuit. An example of this calculation is given in the next chapter.

Circuits and cables for which derating is not required

Where the horizontal distance between adjacent cables exceeds twice their overall diameter, no correction factor for grouping need be applied. When considering this exemption use the larger diameter of two adjacent but dissimilar sized cables, as shown in Figure 5.

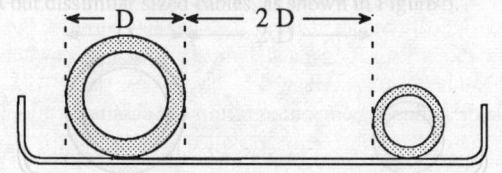

Figure 5 - Obviating the need to derate for grouping

Ambient temperature

Ambient temperature is the temperature of the immediate surroundings of the equipment and cables, before such equipment or cables contribute to the temperature rise, in that location.

Two tables are provided in the regulations: one table (4C2) is for BS 3036 fuses i.e. rewirable fuses, and the other table (4C1) is for all other types of protective device. The derating factors for rewirable fuses are the least onerous, because the rewirable fuse element is not enclosed, enabling it to dissipate heat more easily than the other types of device. Tables CR 1 to 17 in Part 3 contain the ambient temperature derating factors at the bottom of the table. When the protective device is being used for short-circuit protection, including BS 3036 fuses, the derating factor is taken from the table headed 'For other protective devices' in the CR tables in Part 3, or Table 4C1, in the regulations.

Application of ambient temperature factor A

The nominal rating of the protective device is divided by the factor obtained from the table of ambient temperature correction factors. i.e.

$$I_t = \frac{I_n}{A}$$

The value I_t is the calculated current-carrying capacity of the conductor required. The cable rating is selected from the current rating tables, such that the tabulated current-carrying capacity I_{tab} is not less than I_t. The value of 'A' is obtained from the table of derating factors for ambient temperature for the type of protective device being used.

Determination of final conductor temperature

Where a more accurate design is needed, the final conductor temperature is required. This can be obtained by using part of the formula already given in the regulations. However, this formula is not applicable when $t_a < 30\ °C$, or when BS 3036 fuses are used as the protective device.

$$t_f = t_p - \left(G^2 A^2 \cdot \frac{I_b^2}{I_{tab}^2}\right)(t_p - t_a)$$

where: t_f is the final conductor operating temperature, t_p is the maximum conductor temperature allowed and t_a is the ambient temperature; which in the regulations is taken as 30 °C
This formula can be used to determine the initial temperature of a conductor for determining a revised 'k' factor, or the resistance of the conductor at its actual operating temperature.

Correction for mixing cables with different operating temperatures

Where cables having different operating temperatures are grouped together, the current rating has to be based upon the lowest operating temperature of any cable in the group. This requires a derating factor for the higher operating temperature cables. This derating factor can be obtained by using the above formula. By re-arranging the formula we obtain:

$$\frac{I_b}{I_{tab}} = \sqrt{1 - \frac{t_p - t_f}{t_p - t_a}}$$

Using this formula the following correction factors are obtained.

Lowest operating temperature of conductor in group	Correction factor to be applied to the I_{tab} of the higher operating temperature conductor in the group		
	70 °C	85 °C	90 °C
70 °C	1	-	-
85 °C	0.852	1	-
90 °C	0.816	0.957	1

This derating factor is in addition to the derating factor for grouping. The easiest way to use it is to combine it with the derating factor for grouping, when sizing the conductor with the higher operating temperature. For example assume 90 °C cables are being grouped with 70 °C cables and no other derating factors are involved. I_t given by the formula below is now I_{tab} in the 90 °C cable table.

$$I_t = \frac{I_n}{G \times 0.816}$$

Thermal insulation

Where cables are likely to be in contact with thermal insulation, the current-carrying capacity of the cable has to be increased to allow for the thermal insulation.

Tables are provided giving the current-carrying capacities of cables installed in a thermally insulated wall or ceiling, where one side of the cable is in contact with a thermally conductive surface. Where cables are totally enclosed in thermal insulation, the current-carrying capacity of the cables has to be increased by a factor dependent upon the length the conductor is enclosed in the thermal insulation.

Length in mm conductors totally enclosed in thermal insulation	Derating factor
50	0.89
100	0.81
200	0.68
400	0.55
500 or over	0.5

The derating factor is applied to the current-carrying capacity taken from the 'open and clipped direct column' in the table for the type of cable being used.

Application of thermal insulation factor T
Where cables are totally enclosed in thermal insulation, the nominal rating of the protective device I_n is divided by the appropriate derating factor to give the current-carrying capacity required for the cable.

$$I_t = \frac{I_n}{T}$$

Semi-enclosed fuse factor S
The tables in the regulations give the full thermal ratings of the conductors. They are based on the current needed to operate the device, not exceeding 1.45 times the nominal rating of the device. Where the current required to operate a protective device exceeds 1.45, then a calculation is needed to determine the derating factor to be applied to the cable rating.

A typical example of the calculation has been given earlier in this chapter. When a rewirable fuse is used for overload protection, the figure of 0.725 is used to derate the current-carrying capacity of the conductors. It is easier to apply the derating factor to the device's nominal rating as follows.

$$I_t = \frac{I_n}{S}$$

Where I_n is the nominal rating of the protective device, and S is the calculated derating factor.

Application of semi-enclosed fuse derating factor S
When overload protection is required, the nominal rating of the semi-enclosed (rewirable) fuse is divided by 0.725 to give the current-carrying capacity required for the cable

$$I_t = \frac{I_n}{0.725}$$

General rules
The derating factors are only applied to the section of the cable run where they are applicable, because the transmission of heat from the cables will only be affected by Grouping, Ambient and Thermal insulation, over the length of cable where the particular factor is applicable. The factor for semi-enclosed fuses affects the whole length of the cable run since it is installed at the source of the circuit.

188 SIZING CONDUCTORS

Where more than one of the derating factors are going to affect a section of a cable run they can be combined into a general formula as follows:

$$I_t = \frac{I_n}{G \times A \times T \times S}$$

The notes below the tables state that the factors are applicable to groups of one size of cable, the reason being that smaller cables in the group could become overheated. Since no other guidance for the derating of grouped circuits or cables is available, use the derating factors given in the regulations as a guide when small conductors are grouped with larger conductors, but err on the safe side when choosing the conductor size.

Fault current.

Separate chapters are devoted to fault current protection and sizing protective conductors; but since the final circuit temperature has been covered in this chapter, it seems the appropriate place to discuss final conductor temperature in relation to fault currents.

Calculations to determine fault current use the resistance at the average temperature of the initial conductor temperature and the limit temperature for the conductor's insulation. In general, the operating temperature of the conductor is assumed to be the figure given in the current rating tables.

When a conductor is not carrying its rated current-carrying capacity, its operating temperature is less than that given in the tables; this means that the average temperature will be lower, so the conductor resistance will be lower; this improves protection against indirect contact, voltage drop, but increases the fault current.

In large installations, the supply to the final circuit will be through several distribution cables and items of switchgear; each distribution cable getting progressively smaller as it approaches the final circuit. The temperature rise on conductors, caused by a short-circuit or earth fault in the final circuit, will progressively become less, as the distribution cables become larger, as they approach the source of electrical power; even to the point of having negligible effect on conductor temperature.

For calculation purposes, the formula for final conductor temperature, could be used to determine the temperature a distribution's cable conductor would reach, when carrying full load current and the additional current caused by a fault in the final circuit.

The same is true when a fault occurs in a distribution cable; but the temperature rise effect on the other distribution cables, with which it is in series, back to the source of power, is dependent upon its position relative to the source of energy.

Resistance of conductor at another temperature

Having calculated the conductor's final temperature; the resistance of the conductor at the average of operating and insulation limit temperature, can be worked out from the following fromulae:

$$t_1 = \frac{t_f + t_L}{2} \qquad R_{av} = \frac{(230 + t_1) R_{20}}{250}$$

where: t_L is the limit temperature of the conductor's insulation; R_{av} is the resistance of the conductor at the average temperature; R_{20} is the resistance of the conductor material at 20 °C; t_f is the conductor operating temperature; t_1 is the average of conductor operating temperature and conductor's insulation limit temperature.

Part 4 contains tables giving conductor resistances at 20 °C, for copper and aluminium.

5. Examples of sizing conductors

Example 1 - Selecting section board fuse rating.
What size of fuse in Section board A will be required for the load on the final distribution board if the four 415V motors are 4kW and are started together, and take 10 seconds to get up to speed, and the remaining two loads are heating loads which are permanently on? Assume that starting current is seven times full load current.

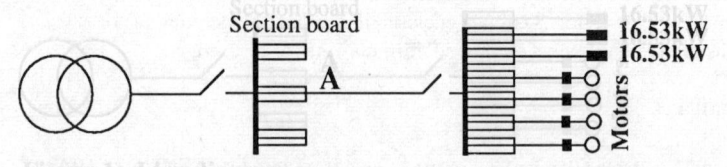

Figure 1 - Line diagram

From table MC1Part 4 the full load current of a 4 kW motor at 415V is 8.4A. Each heating load will have a full load current of:

$$I_{HL} = \frac{16.53 \times 1000}{415 \times 1.732} = 23 \text{ A}$$

The total connected load will be $(23 \times 2 + 8.4 \times 4) = 79.6$ amps, which would require an 80A fuse in section board 'A'. The motors all start together, so the current for which the fuse in the section board 'A' has to cater for upon start-up will be $(23 \times 2 + 8.4 \times 4 \times 7) = 281.2$ amps.
It is now necessary to consult the fuse characteristics for the 80A HRC TIA fuse.

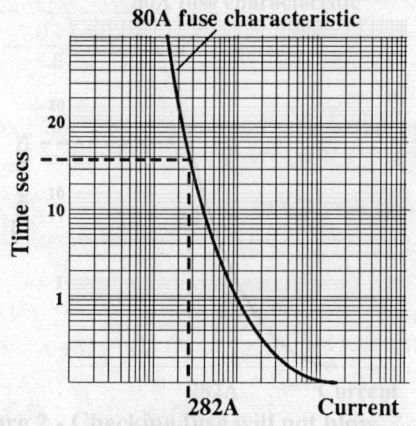

Figure 2 - Checking fuse will not blow

The total start-up load of 282 amps is checked against the fuse characteristic to ensure that the point at which it touches the fuse characteristic is greater than 10 seconds; if it is, then the fuse size is satisfactory: if it isn't, then a larger fuse would have to be selected.

Example 2 - Grouping of cables in conduit, with HRC fuses.

Three circuits, each with a full load current of 32A, are to be wired in p.v.c. enclosed in conduit from HRC BS88 40A fuses. If the ambient temperature is 30°C and the circuits are not in contact with thermal insulation, what size cables will be required.
 a) If the circuits are feeding 3 phase 3 wire BS 4343 socket outlets ?
 b) If the circuits are feeding single-phase heating loads ?

Working problem (a)
The derating factor for ambient temperature and thermal insulation will be 1.
From table GF 1 Part 4, for 3 circuits enclosed in conduit the derating factor is 0.7
The circuits are feeding socket outlets, so formula 1 has to be used.

$$\text{Formula 1} \quad I_t = \frac{I_n}{G} = \frac{40}{0.7} = 57.14 \text{ amps}$$

From column 6 of table CR1 Part 3, 16 mm² cable is required with an I_{tab} of 68A

Working problem (b)
If the loads are heating and simultaneous overloading of the circuits will not occur, formulae 2 and 3 can be used.

$$\text{Formula 2} \quad I_t = \frac{I_b}{G} = \frac{32}{0.7} = 45.71 \text{ amps}$$

$$\text{Formula 3} \quad I_t = \sqrt{I_n^2 + 0.48 I_b^2 \times \left(\frac{1 - G^2}{G^2}\right)}$$

$$\text{therefore} \quad I_t = \sqrt{40^2 + 0.48 \times 32^2 \times \left(\frac{1 - 0.7^2}{0.7^2}\right)} = 45.95 \text{ amps}$$

The worst condition for I_t is formula 3; I_{tab} from CR1 Part 3 must not be less than this value. From column 2 table CR 1 Part 3, a 10mm² cable is required with an I_{tab} of 57A.

Example 3 - Grouping with BS 3036 fuse.

Use the same example as number 2, but circuits are now protected by a semi-enclosed fuse and enclosed in conduit.

Working problem (a)
Derating factor from GF1 for 3 circuits is 0.7; but since a rewirable fuse is being used, formula 4 has to be used

$$\text{Formula 4} \quad I_t = \frac{40}{0.725 \times 0.7} = 78.818 \text{ amps}$$

From Table CR1 column 6, a 25 sq. mm cable is required with an I_{tab} of 89A.

EXAMPLES OF SIZING CONDUCTORS   191

Working problem (b)
This time formulae 5 and 6 have to be used.

Formula 5 $\quad I_t = \dfrac{32A}{0.7} = 45.71$ amps

Formula 6 $\quad I_t = \sqrt{1.9 \times 40^2 + 0.48 \times 32^2 \times \left(\dfrac{1 - 0.7^2}{0.7^2}\right)} = 59.595$ amps

Formula 6 gives the largest value for I_{tab}, and from column 2 table CR 1 Part 3, 16 sq. mm cable required.

Example 4 - Ambient temperature correction with mcb.

A three-phase circuit is to be wired in p.v.c. single core cable (6491X) in its own conduit installed on the surface, and will be protected by a 20 amp mcb when the installation is being carried out in a boiler house where the temperature is normally 45° C. If the cables are not in contact with thermal insulation, what size cable will be required ?

Working
Looking at the question, G,T and S all equal 1. The protective device is an mcb, so the ambient temperature table for all other types of protective device has to be used; the cable size required will be:

$$I_t = \dfrac{I_n}{A}$$

From the bottom of table CR1, A = 0.79 therefore,

I_t required will be: $\dfrac{20A}{0.79} = 25.32$ amps

and from column 6 of table CR1, a 4 mm² cable is required, with an I_{tab} of 28A.

Example 5 - Ambient temperature correction with BS 3036 fuse.

If the mcb is changed to a rewirable fuse in example 4, what size cable would be required?

Working
In this example, both G and T will equal 1, and the cable size is found by:

$$I_t = \dfrac{I_n}{A \times S}$$

The value of S is 0.725, but the value of A is found from table CR1 for rewirable fuses, and is 0.91, since 6491X cable is a general purpose p.v.c cable;

therefore $I_t = \dfrac{20}{0.725 \times 0.91} = 30.314$ amps

From column 6 of table CR1, cable size required is 6 mm² with an I_{tab} of 36A.

Example 6 - Thermal insulation; cable in contact on one side only.

A house is to be fitted with a 7 kW shower heater, using twin & cpc cable for the supply, which will not be grouped with other cables, but will be installed along the side of a joist with thermal insulation installed between the joists. If the ambient temperature is 30° C and the supply voltage is 240V, what current rating will be required for the mcb and what size of cable will be required?

Working
First step: Method of installation: contact on one side with thermal insulation.
Second step: Full-load current of the heater: = 7000W ÷ 240V = 29.17 amps; therefore a 30A mcb can be used.
Third step: No calculation is required, since tables are now given for cables in contact with thermal insulation on one side only.
Fourth step: Table CR8 Part 3 column 2; 6 mm² cable required with an I_{tab} of 32A.
Fifth step: The volts drop/A/m for 6 mm² cable is 7.3 mV from table CR8 Column 3.
The maximum length the circuit can be is:

$$L = \frac{9.6 \text{ V} \times 1000}{29.17 \text{ A} \times 7.3 \text{ mV}} = 45.08 \text{ m}$$

Alternative method using factor from Table CR8 and avoiding the use of 10^{-3} in the formula on page 198.

$$L = \frac{1315}{29.17} = 45.08$$

Sixth step: Part 3 Table INST 1: 30A mcb maximum length to comply with regulations is 45m.

Example 7 - Thermal insulation; cable totally enclosed.

The same problem as example 6, but cable is totally enclosed in thermal insulation; work out the cable size required.

Working
The full load current of the heater will be 29.17 amps; the mcb size 30A
First step: derating factors applicable; G = 1, A = 1, T = ?, S = 1.
In this case the thermal insulation factor is 0.5 from MICS 3 Part 4.
The value of I_t required is found by calculation.

$$I_t = \frac{I_n}{T} = \frac{30A}{0.5} = 60 \text{ amps}$$

The size of cable is now selected from CR7 column 10 for cables 'clipped direct', since they have already been derated by the above calculation. In this case, the cable size required is 10 mm², with an I_{tab} of 63A
From INST 1 Part 3. Maximum cable length allowed 74.8m.

Example 8 - Cables not subject to overload.

A three-phase heater with a full load current of 21A is protected by a 30A HRC BS 88 Part 2 fuse. The circuit is to be wired in p.v.c. single core cable (6491X) which will be installed along with four other motor circuits in conduit. If the cables are not in contact with thermal insulation and the ambient temperature is 50° C, what size cable will be required?

Working
In this example the cables will not be required to carry overload current, since the load is fixed. It cannot be guaranteed that simultaneous overload will not occur, since the heating circuits are mixed with motor circuits, but advantage can be taken of Note 6.3 in Appendix 4.
From table CR1, the ambient temperature factor is 0.71, and from table GF1 the grouping factor for five circuits is 0.6. Therefore:

$$I_t = \frac{I_b}{G \times A} = \frac{21 \text{ A}}{0.6 \times 0.71} = 49.296$$

From column 6 table CR1, cable size required is 10 mm^2, with an I_{tab} of 50A.

Example 9 - Lightly loaded conductors.
This example illustrates the application of the note 2 of Table 4B1/2.
Two three-phase three-wire motor circuits are each protected by a 50A HRC fuse, and are to be installed in the same trunking as 16 single phase control circuits, each of which has a design current I_b of 1.5A, and which is protected by a 2A HRC fuse. What size cables will be required if all other derating factors = 1 ?

Working
Total number of circuits grouped together = 2 + 16 = 18
From table GF1 Part 4, Grouping factor G = 0.39
Consider the lightly loaded conductors first:

$$I_t = \frac{2A}{0.39} = 5.128 \text{ amps}$$

From table CR1 Part 3, 1.0 mm^2 cable could be used having an I_{tab} of 13.5 amps.
If the control cables will not carry more than 30% of their grouped current rating, they can be ignored when considering the three-phase motor circuits.

$$\text{Percentage of grouped rating} = \frac{100 \, I_b}{I_{tab} \times G} = \frac{100 \times 1.5}{13.5 \times 0.39} = 28.49\%$$

Grouping factor for two three-phase circuits = 0.8

$$I_t = \frac{50A}{0.8} = 62.5 \text{ amps}$$

From Table CR1, column 6, 16 mm^2 cable required. Note: without this relaxation 50 mm^2 conductors would have been required.
So far, circuits have been considered where the derating factors affected the whole circuit, but the derating factors are only applicable over that portion of the circuit they affect; this is illustrated in the following example.

Example 10 - Combined factors.
Two three-phase loads 'A' and 'B' in Figure 3, are to be supplied by PVCSWA & PVC.cable, the circuits being fed from a distribution board with BS 3036 fuses; the normal ambient temperature is 30°C. The cables start off clipped to a perforated cable tray with their sheaths touching. The cable for load 'A' passes through a cavity wall where it is totally enclosed in thermal insulation for 50 mm. The cable for load 'B' is installed into a boiler house where the ambient temperature is 60° C. Ignoring voltage drop, what size cables are required ?

EXAMPLES OF SIZING CONDUCTORS

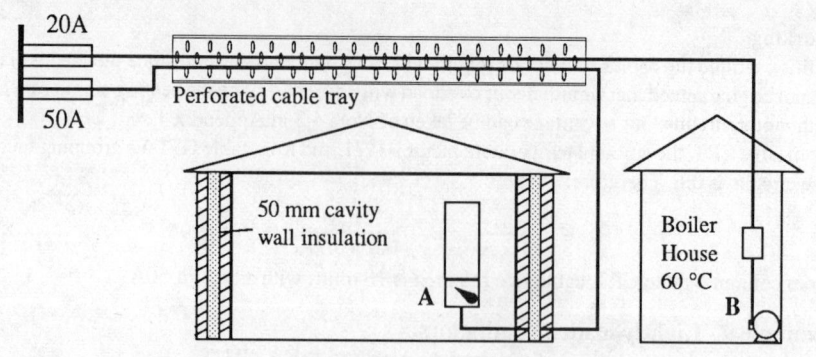

Figure 3 - Combining the derating factors

Working
General Formula :

$$I_t = \frac{I_n}{G \times A \times T \times S}$$

Load 'A' cable
G = grouping = 0..86 from table GF1.
A = ambient = 1 for 30°C.
T = thermal insulation = 0.89 from MISC 3 Part 4.
S = type of protective device = 0.725, from MISC 3 Part 4.

Condition 1 - Length on cable tray

$$I_t = \frac{50}{0.86 \times 1 \times 1 \times 0.725} = 80.19 \text{ amps}$$

Condition 2 - Length in contact with thermal insulation

$$I_t = \frac{50}{1 \times 1 \times 0.89 \times 0.725} = 77.48 \text{ amps}$$

The cable size is selected for the most onerous case, i.e condition 1 - 80.19 A.
From Table CR 9 column 14, 16 mm^2 cable required with an I_{tab} of 83A.

Load 'B' cable
G = grouping = 0.86 (as before).
A = ambient = 0.69 from table CR9.
T = thermal insulation = 1.

Condition 1 - Length on cable tray

$$I_t = \frac{20}{0.86 \times 1 \times 1 \times 0.725} = 32.07 \text{ amps}$$

Condition 2 - Length in boiler house

$$I_t = \frac{20}{1 \times 0.69 \times 1 \times 0.725} = 39.98 \text{ amps}$$

Condition 2 is the worst case so this is used. From Table CR 9 column 14 the cable size required is 6 mm^2.

Comments

It should be noted that the cable is only sized to the worst condition, and the derating factors are only applied to that section of the conductor they affect.

Where long runs of large cables are involved, it may be economic to change the size of cable in each section according to how it is affected by the derating factor; the saving in conductor material and labour costs would have to be weighed against the cost of jointing material and the cost of installing the joints. This arrangement would satisfy the overload regulations, providing the smallest conductor was protected by the overcurrent device; care would be needed with fault current protection.

It should also be noted that in practice (except where conductors are in contact with thermal insulation on one side only) it is not necessary to carry out separate calculations for each section of cable; the cable can be sized by taking the derating factors that will most increase the cable size.

A further point to note is that the ambient temperature derating factor for BS3036 rewirable fuses is only used when the fuse is giving overload protection. When the fuse is giving only short-circuit protection, the ambient temperature derating factor is obtained from the list of derating factors 'for other protective devices' in the CR Tables, or Table 4C1 in the regulations.

6. Voltage drop

An explanation of the regulations
The regulations require that the voltage drop up to the terminals of fixed equipment, shall ensure that the voltage at the equipment, is greater than the lowest operating voltage specified for the equipment in the relevant British Standard. Where no such standard exists, then the voltage drop within the installation should not exceed the value appropriate to the safe functioning of the equipment.

Where the electrical supply complies with the Electricity Supply Regulations, the above requirement can be satisfied by ensuring that the voltage drop from the origin of the installation to the fixed equipment does not exceed 4% of the nominal supply voltage.

When calculating the voltage drop in an installation the starting conditions, or inrush currents can be ignored. For instance, it is the full load current taken by a motor and not the starting current that is used in the voltage drop calculation.

If diversity has been applied, then the current load after diversity has been taken into account is used and not the total full load current. For example, the total connected load on a distribution board may be 100 amps, but after diversity is taken into account, the load could be 90 amps. 90 amps would therefore be used in the voltage drop calculation.

Extra low voltage
The same rule applies to extra low voltage circuits. Care is needed however, when applying the 4% voltage drop, to ensure that the voltage drop in the installation does not exceed a value appropriate to the safe functioning of the associated equipment in normal service.

Additional precautions have to be taken where extra-low voltage lighting is supplied from a transformer. Consideration has also to be given to the regulation of the transformer, and to what would happen if lamps in the circuit blew.

Motor circuits
Although the full load current of the motor is used to calculate voltage drop, consideration has to be given to the voltage drop due to the starting current for motors or similar circuits, otherwise trouble may be encountered.

For example: the supplier of electricity is allowed a tolerance on the supply voltage of ± 6%. If the nominal supply voltage is 415 volts, and is supplied at -6% and a further 4% voltage drop, allowance is made within the installation based on a supply voltage of 415V; then the voltage at the motor terminals when the motor is started will be 415V - 6% - 4%, giving a final voltage of 415V - 24.9V - 15.6V = 374.5 volts. The voltage available for the motor is only 374.5V, before any voltage is dropped due to the starting current. Further reductions in voltage could seriously affect the motor's ability to reach its final speed, or to deliver the necessary driving torque. This could cause a breach of the regulations.

Basis of voltage drop tables
The voltage drop in mV/A/m in the tables in the regulations gives the single-phase voltage drop for single-phase circuits, i.e the value given allows for the voltage drop in the phase and neutral conductors.

The three-phase voltage drop for three-phase three wire or four wire circuits, gives the line voltage drop between phases.

The voltage drops given in the tables are based on the circuit conductors working at the maximum permitted operating temperature, and at a load power factor the same as that for the cable. This leads to a larger voltage drop for the circuit, resulting in shorter lengths of circuit cables. Therefore, the voltage drop tables give the resistance, reactance and impedance of the cables for cable sizes larger than 16 mm^2. The tables in the regulations give the voltage drop per amp per metre, for each size of cable, whether used on single-phase or three-phase.

Correcting for load power factor

Where the load has a power factor less than unity a correction can be made to the mV/A/m value given in the tables to give a more accurate value of voltage drop.

For cables up to 16 mm^2 the figure in the volt drop tables is multiplied by cos Ø, i.e. the power factor. For cables larger than 16 mm^2 the following formula is used:

(cos Ø r + sin Øx) mV/A/m × cable length × full load current × 10^{-3},

where 'r' is the resistance and 'x' the reactance figure taken from the volt drop table for the cable size concerned, and ' Ø' is the power factor angle of the load.

In general, the voltage drop calculation should first be made by using the mV/A/m impedance value given in the tables, the above formula being used only if the calculation gives a voltage drop slightly higher than the desired value, or a conductor length less than that required.

Table MISC 2 in Part 4 gives the sine of the power factor angle, to enable the sine of the power factor to be easily obtained; it can of course, be calculated

Correction for conductor operating temperature

Where a conductor is not carrying its tabulated current rating, its temperature will not reach the operating temperature allowed for the conductor. This means that a correction for the lower temperature can be made, as the voltage drop tables are based on the allowed operating temperature for the conductor.

Since reactance is not affected by temperature, it is only the resistive element that needs correction. This can be done by using the formula given earlier to calculate the actual temperature of the conductor and then to calculate it's resistance; alternatively the correction factor for operating temperature is obtained by the following formula:

$$C_t = \frac{230 + t_p - \left(A^2G^2 - \frac{I_b^2}{I_{tab}^2}\right)(t_p - 30)}{230 + t_p}$$

Where: t_p is the operating temperature given in the tables.
           A is the ambient temperature correction factor.
           G is the correction factor for grouping.
           I_b is the design or full load current of the circuit and,
           I_{tab} is the tabulated current of the cable from the tables.
           C_t is the correction factor, which is equal to the ratio of conductor resistances.

This formula cannot be used where the protective device is a BS 3036 fuse, or where the actual ambient temperature is less than 30 °C.

Where the conductors are 16 mm² or less the resistance given in the volt drop tables is multiplied by C_t to give the revised mV/A/m volt drop; for larger cables, the impedance given in the voltage drop tables is multiplied by C_t.

Combining the factors

Where the load for the circuit has a power factor, then correction for the conductor operating temperature can be combined with the formula for correcting for the loads power factor. Thus for cables up to 16 mm² the revised mV/A/m is obtained from the following formula:

$C_t \cos \emptyset$ (mV/A/m from the tables) × cable length × full load current × 10^{-3}

Where the cables are larger than 16 mm² the following formula is used:

$$(C_t \cos \emptyset \; r + \sin \emptyset x) \times \text{cable length} \times \text{full load current} \times 10^{-3} \text{ volts}$$

The correction factor is not applied to the reactive component, since temperature has no effect on reactance. Thus for very large cables where the reactance is greater than the resistance such that:

$$\frac{x}{r} \geq 3$$

no correction need be made for conductor temperature.

Calculating voltage drop

When corrections for operating temperature or load power factor are not important, there are two ways that the correct size of cable can be checked as being suitable for voltage drop. The first is to look at the mV/A/m for the size of cable for the correct current carrying capacity, and then calculate the voltage drop as follows:

$$\text{Voltage drop} = \text{full load current} \times \text{length} \times \text{mV/A/m} \times 10^{-3}$$

The 10^{-3} converts the mV into volts, (i.e. divide by 1000). The problem with this method is that, if the voltage drop is larger than that allowed for the circuit, the calculation has to be repeated with different cable sizes until the voltage drop is within the amount allowed.

The second method is to re-arrange the voltage drop formula, so that the mV/A/m is the unknown quantity as follows:

$$\text{mV/A/m} = \frac{\text{Voltage drop allowed} \times 1000}{\text{Full load current} \times \text{cable length}}$$

The cable size is then found by looking down the impedance column in the volt drop table until the value is either the same as, or less than, the figure calculated. This method, therefore, only requires one calculation.

There is another method which is to use the Factor from the CR tables in Part 3. This is fully explained in 'How to use the Tables' in Part 3. A brief explanation of how it works is as follows: The load is multiplied by the length of circuit, and a factor that is equal to or larger than the figure calculated for load multiplied by current, is chosen from the appropriate current rating table. This then gives the size of cable which is suitable for 4% voltage drop. It makes calculations quick and easy, and does not involve using 10^{-3}.

Worked examples
Example 1 - Single-phase voltage drop.
A 3 kW fan heater is to be wired in twin and cpc cable clipped direct to a non-metallic surface. If the length of cable run is 25 metres, what size of cable will be required for 9.6V voltage drop, if the supply voltage is 240V ?

Working (first method)

Current taken by fan heater $= \dfrac{3000\ W}{240\ V} = 12.5$ amps

Table CR7 column 10, 1.0mm^2 rated 15A appears large enough. Volt drop/A/m = 44 mV from column 11 of same table.

$$\text{Total voltage drop} = \dfrac{25m \times 12.5A \times 44mV/A/m}{1000} = 13.75V$$

13.75 volts drop is too much. Try 1.5 sq.mm. rated at 19.5A, mV/A/m = 29

$$\text{Total voltage drop} = \dfrac{25 \times 12.5 \times 29}{1000} = 9.06V$$

which is satisfactory. Cable size achieved with a minimum of two calculations.

Working (second method)
Current taken by fan heater as before = 12.5 amps.

$$mV/A/m = \dfrac{9.6V \times 1000}{25 \times 12.5} = 30.72\ mV/A/m$$

Now look in column 11 of table CR 7 for a mV drop equal to, or less than 30.72. The nearest lower rating is 29 mV/A/m for 1.5 mm^2 cable, so this is the size to choose for voltage drop purposes. A check must now be made that this cable size has the correct current-carrying capacity, and whether any derating factors that could affect the cable size are applicable.

Working (third method)
Using the factor method from CR 7, length × load = 25 × 12.5 = 312.5. Now select a factor from column 13 equal to, or larger than 312.5. This gives a cable size of 1.5 mm^2 as before, but is easier to work out.

Working (fourth method)
An even quicker method is to look at Table INST 2 in Part 3, which gives the cable size for a 3 kW load as 1.5 mm^2, irrespective of protective device type. This table also shows that the cable size is also suitable for protection against indirect contact, short-circuit protection, and thermal protection of the cpc, in addition to voltage drop.

Example 2 - Three-phase voltage drop.
A motor with a full load current of 15 amps, is to be wired in pvc single core cable in conduit installed on the surface. If the supply voltage is 415, and the length to the motor is 30 metres, what size cable will be required for voltage drop purposes ?

VOLTAGE DROP

Working
Voltage drop allowed 4% of 415V = 16.6V.

$$mV/A/m = \frac{16.6 \times 1000}{15A \times 30m} = 36.89 \, mV$$

From Table CR 1 column 7, 1.5 mm^2 is required, with a mV/A/m of 25.
Proof: Voltage drop = 15A x 30 m x 25 ÷ 1000 = 11.25 volts.

Working (alternative method)
Using the factor method. Length 30 m × load 15A = 450. Select a factor from table larger than 450.
Column 9 next larger factor = 664 for 1.5 mm^2 cable. This is the cable size required for voltage drop.

Calculation of circuit length
The volt drop formula can also be used to determine how long a circuit cable can be run for a certain voltage drop.

Example 3 - Motor circuit.
Suppose a motor with a full load current of 20 amps, wired in 2.5 mm^2 pvc cable enclosed in conduit, is to be moved to a new location, and suppose the volt drop up to the motor distribution board is 3 volts. How far can we move the motor without having to change the existing 2.5 mm^2 cables, if the supply voltage is 415V, the existing length of run is 24 metres, and the total voltage drop allowed is 16.6 volts?

Working

$$\text{Length of cable allowed} = \frac{\text{Volt drop allowed} \times 1000}{\text{Full load current} \times mV/A/m}$$

From CR1 column 7 mV/A/m for 2.5 mm^2 cable is 15 mV

$$\text{Length allowed} = \frac{(16.6 - 3) \times 1000}{20 \times 15} = 45.33 \text{ metres}$$

Maximum length to which cables can be extended is 45.33 - 24 = 21.33 metres.

Working (alternative method)
Using the factor method explained in Part 3.

$$\frac{\text{Factor}}{I_b} \times \frac{\text{Available Volt drop}}{\text{Allowed Volt drop}} = \text{Actual length allowed}$$

From CR1 for 2.5 mm^2 cable, cable factor = 1107 (Column 9)

$$\frac{1107}{20} \times \frac{(16.6 - 3)}{16.6} = 45.34 \text{ metres. As before the circuit can be extented by 21 m.}$$

Although extending the circuit 21 metres would satisfy the voltage drop requirements, a check would have to be made that the following were still satisfactory, with the added impedance in the circuit:
1. the phase earth loop impedance of the circuit for protection against indirect contact,
2. the protective conductor was still satisfactory,
3. short circuit protection was still given by the protective device,
4. the voltage drop due to the starting current, when the motor was started, would allow the motor to develop its correct starting torque.

Correction factor examples
So far, only simple voltage drop calculations have been considered, but the occasion will arise when the voltage drop is limiting the correct length of cable being installed by just a few metres; under these circumstances it can be economical to carry out the additional calculations.

Example 4 - Power factor in motor circuit.
A three-phase motor with a full load current of 150 A, and a power factor of 0.8, is to be fed by a 150 metre PVCSWA&PVC armoured cable clipped direct to a non-metallic surface. What size cable would be required if the maximum voltage drop allowed in the cable is 16.6 volts?

Working
Trying the simple method first:
From table CR9 Column 6, a 50 mm^2 cable will carry 151 A, and from column 9 the factor is 20494

$$\text{Actual length allowed} = \frac{\text{Factor}}{I_b} = \frac{20494}{150} = 136.6 \text{ metres.}$$

This is less than the 150 metres required, so try compensating for the power factor.
Power factor = 0.8. From Table MISC 2 Part 4, when power factor is 0.8, sine = 0.6.
From Table CR9 column 7 'r' = 0.8 and from column 8 'x' = 0.14.
Substituting these in the formula from Part 3:

$$\text{Actual length allowed} = \frac{\text{Available volt drop} \times 1000}{(Cos\emptyset r + Sin\emptyset x) I_b} =$$

$$\frac{16.6V \times 1000}{(0.8 \times 0.8 + 0.6 \times 0.14) \times 150} = 152.85 \text{ m}$$

Thus by taking the load power factor into account, the cable is suitable as far as voltage drop is concerned.

Example 5 - Correcting for both power factor and conductor temperature.
A PVCSWA&PVC armoured cable is to be installed from an HRC fuse in a distribution board to a 415V 55 kW motor, along with 5 other cables fixed to a perforated metal cable tray, where the cable sheaths will be touching. If the cable length is 100 metres, and the setting of the overloads in the starter are: min 90, mid point 100 A, maximum 110 A, and the power factor of the load is 0.75, what size cable would be required to satisfy voltage drop, if the ambient temperature is 30 °C, the voltage drop in the 3 phase feeder cable up to the distribution board is 9 V, and the total voltage drop allowed is 4%?

VOLTAGE DROP

Working

In this case the cable is grouped, so this will determine the size of cable. From table GF1 Part 4, the derating factor for 6 cables is 0.74. The voltage drop allowed is 415V × 4% = 16.6V. From Table MC1 Part 4, 55 kW at 415V gives motor current of 98 A; but since the overloads can be turned up to the maximum value, this should be used for the voltage drop calculation. Therefore :-

$$I_t = \frac{110}{0.74} = 148.65 \text{ amps}$$

From table CR9 column 14, 50 mm² cable is required. I_{tab} = 163 A, and the factor is 20494, but is this cable size satisfactory? Try the simple calculation first, using formula from example 4.

$$\text{Length allowed} = \frac{20494}{110} \times \frac{(16.6 - 9.0)}{16.6} = 85.3 \text{ metres}$$

This is too much, so try compensating for the load's-power factor.
From CR9, for 50 mm² cable 'r' = 0.8 and 'x' = 0.14
From MISC 2 Part 4 power factor 0.75 = sine 0.6614

$$\text{Length allowed} = \frac{\text{Voltage drop available} \times 1000}{(\cos\emptyset r + \sin\emptyset x) I_b}$$

$$\text{Length allowed} = \frac{7.6 \times 1000}{(0.75 \times 0.8 + 0.6614 \times 0.14)\, 110} = 99.76 \text{ metres}$$

This is still too short, so investigate whether using the actual operating temperature of the conductors will allow a longer cable. (In practice this figure would be acceptable)

$$C_t = \frac{230 + t_p - \left(A^2 G^2 - \dfrac{I_b^2}{I_{tab}^2}\right)(t_p - 30)}{230 + t_p}$$

$$C_t = \frac{230 + 70 - \left(1 \times 0.74^2 - \dfrac{110^2}{163^2}\right)(70 - 30)}{230 + 70} = 0.988$$

Use this value in the formula:

$$\text{Length allowed} = \frac{7.6 \times 1000}{(0.988 \times 0.75 \times 0.8 + 0.6614 \times 0.14)\, 110} = 100.8 \text{ metres}$$

This is now greater than the length required for the cable.

Summary

This chapter has shown various ways of calculating the voltage drop in a circuit. Whichever method is used, a check must always be made that the cable size chosen will have the correct current-carrying capacity for the load and the conditions under which the cable is installed. It is also necessary to check that the circuit complies with the other requirements of the regulations, such as; shock protection, earthing, short circuit current, overloads and current-carrying capacity of the conductors, etc.

7. Protection against electric shock

A person can receive an electric shock in two ways; firstly by coming into contact with live parts (direct contact) and secondly by touching metallic parts that have become live due to a fault (indirect contact). Protection against electric shock is covered by three Sections in Chapter 41 of the regulations.

Section 411 - Protection against both direct and indirect contact.
Section 412 - Protection against direct contact.
Section 413 - Protection against indirect contact.

Protection against electric shock can be achieved by application of the regulations in Section 411, or by the combined application of the regulations in Sections 412 and 413. Looking at each of these in turn;
Section 411 concerns the following methods of protection:
 1. protection by SELV,
 2. protection by limitation of discharge energy.
The rules for item 1 are covered in Part 1 of the handbook, so there is little point in repeating them again in this chapter.

Protection by limitation of discharge energy means that the equipment incorporates a means of limiting the current which can pass through the body of a person or animal to a value lower than the shock current. Any circuits using this method of protection must be separated from other circuits; the separation being equivalent to that specified in the regulations, for SELV circuits.

Functional extra low voltage is the term used, where extra low voltage is used, but compliance with all the SELV regulations is not possible. In these circumstances measures have to be taken to provide protection against electric shock.

Protection against direct contact
Four basic protective measures for protection against direct contact are given in the regulations, and are illustrated by the following diagram:

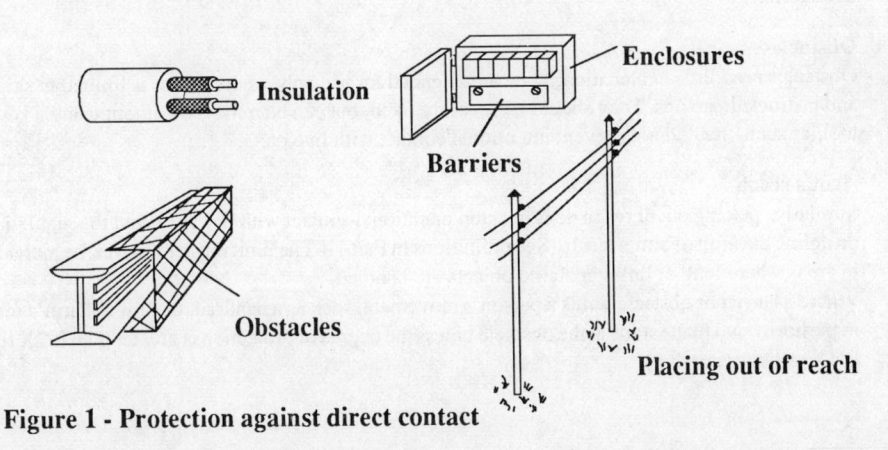

Figure 1 - Protection against direct contact

Insulation

Insulation is the basic insulation of cables and parts required in every installation. The regulations require that the insulation can only be removed by destruction, and that it will withstand any electrical, mechanical, thermal and chemical stresses to which it may be subjected whilst it is in service.

Barriers or enclosures

Where protection by barriers or enclosures is used to protect against direct contact, then the degree of protection must be at least IP2X, which is the Index of Protection for the standard finger 80 mm long and 12 mm in diameter. Where the opening in equipment has to be larger than IP2X to enable maintenance to be carried out, precautions must be taken to ensure that there can be no unintentional touching of live parts, and that persons are warned of the proximity of live parts within the enclosure. Where a top surface of equipment is readily accessible, the degree of protection is more stringent, and must not exceed IP4X.

During the maintenance of an installation, or the carrying out of new works, enclosures may have to be opened and barriers removed. The regulations give four alternative methods of safety against direct contact, these are detailed as follows:

1. the opening of the enclosure, or the removal of a barrier must only be possible by using a key or tool. This means that access is limited to skilled persons who should know the dangers, and take the necessary precautions,
2. opening an enclosure, or removing a barrier can only be carried out after the supply to live parts has been disconnected. The supply may only be restored after the barriers have been replaced or the equipment reclosed. Isolators interlocked with doors is an example of this type of protection,
3. an intermediate barrier having a degree of protection of IP2X is provided to prevent contact with live parts, the barriers being removable only by using a tool.

Care has to be taken when selecting distribution boards to comply with the requirements for protection against direct contact. Quite often, access to the interior of distribution boards is not limited to skilled or instructed persons; quite frequently, barriers are not replaced after work has been completed, and many boards have knurled screws for fixing the door, allowing anyone access to the interior.

Obstacles

Obstacles need little explanation. They are intended for use only where access is limited to skilled and instructed persons. They should be securely fixed, but can be removed without using a key or tool; as such, they do not prevent intentional contact with live parts.

Arm's reach

Similarly, placing out of reach does not stop intentional contact with live parts, but the regulations do define the limit of arm's reach. (See definitions in Part 1.) The limit must, however, be increased in areas where long or bulky metallic objects are handled.

Where a barrier or obstacle limits a person's movement, such as a handrail, the limit of arm's reach in the horizontal plane starts at the obstacle unless the degree of protection is greater than IP2X (i.e., IP3X or larger number)

Residual current devices (RCD)

Although a residual current device can reduce the risk of injury from electric shock, a residual current device cannot be used as the sole means of protection against direct contact. For the risk of injury to be reduced, the live parts must either be insulated, protected by barriers, enclosures, or obstacles, or by placing the conductors out of reach. Additionally, the tripping current of the RCD must not exceed 30 mA, and must operate within 40 msec., with an earth fault current of 150 mA, when used as supplementary protection against direct contact.

Protection against indirect contact

When the primary insulation fails, a fault which can give rise to danger occurs, even though the live electrical parts and conductors cannot be touched. The regulations therefore give requirements that try to limit the degree of danger.

Five methods of protection against indirect contact are given:
1. earthed equipotential bonding and automatic disconnection of the supply,
2. using Class II equipment or equivalent insulation,
3. non-conducting location,
4. earth-free local equipotential bonding,
5. electrical separation.

The last three items in the list would normally be covered by a detailed specification; they do not arise very often, and are therefore beyond the scope of this handbook.

As far as the second item is concerned, a protective conductor has to be installed to all double insulated items used in an installation such as switches and ceiling roses. Other areas where double insulation is used exclusively for protection against indirect contact would be covered by a detailed specification and will not arise very often, so they will not therefore be covered.

One area where double insulation will be encountered is with portable tools and equipment; in this case a protective conductor is not installed to the tool or the equipment.

The most common form of protection against indirect contact used in this country is item one in the list, so this will be looked at in detail.

Earthed equipotential bonding and automatic disconnection

The basic rule requires the characteristics of the protective devices, the earthing arrangements for the installation, and the impedances of the circuits concerned to be co-ordinated, so that during an earth fault, the voltage appearing between simultaneously exposed and extraneous conductive parts occurring anywhere in the installation shall be of such magnitude and duration as not to cause danger.

This basic rule co-ordinates two subjects: equipotential bonding and automatic disconnection. They are more easily understood if considered separately.

Equipotential bonding

To minimise the magnitude of the voltage that can appear between exposed and extraneous conductive parts, the regulations require that main equipotential bonding conductors are installed between the extraneous conductive parts and the main earthing terminal of the installation.

The items to be bonded include, water pipes, central heating pipes, gas installation pipes, other service pipes, metallic ducting, exposed metallic parts of the building structure and the lightning protective system. These are illustrated by 1 to 7 in the following drawing.

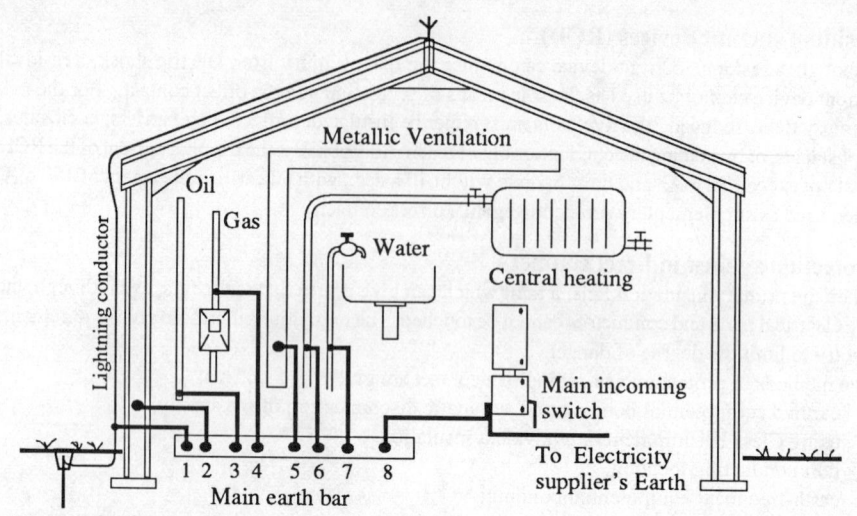

Figure 2 - Main equipotential bonding

The bond to the gas installation pipe, water pipe, or any other metallic service should be made as near as possible to the position the service enters the building. Where a meter is installed, or there is an insulating insert in the service pipe, the connection has to be made on the consumer's side of the insert or meter, before there are any branches off the pipe. It is recommended that the bond should be within 600 mm of the meter outlet, or the entry point of the service into the building.

Where the supply is not PME, the cross-sectional area of the bonding conductors should be half that of the main earthing conductor, the minimum size allowed being 6 mm^2. The maximum size need not exceed 25 mm^2 if the bonding conductor is a copper conductor, or a conductor of a different material, having an equivalent conductance to the copper conductor.

If the supply is PME, then the PME regulations apply, and the minimum size will be as given by the local electricity supplier, as shown in the following table.

Equivalent copper cross-sectional area of supply neutral conductor	Minimum size of copper main equipotential bonding conductor
Up to 35 mm2	10 mm^2
36 mm^2 to 50 mm^2	16 mm^2
51 mm^2 to 95 mm^2	25 mm^2
96 mm^2 to 150 mm^2	35 mm^2
Over 150 mm^2	50 mm^2

Supplementary bonding

If there are any metallic parts going out of the building into the ground, they should have been included in the main equipotential bonding, isolated parts being bonded to extraneous conductive parts connected to the main bonding conductors.

The only area left for supplementary bonding (except for special situations) is where a voltage can appear between an exposed conductive part and an extraneous conductive part.

The principle is that if the current can be disconnected within the times specified in the regulations, then there is no need for a supplementary bond.

Where the disconnection time cannot be met, then either local supplementary equipotential bonding has to be installed, or protection has to be provided by an RCD.

Where supplementary equipotential bonding is provided between exposed and extraneous conductive parts, the bonding conductor is sized in accordance with the following formula:

$$R = \frac{50}{I_a}$$

where R is the resistance of the bonding conductor and I_a is the minimum current that will disconnect the protective device in 5 seconds. The value of I_a can be easily obtained by taking the design value of Z_S for 5 second disconnection time from the ZS tables, and dividing this into 240 V.

A check is then made to ensure that the conductance of the supplementary bonding conductor is at least half that of the protective conductor connected to the exposed conductive part to which the bonding conductor is connected. If not, the larger conductor must be installed. If mechanical protection is not provided for the bonding conductor, its cross-sectional area shall be not less than 4 mm^2.

If it is decided to provide protection by using an RCD instead of providing supplementary equipotential bonding, then the RCD must comply with the condition: $Z_s \times I_{\Delta n} \leq 50$ V

Rooms containing a fixed bath or shower must have local supplementary equipotential bonding. Where plastic baths, plastic waste pipes, plastic shower trays and no exposed conductive parts are present, it will only be necessary to bond all the extraneous conductive parts at one point within the bathroom; this point will however have to be accessible for inspection, especially in domestic premises, where the electricity supplier will inspect the installation before connecting the supply.

Where extraneous conductive parts are bonded together, the minimum size of conductor, if it is sheathed or otherwise mechanically protected, is 2.5 mm^2; if not, the minimum size is 4 mm^2. If one of the extraneous conductive parts is bonded to an exposed conductive part, then the conductor must be sized in accordance with the earlier paragraph on supplementary bonding.

Where supplementary bonding conductors are installed in the roof void over the bathroom or in the floor, they can be considered as being protected against mechanical damage, and the minimum size of conductor specified in the regulations can be used.

Metal framed windows

Metal windows do not need supplementary bonding, if the circuits in the vicinity of the windows can be disconnected within the disconnection time specified in the regulations for the type of circuit installed. Bonding always needs careful consideration, since it is very easy to introduce a danger where none existed previously.

Installing a supplementary bond onto a metal window frame may create a safe situation inside the equipotential zone, but a possible dangerous one outside the zone. Protection using an RCD would be preferable to installing supplementary bonds onto metal parts that are exposed to the outside of the building, such as the metal framed window.

It must be remembered that the object of bonding is to create an equipotential zone, so that a voltage appearing on the main earth bar will also appear on all exposed and extraneous conductive parts within the zone. This does not mean that a person will not receive an electric shock when main equipotential bonding is carried out.

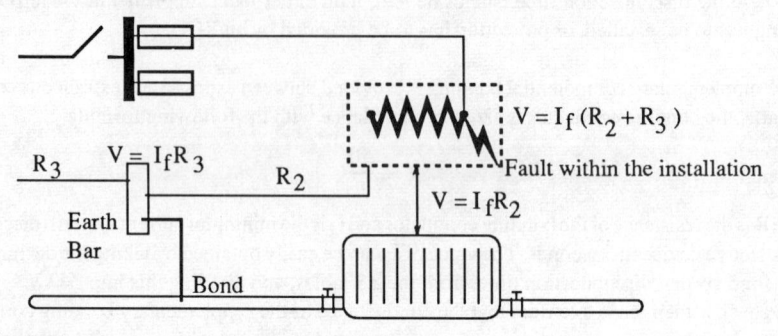

Figure 3 - Shock voltage with bonding

In Figure 3 the voltage $I_f(R_2 + R_3)$ is the voltage appearing on the exposed conductive part with respect to the earthed point at the source of supply when a fault occurs. I_f is the fault current, R_2 the resistance of the cpc to the main earth bar, and R_3 is the resistance of the cpc from the earth bar back to the source of supply. It can also be seen from Figure 3 that even when the extraneous conductive parts are bonded to the main earth bar, a voltage will appear between an exposed conductive part and an extraneous conductive part when a fault occurs. The bond has only reduced the magnitude of the voltage.

If a metal framed patio door is electrically bonded to the main earth bar, a voltage on the main earth bar will also appear on the frame of the patio door. A person touching the frame of the door, would then be touching that voltage. The degree of current flowing through the body would depend upon the type of floor, i.e., its insulation value, and the type of footwear worn by the individual. The resistance of the ground outside the window would be zero; a window cleaner, for instance, would therefore be subjected to a higher shock voltage.

Automatic disconnection

Since a voltage will exist between exposed and extraneous conductive parts when a fault occurs, it is essential to remove this voltage as quickly as possible; this is the object of automatic disconnection.

The actual disconnection time is dependent upon the environmental conditions, and whether a person or livestock is likely to be in contact with exposed conductive parts at the instant of the fault. The basic formula used to ensure that a protective device disconnects in a specified time is:

$$Z_s \leq \frac{U_o}{I_a}$$

Where a circuit supplies a socket outlet, hand-held Class I equipment, or portable equipment intended for manual movement, then the disconnection times allowed are:

PROTECTION AGAINST ELECTRIC SHOCK

Voltage to earth U_o	Disconnection time 't' seconds
120	0.8
220 to 277	0.4
400	0.2
Over 400	0.1

These disconnection times are only suitable for normal environmental conditions. The disconnection times allowed are therefore modified for areas having poor environmental conditions, such as bathrooms and construction sites.

Where a circuit feeds fixed equipment, or when reduced low voltage is used, the disconnection time allowed is 5 seconds, but if the distribution board feeding the fixed equipment also has circuits supplying socket outlets, hand-held class I equipment, or portable equipment intended for manual movement, then either of the two following conditions have to be complied with.
1. The impedance of the protective conductor from the distribution board back to the main earth bar, at which point the main equipotential bonding conductors are connected, should comply with table 41C (Table PCZ 8 Part 3) or,
2. equipotential bonding shall be carried out from the distribution board to all the services that are connected to the main equipotential bonding conductors.

As far as condition '1' is concerned, the impedance of the cpc can be found for any size and type of protective device by using the following formula:

$$Z_{cpc} = \frac{50 \times Z_S}{U_o} \, \Omega$$

where Z_S is the earth loop impedance for 5 seconds disconnection and U_o is the nominal voltage to earth. The value of Z_S can be obtained from the ZS tables in Part 3, when the voltage is 240 V.

The regulations state that the figures given in the tables are for the 'gG' type fuses; for the type 'gM' or other types of fuses, the manufacturer should be consulted.

The types 'gG' and 'gM' are the utilisation category introduced in the IEC 269 standard, and it would perhaps be appropriate if these categories were explained at this point.

Utilisation categories were introduced as some of the fuse links used in some counties will interrupt short circuit fault currents safely, but will not safely interrupt overload currents.

The utilisation classes are based on a two letter code outlined as follows:
First letter indicating the breaking range:
    g full range breaking capacity fuse link,
    a partial range breaking capacity fuse link.
Second letter indicating the utilisation category:
    G General application including motor circuits,
    M Protection of motor circuits.

Confusion often exists as to the rating of a motor fuse due to the label having two numbers as shown below.

Fuse reference

32 M 50

Fuse body size ⎯⎯⎯⎯⎯⎯⎯⎯↑   ↑⎯⎯⎯⎯⎯⎯ Fuse element size

If the Type T characteristics in Part 4 are looked at, it will be seen that the 32M50 fuse has the same characteristic as the 50 A fuse. This means that the 32M50 fuse is a 50 A fuse element in a 32 A fuse size body. The lower number therefore indicates the maximum continuous current rating of the fuse, whilst the larger number indicates the fuse element size for short duration overload.

There is no need to consult the manufacturers as far as gM type fuses are concerned: just use the Z_s of the larger number, which in the above illustration is 50 A When dealing with shock protection, earth fault protection, or short-circuit protection, the fuse element size is used.

Calculations

There is a close relationship between automatic disconnection and the sizing of protective conductors, since the same conductors are involved, i.e. the phase earth loop impedance Z_s. This chapter should therefore be read in conjunction with the chapter on circuit protective conductors.

The calculation of phase earth loop impedance is carried out in the same way as that for sizing the circuit protective conductors.

$$Z_s = Z_E + Z_{inst}$$

where Z_E is considered to be the earth loop impedance external to the circuit under consideration, and Z_{inst} is the impedance of the phase and cpc conductors of the circuit.

Since reactance only needs to be taken into account when the conductor size exceeds 35 mm^2, impedance in the above formula can be replaced by resistance for small cables.

The values of resistance or impedance used should be based on the design temperature of the circuit conductors. The value of Z_s will also have to be worked out for the temperature at which the installation will be tested.

In practice it is better to work out the size of the protective conductor and Z_s. Then check with the tables that Z_s complies with the value required for the protective device being used, and the disconnection time required for the circuit. In certain cases, the size of the cpc is determined by the thermal constraints on the cpc, and not by the limits laid down in the tables for shock protection. This accounts for the difference between the PCZ tables and the values for 5 seconds in the ZS tables.

Temperature effect on Z_s

It must be remembered that automatic disconnection is concerned with phase earth faults. The fault current will therefore raise the temperature of the phase and protective conductors. The calculated value of Z_s must, therefore, take into consideration the resistance of the conductors at the temperature created by the fault current in those conductors. Otherwise, the increase in conductor resistance due to the fault current will reduce the fault current flowing, thus increasing the disconnection time for the circuit. This could be more than that allowed by the regulations as shown in Figure 4.

I_{f1} = Fault current ignoring temperature rise of conductors caused by fault current

I_{f2} = Reduced current due to temperature rise of conductors caused by fault current

$I_f = \dfrac{V \text{ phase}}{Z_s}$

t_1 = Disconnection time required

t_2 = Actual disconnection time due to conductor resistance and fault current

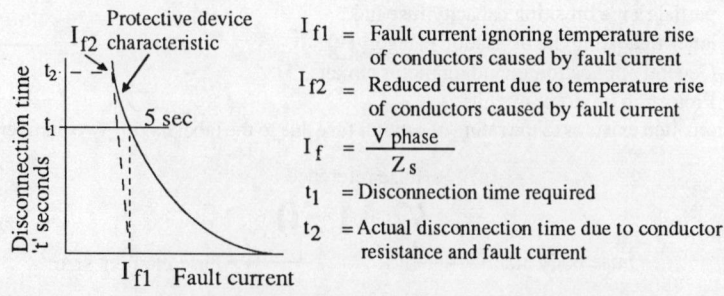

Figure 4 - Effect of conductor temperature rise on fault current

In most cases the final temperature of the conductors due to the fault current will not be known. In these circumstances the impedance of the conductors should be calculated on the average temperature attained by the conductor. The average temperature is obtained by taking the average of the assumed initial temperature and the limit temperature for the conductor insulation.

For example: working temperature of the conductor 70 °C: limit temperature of pvc insulation 160 °C: average temperature is 70 °C + 160 °C divided by 2, giving 115 °C. The impedance of the conductors at 115 °C will be used in calculations.

When conduit, trunking, or steel wired armoured cables are used as the cpc, it is important that the impedance value of the cpc is used, since the reactance has a significant influence on the overall impedance, although the reactance is not affected by temperature.

Socket outlet circuits

There is a relationship between the effects of an electric shock on a person, the magnitude of the shock voltage, and the speed at which the current is disconnected. Thus a person can tolerate a relatively high shock current through the body, providing that the circuit is disconnected fast enough; this is the basis of Regulation 413-02-12.

The principle of Regulation 413-02-12 is to limit the resistance (impedance) of the protective conductor so that the voltage appearing on exposed conductive parts under fault conditions is limited to 50 volts, or if the voltage is a higher value, the circuit is disconnected faster.

The application of Regulation 413-02-12 is achieved by ensuring that the resistance (impedance) of the protective conductor, back to the point at which equipotential bonding is carried out, does not exceed the value given in table PCZ 8 for the appropriate protective device, or the value given in Table 41C in the regulations.

In practice, where the circuit is single-phase and the voltage U_o is 240 volts, it will be found that voltage drop constraints limit the circuit length more than the impedance of the protective conductor.

Worked examples

Example 1 - Protection by limiting impedance of ring circuit cpc.

A ring circuit is protected by a 30 A rewirable fuse, the cable used is 2.5/1.5 twin & cpc, what can the maximum length of the circuit be from the main equipotential bonding?

Working

The resistance of 1.5 mm^2 cpc from table RC1 Part 4 is 16.698 Ω / 1000 m.
From Table PCZ 8 in Part 3, maximum value cpc impedance for 30 A rewirable fuse is 0.58 Ω

$$\text{Maximum circuit length} = \frac{0.58 \times 4}{0.016698} = 138.94$$

From Table INST 4, maximum length allowed is 105 metres; thus, other factors restrict length.

Example 2 - Protection by limiting impedance of radial socket circuit cpc.

A three-phase 415 V radial socket outlet circuit is to be protected by a 32 A BS 88 fuse, and wired in 4 mm^2 PVCSWA&PVC copper cable installed open and clipped direct. What can the maximum length of the circuit be?

Working

From Table PCZ 8, maximum allowed value of impedance for cpc is 0.4 Ω. From RC1 Part 4, the impedance of the phase conductor is 6.362 Ω per 1000 metres. From ELI 1 Part 4, the phase earth loop impedance of the cable is 15.22 Ω per 1000 metres. The impedance of the cable armour will be 15.22 - 6.362 = 8.858 Ω / 1000 m. = 0.008858 Ω / m. This result is the same as given in ELI 1 for the armour; the calculation was, therefore, unnecessary.

Maximum length for circuit for shock protection = $\dfrac{0.4}{0.008858}$ = 45 metres.

From CR9, Factor = 1747. Socket circuit full load current will therefore be the rating of the protective device.

Maximum length for voltage drop = $\dfrac{1747}{32}$ = 54.59 metres.

Shock protection constraint limits circuit length more than voltage drop.

Example 3 - Protection by limiting phase earth loop impedance.
A 6 mm^2 twin & cpc cable is to be installed to a cooker unit containing a socket outlet; the protective device is a 30 A rewirable fuse, the circuit length is 28 m, and Z_E up to the distribution board is 0.35 Ω. Will it satisfy the regulations for indirect contact ?

Working
A 6 mm^2 cable has a cpc of 2.5 mm^2 (Table MISC 1 Part 4)
The resistance per metre = 0.00425 + 0.010226 Ω at 115 °C
Z_S = Z_E + Z_{inst}
Z_S = 0.35 + 28(0.00425 + 0.010226) = 0.755 Ω
From Table ZS 3 Part 3, for 30 A fuse 0.4 second disconnection Z_S allowed = 1.14 Ω.
Therefore circuit is suitable for shock protection.

Example 4 - Protection by limiting impedance of cpc.
Same problem as example 3.

Working
From PCZ8, maximum impedance for cpc and 30 A rewirable fuse is 0.58. From MISC 1, size of cpc is 2.5 mm^2. From RC1, resistance of 2.5 mm^2 cable is 0.010226 Ω /m.

Maximum length of circuit from equipotential bonding = $\dfrac{0.58}{0.010226}$ = 56.72 metres.

Maximum length from INST 1 for 7 kW load is 45 metres. 28 m therefore satisfactory.

Example 5 - Protection by limiting phase earth loop impedance.
A 2.5 mm^2 circuit is to be installed in 20 mm heavy gauge conduit to fixed equipment from a 20 A, HRC fuse. If it is 30 m long, will it comply with the shock protection regulations if Z_E up to the distribution board is 0.8 Ω ?

Working
The resistance per metre of 2.5 mm^2 cable at 115 °C is 0.010226 Ω from Table RC1 Part 4
The impedance per metre of 20 mm conduit is 0.005875 Ω (Table ZCT1 Part 4)
Z_S = Z_E + Z_{inst}
Z_S = 0.8 + 30(0.010226 + 0.005875) = 1.283 Ω
From Table ZS1A, Z_S allowed for 20 A HRC fuse disconnection 5 secs is 3.04 Ω; therefore circuit complies.

General

It does not matter whether the cable is micc, PVCSWA&PVC, or whether trunking and conduit are used, the calculation is always worked out the same way. It saves time if the cpc is sized first, then the value of Z_S obtained for sizing the protective conductor is checked against the tables for protection against indirect contact.

Using the operating temperature given in the current-carrying capacity tables to obtain the average of operating temperature and the conductor's insulation limit temperature, will give a lower fault current than will occur if the conductor is not carrying its rated current-carrying capacity.

Calculations based on the operating temperature given in the current rating tables, will err on the safe side. The formulae given earlier in Chapter 4 Sizing Conductors can be used to determine the resistance of a conductor, when the operating temperature is less than that given in the current rating tables.

Where distribution cables are concerned, the fault current of the final circuit should be added to the full load current carried by the cable; this value of current is then used to determine the operating temperature of the conductor, used in the phase earth loop impedance calculations.

8. Circuit protective conductors

Options available
The regulations give two options for the determination of the size of the circuit protective conductor; it can either be calculated or determined from table 54G.

Phase earth loop
It is important to understand what an earth fault is, and the path the current takes when such a fault occurs. The following diagram shows a TN-S system, starting at the source of energy, with the feeder cables up to the origin, and from the origin to the end of a final circuit. Meters and fuses etc., have been omitted for clarity.

Phase earth loop = $Z_A + Z_B + Z_1 + Z_2 + Z_C + Z_D$

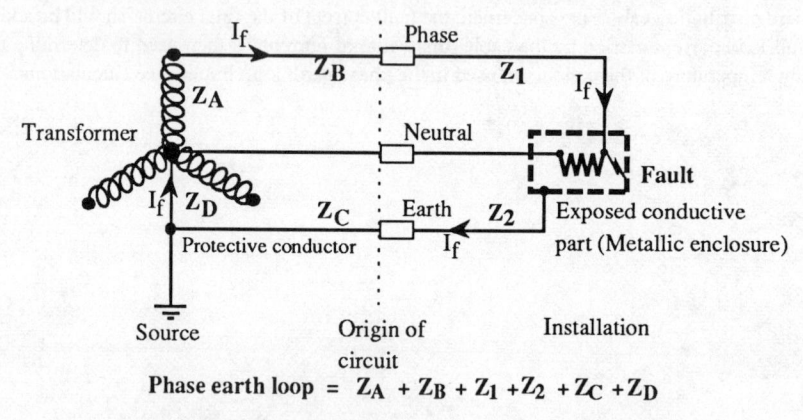

Figure 1 - Phase earth loop impedance

The equipment is shown with a phase to earth fault, which from the point of calculations is always assumed to be of negligible impedance, i.e., there is no resistance between the phase conductor and the exposed conductive part.

At the instant of the fault, current will flow through the phase winding of the transformer, along the phase conductor up to the fault, and then along the protective conductor back to the transformer. This is called the phase earth loop, and the magnitude of the current is only limited by the impedance of the transformer phase winding, the phase conductor, and the impedance of the protective conductor.

These are represented in the diagram by Z_A, Z_B, Z_1, Z_2, Z_C, and Z_D. The total of all these impedances is called the system impedance Z_s. The phase earth loop impedance Z_s (sometimes just referred to as the earth loop impedance) is obtained from the addition of all the individual impedances in the loop; i.e.

$$Z_s = Z_A + Z_B + Z_1 + Z_2 + Z_C + Z_D$$

In the diagram, impedances Z_A, Z_B, Z_C, and Z_D are all external to the installation and are referred to as Z_E.

Small conductors

If the conductors in the installation do not exceed 35 mm², reactance of the conductors can be ignored, so only the resistance of the conductors is used. For conductors that do not exceed 35 mm², the system impedance Z_S is therefore equal to $Z_E + R_1 + R_2$, where R_1 is the resistance of the installations phase conductor, and R_2 is the resistance of the installation's circuit protective conductor. Where the conductors exceed 35 mm², then $Z_S = Z_1 + Z_2$.

The installation impedances Z_1 and Z_2, or the resistances R_1 and R_2, can be referred to as Z_{inst}, representing the phase earth loop within the circuit or installation. The system impedance is then: $Z_S = Z_E + Z_{inst}$.

The regulations talk of protective conductors, but where the conductor is used for a particular circuit, it is then known as the circuit protective conductor, which is abbreviated to cpc.

Relationship between earth fault and short-circuit

There may appear to be no connection between an earth fault and a short-circuit fault, but a relationship does exist between them. When a fault to earth occurs, the fault current not only flows down the protective conductor, it also flows in the phase conductor. When the phase conductor has been sized to allow for the prospective short-circuit current, it is possible for the earth fault current to be less than the prospective short-circuit current. In these circumstances, the protective conductor has to be sized by using the formula. Additionally, since the protection of the phase conductor is based on the fault current being not less than a certain minimum value, as explained in the Chapter on Fault Currents, a check must therefore be made to ensure the phase conductor is still protected with the smaller earth fault current.

The formula used for calculating that the thermal capacity of the protective conductor is satisfactory, is the same as the formula used for checking the thermal capacity of the live conductors. Although at first sight the formula for sizing the protective conductor looks different, it is in fact the same as that used for checking that the live conductors are protected against short-circuit current. The formula has just been re-arranged so that the area of the conductor is the unknown quantity. Consequently, where the disconnection time obtained from the protective device characteristic for a given value of I_f, is 0.1 seconds or less, then the $I^2 t$ (not $I_f \times$ disconnection time) from the manufacturer's $I^2 t$ characteristics for the type of protective device being used, must not be less than $k^2 S^2$. In this case the protective conductor can be sized from $I^2 t \leq k^2 S^2$

Calculating the size of circuit protective conductor

To use the formula for determining the size of the circuit protective conductor (abbreviated to cpc), the values of I_f, t and k have to be known.

$$S \geq \frac{\sqrt{I^2 t}}{k}$$

Look at each one of these values in turn, starting with k. The value of k is determined from the initial temperature of the conductor at the start of the fault and the limit temperature of the insulation with which the conductor is in contact. It is important to check that the assumed initial temperature is correct for the protective conductor that is to be used. For instance, if the protective conductor is to be a pvc insulated copper conductor, installed in plastics trunking along with other circuits, then its initial temperature will rise to the working temperature of the other conductors in the trunking, and in the absence of more detailed information, it has to be assumed that the working temperature of the other conductors will be 70°C. Therefore, the assumed initial temperature of the cpc is 70°C, even though it is not carrying current.

In the case of armoured cable, the armouring is in the temperature gradient from the core of the cable running at 70° C, to the ambient temperature at 30° C; thus its initial temperature is assumed to be 60° C. The final temperature of the conductor is taken as the limit temperature of the conductor's insulation, which for pvc is 160 °C. Tables are provided in the regulations giving the value of k in different situations: they are also given in Table K1 and K1A in Part 4.

The fault current 'I' in phase to earth faults is called I_f, and to determine I_f the phase to earth voltage is divided by Z_S.

$$I_f = \frac{U_o}{Z_S}$$

The difficulty here is that Z_S includes the impedance of the cpc, which of course, in the majority of cases has not yet been sized. This problem does not occur on the very small installations, such as house wiring, since the electricity supplier will declare Z_E at the origin of the installation. The small installation will use twin &cpc cables, so the size of the cpc is known and since the conductors will not exceed 35 mm^2, all that is required is the resistance of the phase and circuit protective conductors. This resistance should be at the average of the working temperature of the conductor and the limit temperature for the conductor insulation.

Determination of 't'

The value of 't' in the formula, is the time it takes the protective device to disconnect a circuit with a given earth fault current flowing, as illustrated in Figure 2. A vertical line is drawn upwards from the current axis on the protective devices characteristic for the value of I_f, and where it touches the characteristic for the protective device being used, a horizontal line is then drawn from this point to the time axis. The disconnection time is read off the time axis at the point where the horizontal line crosses it. This is the value used in the formula. Of course, if the time is 0.1 seconds or less, then the I^2t in the formula becomes the value from the manufacturer's I^2t characteristics.

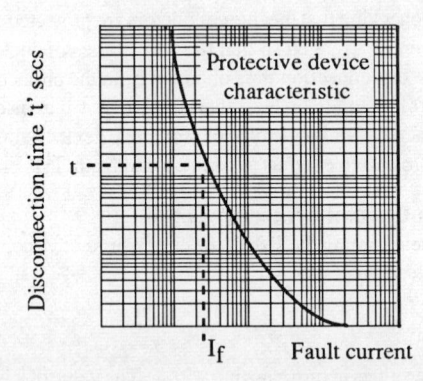

Figure 2 - Finding disconnection time 't'

Determination of 'k'

The value of 'k' is obtained either from table K1 or K1A in Part 4, or from the tables in Chapter 54 in the regulations. Select the initial temperature of the conductor at the start of the fault for the type of conductor insulation. Then select the value of 'k' from the appropriate column, depending

CIRCUIT PROTECTIVE CONDUCTORS

upon whether the protective conductor is bunched with live conductors, installed separately, is a sheath or armour of a cable, or is conduit or trunking.

Resultant size 'S'

The value of 'S' obtained will rarely be a standard cable size, so the next largest cable has to be used.

Where one cpc is used for several circuits, the values of I_f and t have to be worked out for each circuit, and the values that will give the largest size S are then used in the calculation.

Impedance tables

Part 4 of the handbook contains three tables giving the resistance, reactance, and impedance of cables up to 300 mm^2. The tables give this information at the design temperature for calculations. The design temperature is based on the average of the normal operating temperature of the conductor and the limit temperature for the conductor's insulation. Information is also given for an ambient temperature at 20 °C, 30 °C, and for the temperature when the conductor is only carrying 80% of its rated current.

The three tables cover aluminium conductors with either pvc or xlpe insulation (RA1), copper conductors with pvc insulation (RC1), and copper conductors with xlpe insulation (XLPE1).

It is therefore unnecessary to have to calculate the impedance of the conductors at the design temperature, since this has already been done in the tables. Figures are also given at 20 °C to enable the maximum fault current to be determined, when sizing switchgear etc.

Calculations for larger installations

For the larger installations, the exact size of the protective conductor will not be known. It is therefore necessary to start with an assumed cpc size. One way to limit the calculations to a minimum is to assume the cpc will be 10% of the phase conductor size rounded up to the next largest cable available. For example, if the phase conductor is 95 mm^2, then 10% equals 9.5, so assume the cpc will be 10 mm^2, and use the resistance of this conductor in determining the phase earth loop impedance Z_S. The calculation will then prove whether this size of cpc is suitable.

Where the circuit conductor exceeds 35 mm^2 and is close to the substation in a factory, or other type of premises, the impedance of the supply and of the conductors has to be used. For example, using the characters in figure 1.

$$Z_S = \sqrt{(R_A + R_B + R_1 + R_2 + R_C + R_D)^2 + (X_A + X_B + X_1 + X_2 + X_C + X_D)^2}$$

where R is the resistance of each conductor in the phase earth loop, and X is the reactance. Having determined Z_S, I_f is found by dividing the phase to earth voltage by Z_S i.e.:

$$I_f = \frac{U_o}{Z_S}$$

Table 54G

Calculating the size of cpc can be avoided by using table 54G. It cannot, however, be used for twin & cpc cables larger than 1 mm^2, since the cpcs in the larger cables are smaller than the phase conductor.

Where the phase conductor exceeds 35 mm^2, the table will invariably give a non standard size for the cpc, in which case the next largest standard size conductor must be used.

The size of the cpc given in the second column of the table is based on the protective conductor being made from the same material as the phase conductor. Where the protective conductor is made from a different material from that of the phase conductor, it's size is determined by the ratio of the k factor for the live conductors divided by the k factor for the protective conductor material. as detailed in the third column.

Area of phase conductor in mm² S	Area of protective conductor in mm² being the same material as the phase conductor S_p	Area of protective conductor in mm² for material different to phase conductor S_p
S not exceeding 16	S	$\dfrac{k_1}{k_2} \times S$
S from 16 to 35	16	$\dfrac{k_1}{k_2} \times 16$
S exceeding 35	$\dfrac{S}{2}$	$\dfrac{k_1}{k_2} \times \dfrac{S}{2}$

MINIMUM SIZE OF PROTECTIVE CONDUCTORS

The factor k_1 is the factor for live conductors, applicable to the type of conductor and insulation being used. The factor k_2 is the factor for the protective conductor based on the conductor material, conductor insulation, and type of conductor: i.e., whether it is enclosed with live conductors, installed separately, is the armour of sheath of a cable, or conduit and trunking.

Values of k_1/k_2 are given in Table M54G in Part 4 for aluminium, steel and lead, based on the worst values for k_2.

Using table 54G will in most cases be uneconomic, but it can be used to advantage in certain cases, such as:
1. aluminium cables with an aluminium strip armour, since the armour has an area, in most cases, at least equal to the area of the phase conductors.
2. micc multicore cable, because the sheath has an area larger than the phase conductor.
3. steel conduit, since the cross-sectional area of the conduit is so large relative to the size of conductor that can be installed through it.

One protective conductor can be used for several circuits; its size is then determined by the largest conductor of the circuits to be protected. Thus a number of circuits with conductors of 2.5 mm², 4 mm², and 10 mm² can be protected by one cpc; in this case the cpc size is based on the largest conductor, which is 10 mm².

What can be used as a protective conductor

The most common items that are used as protective conductors are: the bare or insulated conductors contained in a cable, the metal sheath, screen or armouring of a cable, and steel enclosures such as conduit or trunking.

The protective conductor can be a single core cable installed on its own, or installed along with live conductors, either in plastic or steel conduit or trunking. It can be the metal enclosure of distribution equipment, or busbar trunking, providing it complies with the following paragraph.

CIRCUIT PROTECTIVE CONDUCTORS 219

Protection of conductors
Protective conductors must be protected against mechanical damage, and against chemical and electrochemical deterioration. With the exception of ducting, conduit, trunking, or underground cables, all joints must be accessible.

Where the metal enclosure is switchgear, distribution boards, controlgear, or busbar trunking used as a protective conductor, it is essential that its electrical continuity is assured, and that its cross-sectional area is suitable for use as a protective conductor.

Protective conductor sizes in twin&cpc cable
The following Table details the size of cpc in twin & cpc cables., along with the conductor resistance at 115°C

Cable size (R_1) mm².	Cpc size (R_2) mm²	Ohms per 1000 metres at 115 °C		
		R_1	R_2	$R_1 + R_2$
1.0	1.0	24.978	24.978	49.956
1.5	1.0	16.698	24.978	41.676
2.5	1.5	10.226	16.698	26.924
4.0	1.5	6.362	16.698	23.060
6.0	2.5	4.250	10.226	14.476
10.0	4.0	2.525	6.362	8.887
16.0	6.0	1.587	4.250	5.837

Sizing cpcs. by using the PCZ tables from part 3
Where the circuit protective conductor size is unknown, it can be found by using the following formula and the PCZ tables:

$$R_2 = Z_s - (Z_E + R_1)$$

Impedance values Z_1 and Z_2 can be used in place of R_1 and R_2.

Determination of Z_s in a large installation
The value Z_E is referred to as the external phase earth loop impedance. This is all right for a very small installation, such as a house, but is inconvenient when considering large installations, since there can be many distribution boards between the origin of the installation and the circuit for which calculations are required.

It is therefore more convenient to consider Z_E as being the earth loop impedance external to the circuit under consideration. This makes calculations easier, since only three components are involved in the calculation, i.e., Z_E, Z_S and Z_{inst}. Knowing any two of these items will enable the third to be calculated. A convenient way of remembering the above is given in the following diagram:

CIRCUIT PROTECTIVE CONDUCTORS

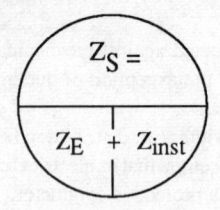

$$Z_S = Z_E + Z_{inst}$$

Figure 3 - Calculation diagram

The three calculations resulting from the diagram are:

$Z_S = Z_E + Z_{inst}$
$Z_E = Z_S - Z_{inst}$
$Z_{inst} = Z_S - Z_E$

Two values of Z_S

When carrying out calculations, it must be remembered that there are two values of Z_S, one that is the maximum allowed for the circuit, (as given in the tables) whilst the other value is the actual Z_S of the circuit.

When using the notation above to carry out a calculation for a circuit from a sub-distribution board, the Z_E for the circuit will be the Z_S at the busbars of the distribution board, as illustrated in Figure 4.

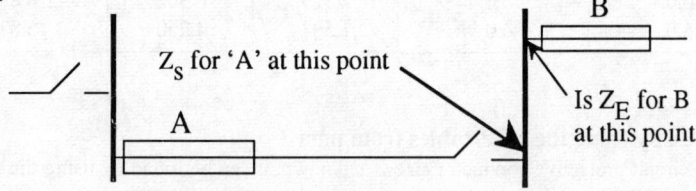

Z_S for 'A' at this point

A

B

Is Z_E for B at this point

Figure 4 - Illustrating how Z_S becomes Z_E

When using the PCZ tables for a small installation, the maximum length of the circuit for the cpc can be found by using the above formula. As Z_S is obtained from the tables for the type and size of protective device and size of protective conductor, the value of Z_E will be known, so Z_{inst} can be found. The maximum length of the circuit as far as the size of protective conductor is concerned, will be;

$$\text{Maximum length of circuit} = \frac{Z_{inst}}{R_1 + R_2 \text{ per metre at temperature T}}$$

where the temperature 'T' is the average of the conductor operating temperature and the limit temperature for the conductor's insulation, which for pvc is 115 °C, R_1 is the resistance of the phase conductor per metre, and R_2 is the resistance of the cpc per metre.

The maximum length obtainedonly satisfies the requirement that the thermal capacity of the protective conductor is satisfactory. Protection against indirect contact, overload and short-circuit regulations will have to be complied with, and calculations will still be needed for voltage drop.

Conduit and trunking

The problem all engineers have when designing an installation using conduit or trunking as a protective conductor, is what values should they use in the calculations.

Table ZCT 1 in Part 4, which gives the impedance of conduit and trunking, has been provided to assist with this problem. Similarly tables ELI 1, 2, 3, 4 and 5 have been provided giving the phase earth loop impedance of armoured cables and m.i.c.c. cables. The tables cover both copper and aluminium conductors with pvc and x.l.p.e insulation with steel wire armouring as well as strip aluminium armouring, and take into account the reactance of the armour. Both light duty and heavy duty micc cables are included in ELI 5.

When considering in calculations the temperature of conduit with fault current flowing, the reactance of the conduit has to be considered. The reactance is quite high relative to the conduit's resistance, that during the short time the fault current is flowing, there will be little, if no increase in the temperature of the conduit. By calculation it can be shown that if a 20 mm heavy gauge conduit has its temperature increased from 20 °C to 115 °C by a fault current the increase in impedance is only 0.00036 Ω/m. The increase is only some 0.4Ω per 1000 metres. Conduit when installed forms a network, so that the fault current can divide and take several routes back to the origin of the circuit. The only conduit that will take the full fault current is the length that feeds the actual piece of equipment in which the fault develops, or where a single run of conduit is installed from a distribution board to a piece of equipment.

What designers are concerned with is the thermal capacity of the protective conductor. As far as conduit is concerned, it can be shown that its thermal capacity is larger than the the pvc insulated copper cables that can be installed through it in accordance with the regulations. Bearing this in mind the values given in ZCT 1 can be applied without alteration due to temperature rise by a fault current.

Table ZCT 1 gives the impedance of conduit with joints. Strictly speaking, conduit joints do not increase the impedance of the conduit. The cross-sectional area of the conduit box or coupler is such that it has a lower impedance than that of the conduit. The value of conduit impedance with joints has been given, since it was felt that the IEE, when they eventually publish a table of conduit impedances, will add a percentage to the conduit impedance for joints, to allow for any deterioration that may occur during the life of the installation.

Armoured cables

As far as the armouring of cables is concerned, the same comments as those made for conduit apply. In this case, the armour does not have the same relative cross-sectional area, so the effects of reactance are not as dramatic as with conduit. However, it should still be taken into account when carrying out calculations.

9. Circuit protective conductors examples of calculations

In the following examples, the values of resistance at the design temperature have been taken from the tables in Part 4

Example 1 - Conduit as cpc by calculation.
A three-phase circuit feeds a 415V 4 HP motor and is wired in 1 mm^2 pvc single core cable enclosed in 20 mm light gauge conduit; the length of the circuit is 35.84 metres, and the protective device in the distribution board is a 20A BS88 HRC fuse. The value of Z_S up to the distribution board is 0.86 Ω. Is the conduit satisfactory as a cpc?

Working.
From Table RC 1 Part 4: resistance of 1 cable at 115 °C = 0.024978 Ω
From Table ZCT 1 Part 4: impedance of conduit = 0.00675 Ω
The first task is to determine Z_S for the circuit.
Z_{inst} = 35.84 (0.024978+ 0.00675) = 1.137 Ω.
Z_S = $Z_E + Z_{inst}$ = 0.86 + 1.137 = 1.997 Ω.

$$I_f = \frac{U_o}{Z_S} = \frac{240V}{1.997\Omega} = 120.18 \text{ A}$$

The disconnection time for the fault current I_f of (say)120 amps is now obtained from the 20A HRC fuse characteristic as shown in Figure 1.

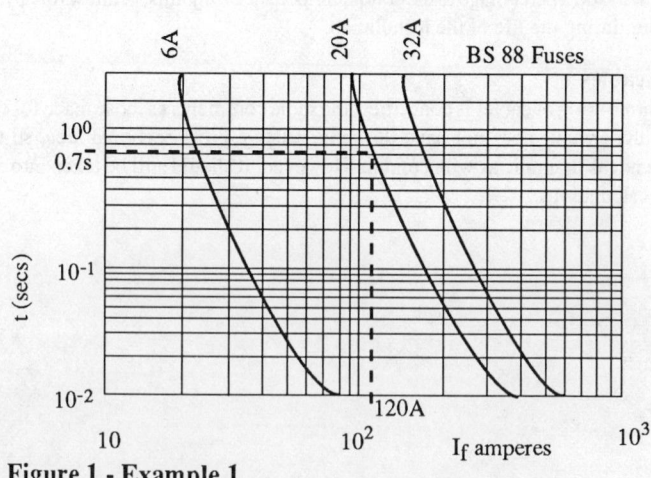

Figure 1 - Example 1

CPC EXAMPLES OF CALCULATIONS 223

From the characteristic, when $I_f = 120$ amps, $t = 0.7$s. It is now necessary to determine the value of k from Table K1A (Part 4). The conduit is in the temperature gradient between the conductors at 70 °C, and the ambient temperature of 30 °C. This means that its initial temperature will be 50 °C. It is however in contact with the pvc insulation of the cables, whose temperature must not exceed 160° C i.e. the final temperature is 160° C. The value of k therefore is 47.

All the necessary information is now available for the calculation:

$$S = \frac{\sqrt{I^2 t}}{k} = \frac{\sqrt{120^2 \times 0.7}}{47} \quad 2.136 \text{ mm}^2$$

From MISC 1 Part 4: The cross sectional area of the conduit is 59 mm², therefore it is satisfactory as a circuit protective conductor.

Example 2 - Conduit as a cpc using Table 54G.

The same problem as example 1, but conduit sized using Table 54G.

Working

From Table K1A Part 4: k for copper/pvc live conductor is 115
k for conduit for same type of cable is 47
Size of phase conductor is 1 mm². From 54G: cross-sectional area of cpc must be :

$$\frac{k_1}{k_2} \times S = \frac{115}{47} \times 1 \text{ mm}^2 = 2.45 \text{ mm}^2.$$ The conduit area is 59 mm², so it is suitable

Example 3 - Armoured cable using Table M54G, Part 4.

Same problem as example 1, but supply to motor is now a 1.5 mm² pvc steel wired armoured pvc cable (PVCSWA & PVC sheathed cable).

From Table M54G in MISC1 Part 4, cross sectional area required for steel armour is 2.25 times area of phase conductor; therefore area of armour = 1.5 mm² × 2.25 = 3.375 mm²
From Table ELI 1 Part 4, area of 3 x 1.5 mm² PVCSWA&PVC cable armour = 16 mm², so it is satisfactory.

Note : Smallest armoured cable available is 1.5 mm²

Example 4 - Twin and cpc cable by calculation.

A cooker circuit is to be installed from a 30A rewirable fuse in a consumer unit using 6 mm² twin and cpc cable to a cooker control containing a socket outlet, the length of run being 28 metres, if Z_E to the consumer unit is 0.35 Ω, check that the cpc will be suitable as a circuit protective conductor.

Working (using formula)

From Table MISC 1 Part 4: size of cpc for 6 mm² twin&cpc cable is 2.5 mm².
From table RC1 Part 4: resistance at 115° C for 6 mm² = 0.00425Ω & 2.5 mm² = 0.010226 Ω per metre.

Z_{inst} = 28 (0.00425 + 0.010226) = 0.4053 Ω
Z_S = Z_E + Z_{inst} = 0.35 + 0.4053 = 0.7553 Ω

$$I_f = \frac{U_o}{Z_s} = \frac{240V}{0.7553} = 317.75 \text{ amps}$$

The disconnection time for a fault current of (say) 318A has to be found from the 30A rewirable fuse characteristic, as shown in Figure 2.

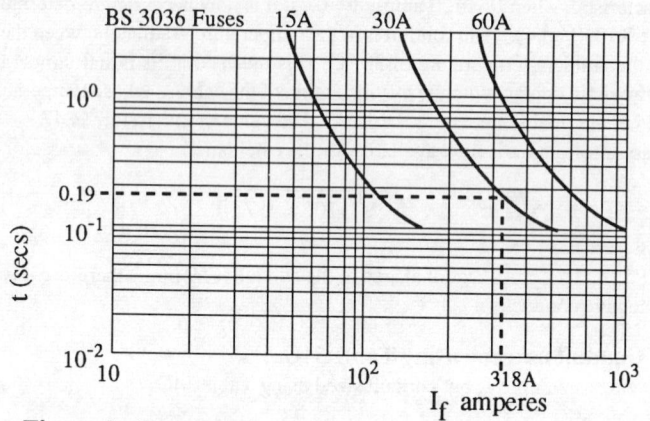

Figure 2 - Example 4

From the characteristic, when I_f is 318A, t = 0.19 secs.
The cpc is enclosed in the cable: therefore its initial temperature will be the same as the live conductors i.e. 70° C, and because the cable is pvc, the final temperature must not exceed 160° C. The value of k from table K1A Part 4 is 115. Substituting this in the formula:

$$S = \frac{\sqrt{318^2 \times 0.19}}{115} = 1.21 \text{ mm}^2$$

The area of the cpc is therefore satisfactory.
Note: Table 54G cannot be used since the cpc is smaller than the phase conductor.

Example 5 - Using PCZ tables from Part 3.

Same problem as example 4, but using PCZ tables from Handbook for checking cpc.

Working

The impedance Z_s for the circuit would be calculated as in the previous working, giving a total Z_s of 0.7553 Ω.

The maximum Z_s allowed for the cpc is obtained from the PCZ table for rewirable fuses, by selecting the value for the fuse size and cpc size.
From Table PCZ 3, the value of Z_s allowed for 2.5 mm² cpc and 30A fuse is 2.76 Ω, so the cpc is satisfactory. From Table ZS 3 Part 3 the Z_s allowed is 1.14Ω, so the circuit also complies with shock protection requirements.

General

Where the sheath of multicore micc cables and the strip aluminium armouring of aluminium cables is used as the protective conductor, no calculations will be required to ensure that the sheath or armour has the correct thermal capacity. The sheath and armour quite easily comply with table 54G. The earth loop impedance for such cables will, however, still be required, to enable protection against indirect contact to be worked out.

10. Fault current regulations

Section 434 of the regulations is only concerned with fault currents flowing in the live conductors belonging to the same circuit. This would comprise a short between two phases, between all three phases on a three-phase circuit, and between all three phases and neutral, or a short between phase and neutral on a single-phase circuit. Where there is a phase to earth fault, Section 434 is concerned with the protection of the phase conductor, and Section 543 with the protection of the protective conductor. For the purposes of calculation, the fault is considered to be a bolted short of negligible impedance on the load side of the device installed.

The fundamental requirements for safety in the regulations, call for every installation and circuit to be protected against overcurrent. The device used must operate automatically at a safe current, related to the current rating of the circuit, and must have an adequate breaking capacity. Where circuit breakers are used for protection, they must also have adequate making capacity. The protective device must be located so that there is no danger from overheating, arcing, or scattering of hot particles.

The regulations stipulate that a protective device must be provided to break a fault current occurring in the conductors of a circuit, before the fault current can cause danger from thermal and mechanical effects. The thermal effect referred to, is the heating effect on the conductors created by the fault current (I^2t). If the fault current is not interrupted, the heating effect can be such as to melt the conductors through which it is flowing. The mechanical effect is the magnetic field set up round the conductor due to the fault current (I^2), which exerts a mechanical stress on the conductors. The regulations also state that the nominal current rating of the protective device 'may be greater than' the current-carrying capacity of the conductor being protected.

There is another stipulation, that where a protective device is used for protection against fault currents, the time taken by the protective device to clear the fault must not allow the limit temperature of the conductor's insulation to be exceeded.

Cable size independent of protective device rating

Protecting against fault current, is only protecting against the thermal and mechanical effects that can occur with a fault. Consequently, the cable conductors do not need to be sized to the nominal rating of the protective device. The relationship that I_b must not be greater than I_n, which in turn must not be greater than I_z, is related only to overloads, and has nothing to do with short circuit protection.

This allows the use of cables with a lower current-carrying capacity than the nominal rating of the protective device when feeding a motor circuit. But the circuit still has to be protected against overload in accordance with Sections 433 and 473 by another device, i.e. an overload incorporated in a motor starter. Similarly, circuits unlikely to carry overload current do not need to be sized to the size of protective device, but to the load, the protection provided being against fault current.

Determination of prospective short-circuit current at relevant point

The regulations call for the prospective short-circuit current and earth fault current at every relevant point of the installation to be determined. Additionally, (with a few exceptions covered later) Section 473 calls for the installation of a short-circuit protective device where there is a reduction in the current-carrying capacity of the conductors of the installation.

Such reductions are caused by changes in cross-sectional area, method of installation, type of cable, or environmental conditions.

Breaking capacity lower than short-circuit current

Section 434 calls for the breaking capacity of the protective device to be not less than the prospective fault current at the point at which it is installed. A protective device that has a breaking capacity lower than the prospective fault current available at the point it is installed, can be used, provided that:
1. there is a protective device on the supply side of the lower rated protective device which has the necessary breaking capacity; and
2. the characteristics of both protective devices are co-ordinated, so that the energy let-through the supply side device will not damage the lower rated device, or the cable conductors protected by both devices. (see examples for details of how co-ordination is achieved).

Protective device allowed at different position than the relevant point

A protective device can be installed at a point other than at the point of change in current-carrying capacity, (473-02-02) provided that:
1. the distance from the point of change in current-carrying capacity does not exceed 3 m, and
2. the conductors are so erected that the risk of short-circuit, earth fault, and the risk of fire or danger to persons is reduced to a minimum.

In general, the conductors will, however, have to comply with the overload requirements for the circuit concerned.

This relaxation allows the conductors which tap off a busbar chamber, feeding a switch fuse mounted on the busbar chamber, to be smaller than the rating of the busbars. However, the conductors between the busbars and the switch fuse must comply with the overload requirements for the circuit under consideration.

The overload regulations allow the device protecting a conductor against overload to be placed along the run of that conductor, provided that the part of the run between the point where the value of the current-carrying capacity is reduced, and the position of the protective device, has no branch circuits or outlets for the connection of current using equipment.

The conductors in the busbar chamber should be installed carefully, so that a short-circuit will not occur between them, or between the conductors and any other conductor or earth. In practice it is better to connect the tails from the switch further along the busbar chamber, rather than directly below the switch, and preferably at the rear of the busbars so that the cables can be set to the back of the chamber away from the other busbars, eliminating the risk of a short-circuit. The actual connection is then in line with the busbar.

The insulation or sheath on the conductors must be suitable for the maximum operating temperature of the busbars. When determining the size of the conductor to install from the busbar to the switch fuse for overload protection purposes, the derating factors should be taken into account.

It must be remembered that the maximum working temperature of the busbar enclosure becomes the ambient temperature for the conductors to the switch fuse. The nominal rating of the fuse in the switch fuse is taken as I_n in the calculation:

$$I_t = \frac{I_n}{G \times A \times T \times S}$$

There is a further regulation, (473-02-03) which allows the short-circuit protective device to be installed at a point other than at the point where the current-carrying capacity of the conductors change, provided that there is a protective device on the supply side of the change in current-carrying capacity; which will protect the conductors against short-circuit or an earth fault.

In this case, a calculation is required to prove that the conductors are indeed protected, but there is no restriction on the length of the conductor installed.

A typical example of the use of this regulation is in a main switchroom, where there is spare capacity in the main switchgear, but the busbar chamber is physically full of switch fuses. In such a situation a cable can be tapped off the busbars and installed to a switch fuse mounted on the switchroom wall. A calculation must then be made to ensure that the main incoming protective device to the switchgear will protect the cable against short-circuit current.

Omission of protective devices

The omission of a device for protecting against a fault current is permitted where conductors connect a generator, transformer, rectifier or battery to a control panel, when a protective device is installed in the panel (473-02-04). One such example of this relaxation, is when the conductors from a transformer are installed to a main l.v. board which incorporates protective devices in the l.v.board.

In this case it is important to remember that the cables are only protected by the protective device on the H.V. side of the transformer; it is for this reason that the regulation specifies that the risk of short-circuit, fire, and danger to persons must be a minimum. This is achieved by installing the transformer tails in smooth ducts between the transformer and the main l.v. switchboard, open type cable ducts are usually provided with covers.

Calculations redundant

One of the most important regulations in Section 434 is 434-03-02, which states that where a protective device is used for both overload and fault-current protection (as in the case, say of a 5 A fuse protecting a lighting circuit), the conductors on the load side of the protective device will have been sized to the nominal current rating of the protective device in accordance with Section 433. If the protective device has a breaking capacity not less than the prospective fault current available at the point at which it is installed, it can be assumed that the conductors on the load side of the device are protected against fault current. However, where conductors are installed in parallel, or the protection is by a non-current limiting type of circuit breaker, the circuit has to be checked by calculation.

The non-current limiting type of circuit breaker to which the regulations are referring are the zero point circuit breakers. These circuit breakers, depending upon the circuit conditions, may not protect the cables. This is because they let through all the energy for at least the first half cycle, and can let it through for several cycles.

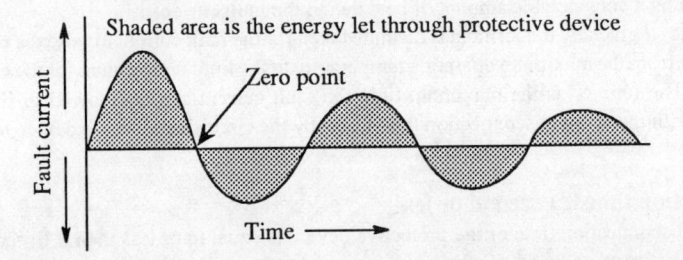

Figure 1. Energy let through a non-current limiting circuit breaker

If a circuit breaker does not have a characteristic similar to those shown in Part 4, it must be assumed to be a non-current limiting type. This means that a calculation is required to ensure the conductors are protected against fault current.

In the majority of cases Regulation 434-03-02 will apply, and it will generally be necessary only to carry out calculations by using the formula of Regulation 434-03-03, where the cable conductors are not sized to the nominal rating of the protective device, as in the case of cables feeding a motor, or where Regulation 473-02-03 is used.

Calculating disconnection time

Where Regulation 434-03-02 is not applicable then Regulation 434-03-03 has to be used to ensure that the live conductors are protected against short-circuit current. This latter regulation gives a formula for determining whether a fault current occurring at any point of a circuit will be interrupted in a time which will not allow the cable conductors to exceed their limiting temperature. The formula given in the regulations is:

$$t = \frac{k^2 S^2}{I^2}$$

What the formula means

In this form, the formula does not convey any particular meaning, but if this formula is re-arranged it becomes:

$$I^2 t = k^2 S^2$$

Now energy is equal to watts × time = I^2 Rt. If R remains constant, or is minute compared with I, then energy becomes proportional to I^2 t, and this information can be inserted into the above formula,

I^2 t	=	$k^2 S^2$
Thermal energy let-through the protective device	=	Thermal capacity of conductor

It follows that if the conductor is not going to be damaged, then I^2 t must never exceed $k^2 S^2$. This is important because the heat produced by a short-circuit current is considered to be contained within the conductor core of the cable for fault durations up to five seconds.

The equation is commonly called adiabatic (without gain or loss of heat). It gets this name because the equation is considered to remain adiabatic for fault current durations up to five seconds, i.e., it ignores heat loss. This is not strictly correct, since the conductor will be losing some heat, and will be gaining a considerable amount of heat due to the fault current.

The time 't' given by the formula is the time taken for the fault current to raise the conductor's temperature from the maximum operating temperature to the limit temperature for the conductor's insulation. The time 't' is the maximum time the fault current can be allowed to flow, and is compared with the actual disconnection time taken by the circuit's protective device for the fault current I.

Disconnection time 0.1 second or less

Where the disconnection time of the protective device is equal to or less than 0.1 seconds, then I^2 t should be less than $k^2 S^2$ for the conductor.

The I^2t referred to, is the energy let-through of the protective device, and is obtained from the manufacturer's characteristics for the protective device protecting the circuit.

It is not the square of the fault current multiplied by the disconnection time. (An example is given in the chapter on examples of short-circuit current calculations).

The value of k used in the formula is obtained from the table 43A (Tables K1A and K1B in Part 4). The values are based on the initial temperature at the start of the fault, this being the conductor's normal operating temperature, and the final temperature being the limit temperature of the conductor's insulation.

The value of 'S' in the formula is the conductor's area in square millimetres.

Application of the fault current regulations

The application of the short-circuit regulations are illustrated in the following diagram. The diagram shows a transformer feeding a main switchboard, which in turn feeds two distribution boards.

Regulation 473-02-04 allows the cables to be installed from the transformer to the switchgear without protective devices being installed at the transformer.

Regulation 473-02-02 allows the tails to switch I_{n2} to have a current-carrying capacity less than the busbars, providing they comply with Section 433 for overloads, and do not exceed 3 m in length. No calculations are required for the cable from I_{n2} to DB1, since it complies with 434-03-02: i.e. its current-carrying capacity is not less than the nominal rating of the fuse in I_{n2}, and the fuse has a breaking capacity greater than the 14 kA at the busbars. (Note BS 88 fuses have a breaking capacity of 80 kA.)

The characteristics of the mcbs I_{n6} to I_{n9} must be co-ordinated with I_{n2}, so that the energy let-through I_{n2} is within the withstand capacity of the mcbs, and that both devices protect the circuit cables L1 to L4, in compliance with 434-03-01.

The other distribution board DB2 is fed through an isolating switch and the cable exceeds 3 m in length, so regulation 473-02-03, along with 434-03-03, has to be used to make certain that the cable feeding the distribution board is protected against short-circuit current by I_{n1}.

Regulation 434-03-02 can again be used for the lighting cable fed from I_{n4}, but a calculation will be needed for the cable to the motor, since its current-carrying capacity is less than the nominal rating of the fuse I_{n5}; so 434-03-03 has to be used.

Regulation 435-01-01 requires that the characteristics of the starter are co-ordinated with its protective device I_{n5}; this means that the withstand capacity of the starter must be suitable for the energy let-through I_{n5}.

ILLUSTRATING WHERE FAULT CURRENT REGULATIONS APPLY

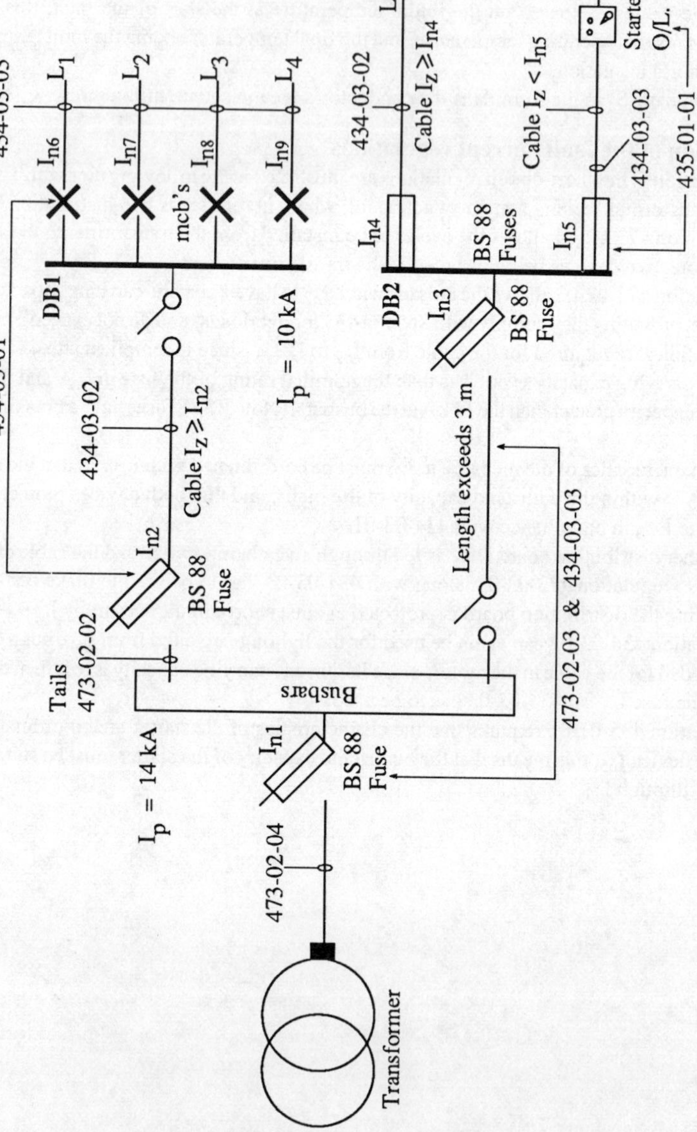

Transformer
473-02-04
I_p = 14 kA
BS 88 Fuse
I_{n1}
Busbars
Tails
473-02-02
BS 88 Fuse
I_{n2}
473-02-03 & 434-03-03
Length exceeds 3 m
434-03-01
434-03-02
Cable $I_z \geq I_{n2}$
I_p = 10 kA
DB1
mcb's
I_{n6}
I_{n7}
I_{n8}
I_{n9}
434-03-03
L_1
L_2
L_3
L_4
BS 88 Fuse
I_{n3}
DB2
BS 88 Fuses
I_{n4}
434-03-02
Cable $I_z \geq I_{n4}$
Lighting
I_{n5}
434-03-03
Cable $I_z < I_{n5}$
435-01-01
Starter, O/L.
Motor

REGULATIONS 130-03-01 & 434-01-01 APPLY TO THE WHOLE SYSTEM

11. Maximum fault current lines

The adiabatic equation can be used to construct what have become known as adiabatic lines. A better name for these lines is fault current lines, since they indicate the maximum fault current a conductor can withstand for a given disconnection time.

Such lines only indicate the withstand capability of the conductor based on the conditions used to obtain the value of k used in the equation. In other words, for a given conductor material, the value of k is dependant upon the initial and final temperature of the conductor.

The values of k given in the tables in the regulations is based on the initial temperature being the operating temperature of the conductor when carrying its rated current, as given in the current rating tables. The final temperature is the limit temperature of the conductor's insulation, which for PVC is 160 °C.

Producing fault current lines

To produce a fault current line, it is only necessary to substitute values in the equation:

$$t = \frac{k^2 S^2}{I^2}$$

The size and type of conductor is chosen - say copper 4 mm^2 pvc insulated cable. The type of conductor and its insulation determines the value of k, which for the above cable is 115 (Table K1A). It is now only necessary to substitute three values for I to give three values of t.

The values of t are then plotted against I on log-log graph paper when a straight line is produced. The following diagram shows a fault current line produced for a 4 mm^2 pvc insulated conductor.

The fault current line was produced by substituting the values of 115 for k, 4 for S, and then using three values of I, namely 2000 A, 1000 A, and 500 A. Each of the values used for I gave three values of t, viz: 0.0529 sec, 0.21 sec, and 0.846 sec. Three points are required to ensure the correct straight line is produced; two for a straight line, three to confirm the line's direction.

To determine the maximum disconnection time for a fault current, draw a line vertically from the fault current axis for the value of fault current, up to the fault current line; where it touches the fault current line, draw a horizontal line to cut the axis t; this is the maximum time allowed for the circuit to be disconnected.

A family of such fault current lines can be produced for a particular type of cable. In the case above, a group of lines could be produced for 1 mm^2 cable up to the maximum size made for that type of conductor.

The fault current lines are more useful if they are superimposed onto the characteristics of protective devices, since this shows at a glance whether the protective device will effect disconnection (of the faulty circuit) before the fault current reaches the fault current line.

Using clear acetate sheet for a set of lines will enable them to be used with various protective device characteristic, providing the characteristics have been produced on the same size of log-log graph paper. Marking the corners of the log-log paper on the acetate will ensure that the fault current lines are correctly positioned.

MAXIMUM FAULT CURRENT LINES

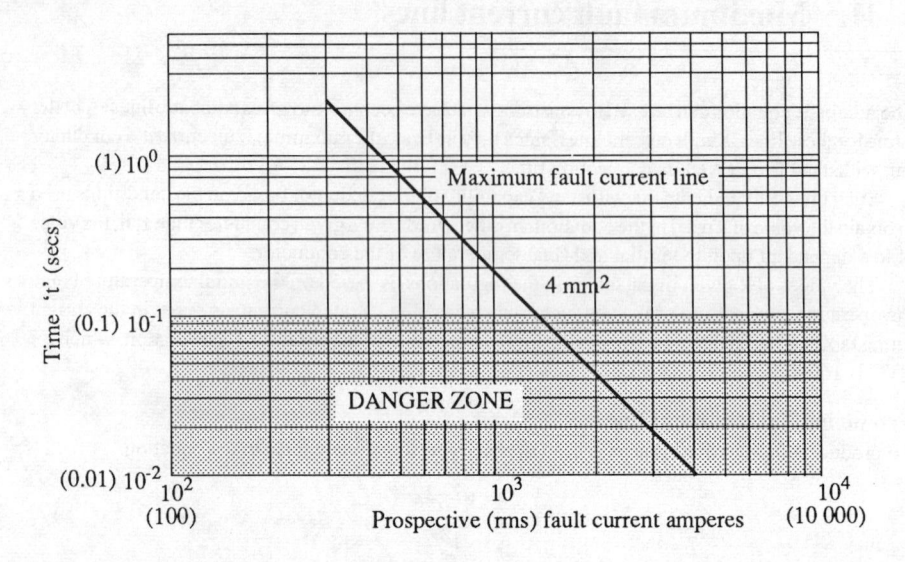

Figure 1 - Maximum fault current line

Dangers in using fault current lines

There are two hidden dangers when using fault current lines for determining whether a circuit is protected. The first is due to the scale of the characteristics; this makes it impossible to determine the exact position of an odd numbered fault current. The danger occurs when the fault current is very close to the intersection of the fault current line with the protective device characteristic.

The second and most important danger (shown shaded on the previous diagram) is where the disconnection time is equal to, or less than 0.1 seconds. It is possible for the fault current line to be above the protective device characteristic in the time range of 0 to 0.1 seconds, visually indicating that the conductor is protected, whereas by checking the I^2t from the manufacturer's characteristics and comparing it with k^2S^2 for the conductor, it can be shown not to be protected. It is for this reason when using a fault current line to check that a conductor is protected against short-circuit current, or an earth fault current, that care has to be taken to ensure that the disconnection time is not less than 0.1 seconds; if it is, then k^2S^2 must be compared with the I^2t let-through the protective device, as given in the manufacturer's literature. (See last page in Part 4.)

12. Short-circuit current calculations

Information required before starting
Before commencing any calculations, there are four fundamental questions that have to be answered, as called for in the assessment of general characteristics:
1. the prospective short-circuit current at the origin of the installation (I_p) which will usually be at the supply company's fuse;
2. the external earth loop impedance (Z_E);
3. the type of protective device installed at the origin (usually the supply company's fuse); and
4. the type of system to be provided.

The type of supply required, and the system of which the installation will form part, will affect the approach to the calculations. If a three-phase four-wire supply is required, then the I_p given by the supply company will be the three-phase prospective short-circuit current.

The Z_E quoted will only give the impedance of the phase and neutral conductors in the TN-C type system, because the protective conductor is combined with the neutral up to the origin of the installation. If the system is TN-S, then the supply company will have to be asked for the phase-neutral impedance external to the installation, since the neutral conductor is a separate conductor from the protective conductor.

Two calculations required
Two calculations will be needed, one to determine whether the equipment (such as switchgear, starters, control panels and distribution boards) have sufficient withstand capacity. The second calculation is to determine whether the circuit conductors are protected.

When calculating symmetrical three-phase short circuit currents, the impedance of the neutral and earth conductors are ignored.

The three-phase symmetrical short-circuit current I_p can be determined by the formula:

$$I_p = \frac{V_L}{\sqrt{3} \times Z_p}$$

where V_L is the line voltage, Z_p is the impedance of one phase conductor, and $\sqrt{3}$ is 1.732.

To determine the single-phase short-circuit current on a TN-C type system, the phase voltage is divided by Z_E, but if the system is TN-S, then the following formula should be used:

$$I_{pn} = \frac{V_{ph}}{Z_{pn}}$$

where Z_{pn} is the impedance of the phase and neutral conductor, and I_{pn} is the prospective single-phase short-circuit current.

To determine the impedance to the short-circuit current:

Single-phase

$$Z_{pn} = \frac{V_{ph}}{I_{pn}}$$

Three-phase

$$Z_p = \frac{V_{ph}}{I_p}$$

The formula for single-phase will give the impedance of the phase and neutral conductor. The three-phase formula will give the impedance of one phase conductor only, where I_p is the three-phase symmetrical short-circuit current. Where the fault is only between two phase conductors then I_{pp} will give the impedance of both phase conductors as indicated in the following formula:

$$Z_{pp} = \frac{V_L}{I_{pp}}$$

The main difference between this formula and the previous formula is that the line voltage is divided by the fault current; instead of the phase voltage.

In all the above cases the impedance must be the total for the circuit from the source of electrical energy (i.e. the generator) to the end of the final circuit, or to any intermediate point at which the prospective short-circuit current is required to be calculated.

Although impedance is always referred to when discussing short-circuit currents, only the resistance needs to be taken into account, if the conductors are 35 mm^2 or less, the effect of reactance on these smaller conductors being considered negligible. In fact, the reactance has a decreasing effect on the prospective short-circuit current as the distance from the origin of the installation increases.

Where short runs of very large conductors are involved, the reactance of the conductor can exceed the resistance by a considerable amount; in such circumstances, the resistance of the conductor can be ignored, only the reactance being used in the calculation.

When calculations are made where the reactance is given separately, or is known, the impedance is obtained from the equation:

$$Z = \sqrt{R^2 + X^2}$$

When considering whether the equipment will be suitable for the prospective short-circuit current that might flow, there are two considerations that have to be made. The first concerns the actual switchgear or equipment, and the second concerns the conductors of the installation.

As far as the equipment is concerned, the maximum fault current that may flow is required, so that switchgear or equipment with the correct withstand capacity can be selected.

On the other hand, to protect the conductors of the installation, the minimum fault current that will flow is required.

Determination of maximum short-circuit current

To determine the maximum prospective short-circuit current that will flow in a three-phase installation, the three-phase symmetrical short-circuit current should be calculated, using the resistance of the conductors at the temperature at which they could possibly be when a fault occurs. For instance, if a cable was installed between buildings, and it was the depth of winter, then the conductor temperature could be the same as the outside temperature. The actual conductor temperature would depend upon whether the conductor was just being energised, or whether it had been loaded for some time. When in doubt as to the temperature, take the lowest value that could occur. The resistance obtained at the selected temperature is then used in the impedance formula given above.

Determination of minimum short-circuit current

To determine the minimum fault current in a three-phase and neutral installation, the phase to neutral short-circuit current should be calculated. The phase to neutral fault current is calculated since this value will be less than the three-phase symmetrical short circuit current.

Where the supply is three-phase three-wire, the minimum fault current will be with a fault between two phases. In this case the line voltage is divided by the impedance of two phases: from the generator up to the end of the circuit, or to any other point at which the calculation is being carried out.

$$I_{pp} = \frac{V_L}{Z_{pp}} = \frac{V_L}{2Z_p}$$

Reason for minimum short-circuit current being used for conductors

The reason for the minimum prospective short-circuit current being taken for the conductors is more clearly seen if the maximum fault current line is superimposed onto protective device characteristics. Figure 1 shows the fault current line superimposed onto a fuse characteristic.

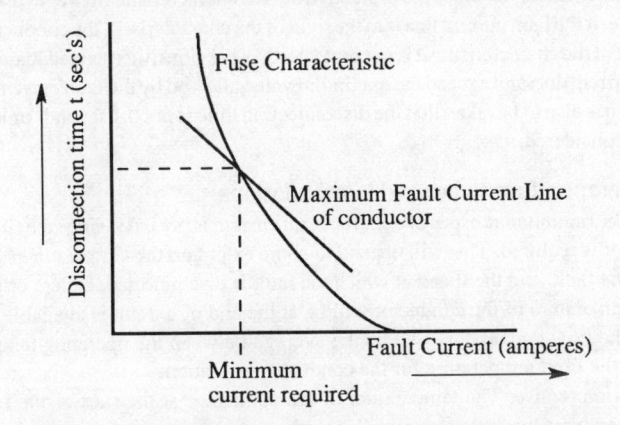

Figure 1 - Fault current line and fuse characteristic

Figure 1 shows the fault current line crossing the fuse characteristic. The minimum fault current and maximum disconnection time for the conductor is determined by the point at which the fault current line crosses the fuse characteristic.

The point of intersection, therefore, gives the minimum fault current that must flow in the circuit for the conductor to be protected. The fact that a smaller fault current will damage the conductor can be seen by drawing a vertical line to the left of the crossover point. As the line crosses the fault current line before it reaches the fuse characteristic, the current would therefore be allowed to flow too long, and the conductor would be damaged.

If the fault current line is superimposed onto a circuit breaker characteristic, as in Figure 2, it can be seen that the fault line crosses the characteristic at two places, corresponding to a minimum and maximum fault current. The fault current in the circuit must therefore lie between these two points if the conductor is to be protected.

236 SHORT-CIRCUIT CALCULATION THEORY

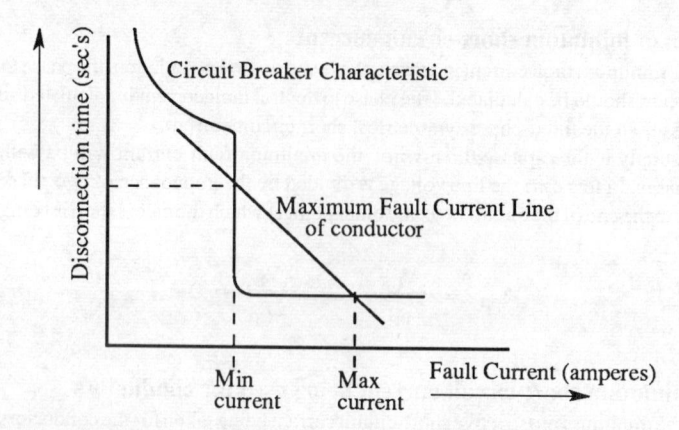

Figure 2 - Fault current line and a circuit breaker characteristic

The fault current line will not cross the protective device characteristic in every case, and therefore as a general rule, if the fault current line is to the right of the characteristic, the conductor is protected; if it is to the left of the characteristic, it is not protected. As previously explained, care has to be taken that the fault current does not exceed the maximum value allowed by the fault current line for circuit breakers. Care has also to be taken that the disconnection time is not 0.1 seconds or less for the fault current being considered.

Conductor temperature to be used in calculations

To determine the minimum prospective short circuit current is not easy, since the final temperature of the conductor is required. This will depend to some extent on the temperature of the conductor at the start of the fault, and the speed at which the fault is disconnected. Unless other information on what the temperature of the conductor will be at the end of a fault is available, it can only be assumed that the final temperature will be the average between the operating temperature of the conductor and the limit temperature for the conductor insulation.

This is a compromise. The temperature of the conductor at the start of the fault will be its operating temperature; the fault current will raise the conductor temperature, but in all probability, the circuit will be disconnected before the limit temperature of the insulation is reached, due to the time taken by the fault current to raise the conductor's temperature.

The temperature of the conductor will rarely be running at the operating temperature given in the tables, and therefore if a more precise determination is required, a new value of k will need to be worked out, based on the conductor's initial temperature; at the same time, the average temperature used for the conductor's resistance will need recalculating. It should be mentioned at this point that temperature has no effect upon reactance: additionally, the conductor's resistance will be required at i's actual operating temperature.

Tables are given in Part 4 for the resistance, reactance and impedance of cables from 1 mm^2 up to 300 mm^2. These tables give the impedance at various average temperatures as well as at 20 °C.

The resistance of a conductor at other temperatures can be calculated by working out the operating temperature of the conductor and then calculating the change in resistance from 20 °C; by using the formulae given in earlier chapters. The following formulae only apply for $t_a \geq 30$ °C, and cannot be used when the protective device is a BS 3036 fuse.

The formula for calculating the actual operating temperature of the conductor is as follows:

$$t_f = t_p - \left(G^2 A^2 - \frac{I_b^2}{I_{tab}^2}\right)(t_p - t_a)$$

The formulae for converting the resistance of the conductor from 20 °C to the average of operating temperature and limit temperature of the conductor's insulation are as follows:

$$t_1 = \frac{t_f + t_L}{2} \qquad R_{av} = \frac{(230 + t_1) R_{20}}{250}$$

It is important to remember that the danger with the above method of determining the prospective short-circuit current, when used to check whether switchgear or equipment has the correct withstand capacity, is that the values obtained are lower than the actual short-circuit current at the start of the fault.

Checking withstand capacity of switchgear etc.

Two calculations are therefore needed, one to determine whether the equipment installed has sufficient withstand capacity, and the other to determine whether the conductors are protected against short-circuit current.

Looking first at checking whether the withstand capacity of equipment is suitable, if the supply is three-phase, the maximum prospective short-circuit current will occur if the fault is across all three phases, since the fault current is only limited by the impedance of the source and one phase conductor. If the source impedance is expressed as $R_1 + jX_1$ and the phase conductor $R_2 + jX_2$ then the prospective short-circuit current will be:

$$I_p = \frac{V_{ph}}{\sqrt{(R_1 + R_2)^2 + (X_1 + X_2)^2}}$$

The damage to the cable or switchgear could occur when the circuit was switched out of service, or when the circuit is lightly loaded; in either event, the conductors could be at ambient temperature, so the resistance values in the calculation ought to be taken at the lowest temperature conditions that can occur, to give the maximum symmetrical short circuit current. This value is then used to check whether the equipment will withstand the prospective short circuit current.

Checking conductor withstand capacity

Checking whether the conductors are protected is done differently. As already explained, it is the minimum fault current that is required, which means taking the resistance of the conductors at the average of the working temperature and the limit temperature for the insulation. This is, however, not the only criteria; the minimum short circuit current is also dependant upon the type of circuit and type of fault.

In diagram 3, a three-phase four wire supply has a short-circuit across all the conductors, so the maximum short-circuit current will be:

$$I_p = \frac{V_{ph}}{Z_{p1} + Z_{p2}} \qquad \text{but the minimum short-circuit current will be:}$$

$$I_{pn} = \frac{V_{ph}}{Z_{p1} + Z_{p2} + Z_n}$$

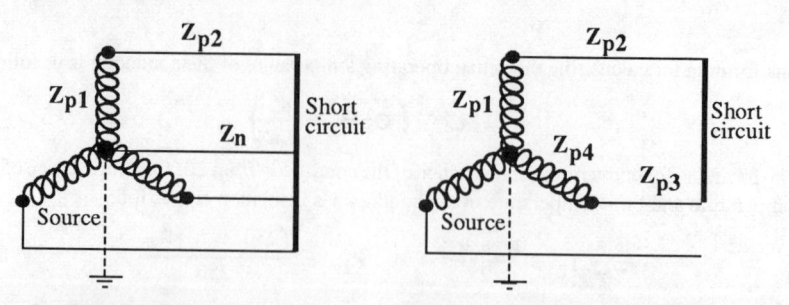

Figure 3 - Three phase and neutral short circuit

Figure 4 - Two phase short circuit

In diagram 4, the fault is shown across two phases, but this time it is a three wire supply: i.e. there is no neutral. In this case the maximum short-circuit current will be the same as in Figure 3 above, but the minimum short-circuit current will occur when there is a fault between two phases; the line voltage is therefore divided by the impedance of all the conductors involved, as given in the following formula:

$$I_{pp} = \frac{V_L}{Z_{p1} + Z_{p2} + Z_{p3} + Z_{p4}}$$

This formula can be simplified if Z_p is taken as the impedance of one phase up to the point of fault. The formula now becomes:

$$I_{pp} = \frac{V_L}{2 Z_p}$$

In each formula, the impedances have been shown added together for simplicity of understanding the conductors involved in the circuit. In practice, the resistance of each item and the reactance of each item would be added together and put in the impedance formula. Using figure 4 as an example:

$$I_{pp} = \frac{V_L}{\sqrt{(R_1 + R_2 + R_3 + R_4)^2 + (X_1 + X_2 + X_3 + X_4)^2}}$$

Calculation sheet

When calculations have to be carried out using R and X, it is advisable to use a table for listing the values of R and X. In this way neutrals and cpcs do not get omitted. The following table is given for guidance. Z_p only has the values of one phase conductor placed in it. The Z_{pn} column has the values of the phase conductor and the neutral conductor placed in it, and the Z_s column has the phase conductor and the cpc conductor values placed in it. The impedance Z does not have to be worked out on each line; it is only required at the point the impedance is required for determining fault current. This will usually be at switchgear and distribution boards, or at the ends of final circuits.

Equipment or location	Z_p			Z_{pn}			Z_S		
	R	X	Z	R	X	Z	R	X	Z

Figure 5 - Calculation sheet

Co-ordination between protective devices

A protective device can have a breaking capacity less than the prospective short-circuit current at the point it is installed, providided that on the supply side of that device there is a second further protective device which has the necessary breaking capacity and the characteristics of the two devices are co-ordinated. This is best explained with an illustration.

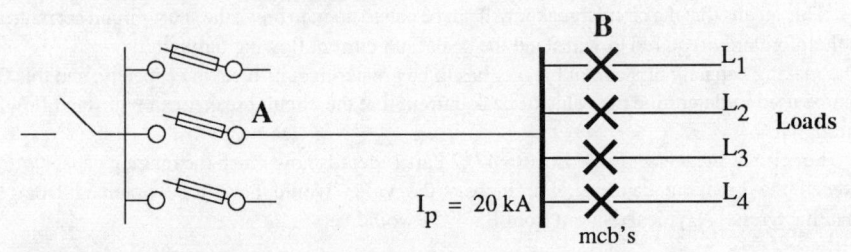

Figure 6 - Co-ordination of protective devices

Where it is required to install lower breaking capacity devices at 'B', with a protecting fuse at 'A', the characteristics of the protective devices have to be co-ordinated. The principle is to select a fuse whose pre-arcing time/current characteristic crosses the circuit breaker time/current characteristic at a point which allows the circuit breaker to operate under all conditions of overcurrent, up to it's short-circuit capacity. This leaves the fuse to take over operation at the short-circuit current values in excess of the circuit-breakers breaking capacity. This involves superimposing the fuse characteristic onto the circuit breaker characteristic to determine the change-over point as shown in the Figure 7.

It must be remembered that all protective devices have a tolerance. The dashed line in Figure 7 shows the envelope for both the circuit breaker and the fuse; it is important that the two characteristics do not overlap.

As can be seen from figure 7, the circuit breaker operates for all fault current values up to the point 'X' on the fuse characteristic.

SHORT-CIRCUIT CALCULATION THEORY

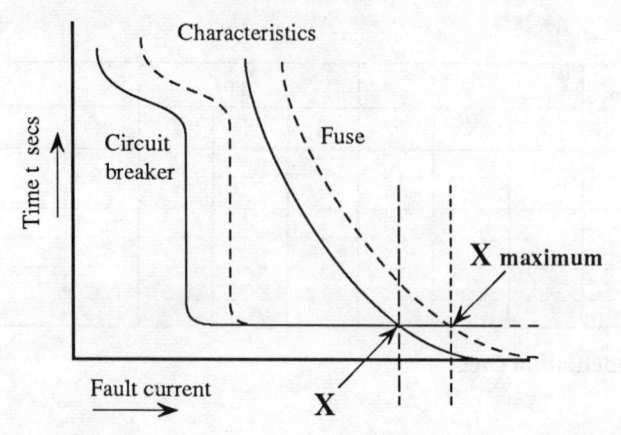

Figure 7 - Fuse and circuit breaker characteristics combined

For fault currents in excess of the point 'X' the fuse will operate first, but both devices will recognise the fault, and both devices will start to operate; but because the fuse characteristic goes down to the current axis (i.e. its operating time is less than the circuit breakers) the fuse will operate first.

This means that the circuit breaker will not be called upon to break the short-circuit current, and is therefore only required to withstand the peak fault current flowing through it.

The making capacity of a circuit breaker has to be greater than its breaking capacity, and this fact can be used to determine the value of peak current that the circuit breaker can withstand flowing through it.

For circuit breakers, Table 1 of BS 4752 Part 1 details how much the making capacity must exceed the breaking capacity. For mcbs, the value would have to be obtained from the manufacturers. A typical column from BS 4752 would be:

Rated short-circuit breaking capacity	Minimum short-circuit making capacity
I_{cn} = 4,500 to 6000	1.53 × I_{cn}

The value of making capacity for the circuit breaker can be used in conjunction with the fuse cut-off characteristic which gives the peak current let-through the fuse for various r.m.s. values of short-circuit current.

The method is to find the peak current let-through the fuse for a given fault current from the fuse cut-off characteristic. The peak current can then be divided by the factor by which the making capacity of the breaker must exceed its breaking capacity. The result must not be more than the breaking capacity of the breaker.

For instance, if the peak current obtained from the fuse cut-off characteristic were 7,500 amps and the factor by which a circuit breaker's making capacity must exceed its breaking capacity were 1.53, then the circuit breaker's minimum breaking capacity would need to be 7,500 ÷ 1.53 = 4,900A.

13. Short-circuit current calculations - examples

The examples given have been kept simple so as to illustrate the method of carrying out the calculations. In practice, it may be more economic to take into account the actual operating temperature of the conductors. In such a case, the following formulae can be used:

$$t_f = t_p - \left(G^2 A^2 - \frac{I_b^2}{I_{tab}^2}\right)(t_p - t_a) \qquad t_1 = \frac{t_f + t_L}{2} \qquad R_{av} = \frac{(230 + t_1)R_{20}}{250}$$

where t_f is the operating temperature of the conductor; t_L is the limit temperature for the conductor's insulation; t_1 is the average, of the initial conductor temperature and limit temperature of the conductor's insulation; R_{av} is the resistance of the conductor at the average temperature; the conductor impedance would then be calculated using R_{av} and the reactance for the conductor. All three formulae could be combined to give just one formula for R_{av}.

Alternatively, having found t_1, the resistance can be found by using the conversion factors in MISC 4 in Part 4. This figure would be used with the reactance taken from either RA 1, RC 1, or RXLPE 1 in Part 4; to calculate the impedance. Additionally, the value of k used can also be based on the lower initial temperature of the conductor.

For those instances where the conductor is only carrying 80% of its current-carrying capacity, tables provided in Part 4 give the impedance of the conductor at the average of the actual operating temperature and the limit temperature for the conductor's insulation. Table K1A and K1B give the k factor for the conductor when it is only carrying 80% of current-carrying capacity.

Example 1 - Co-ordination between protective devices.

Consider the following diagram representing a situation in which it is intended to install at 'B' circuit breaker devices which have lower breaking capacities than the 20 kA prospective short-circuit current available at 'B'. Let the nominal rating of the protective device (I_n) at 'A' be a 63 A BS88 Part 2 fuse. Ignoring any reduction in I_p due to the cable between 'A' and 'B', will the 63 A fuse protect the circuit breakers and what breaking capacity will the circuit breakers have to be ?

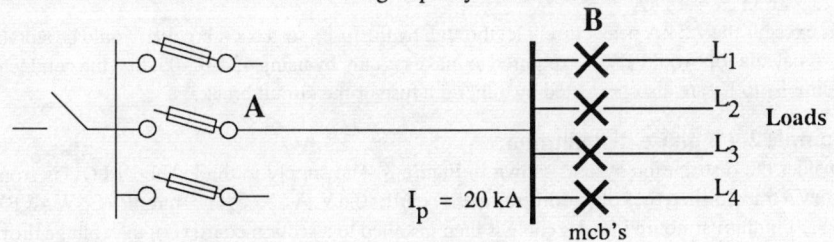

Figure 1 - Co-ordination of protective devices

Working

The breaking capacity of the 63 A fuse is 80 kA, so this exceeds the 20 kA at the point at which device 'A' is installed. Now look at the 63 A cut-off characteristic curves in Part 4, as shown in Figure 2. When the prospective short circuit current is 20 kA, the characteristic shows that the peak cut-off current is 7.5 kA (see dotted line in Figure 2).

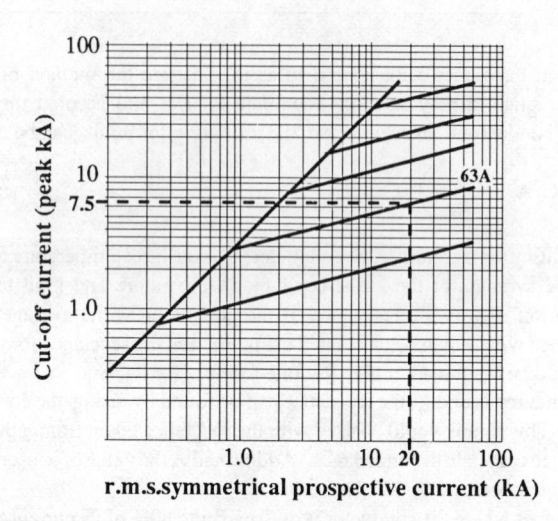

Figure 2 - Fuse cut-off characteristic

The next step is to refer to Table 1 in BS 4752 Part 1. This specifies that the making capacity of the circuit breaker should be greater than its breaking capacity, by the factor given in Table 1. For a breaking capacity of 4.5 kA to 6 kA, the minimum making capacity is 1.53 times the breaking capacity. Since a large fault current will cause the fuse to operate before the circuit breaker, the circuit breaker is not having to break the fault current, but has only to withstand the peak value of fault current flowing through it. As the making capacity of the breaker is greater than its breaking capacity, this can then be used to determine the withstand capacity of the breaker. i.e:
6 kA × 1.53 = 9.18 kA
This exceeds the 7.5 kA peak current let through by the fuse, so a 6 kA breaker would be suitable.

A calculation would also be required to make certain, by using 434-03-03, that the conductors feeding L_1 to L_4 are also protected by either the fuse or the circuit breakers.

Example 2 - 3 phase distribution.

Consider the distribution system shown in Figure 3. The supply to the L.V board (A) is from a 500 kVA transformer; the voltage drop up to the board is 0.6 V. A 4 core × 300 mm² PVCSWA&PVC strip aluminium armoured feeder cable is then installed to a section board (D); the voltage drop in the cable is 10.5 V. A final circuit to a 5.5 kW three-phase motor (H) is carried out, using 3 core 4 mm² cable. The following information is available:

at 'A' $Z_p = 0.0055219 + j0.01786$ and $Z_{pn} = 0.0059538 + j0.01862$.

The Cable from A to D is 161 metres long and is clipped direct. The length of the circuit from board D to motor H is 50 metres, the cable being clipped direct. The distribution boards are suitable for GEC Type T Gg fuses. The object is to check whether the fuses protect the cables against short-circuit current.

SHORT-CIRCUIT CALCULATIONS - EXAMPLES 243

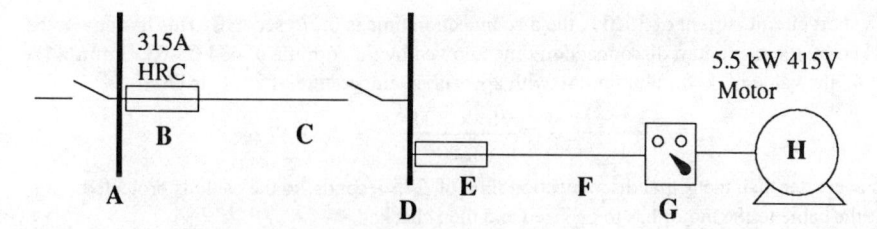

Figure 3 - Line diagram for example 2

Working

In this type of problem it is convenient to log the cable parameters as they are worked out on a calculation sheet, or on the drawing; this enables quick reference to their values to be made.

The first thing to remember is that it is the minimum short-circuit current that is required to check that the cables are protected. All the cables are pvc insulated, so their impedances will be taken at 115 °C.

To check the feeder cable 'C', the total impedance up to board 'D' is required. From table RA1 in Part 4, the values per 1000 m for 300 mm^2 cable are 0.138 + j 0.076

Value for cable 'C' phase = 0.161 (0.138 + j0.076)	0.022218 + j0.012236
Value for cable 'C' neutral =	0.022218 + j0.012236
To this must be added the impedance at the source	
Value of Z_{pn} at Main L.V. board =	0.0059538 + j0.018620
Total	0.0503898 + j0.043092

$$\text{Calculate } Z_{pn} \text{ at D} = \sqrt{0.0503898^2 + 0.043092^2} = 0.0663023 \, \Omega$$

$$\text{Calculate } I_{pn} = \frac{240V}{0.0663023} = 3{,}619.8 \text{ A}.$$

From the fuse characteristic in Part 4, look up the disconnection time for 3,620 amps for the 315 A fuse, as shown in Figure 4.

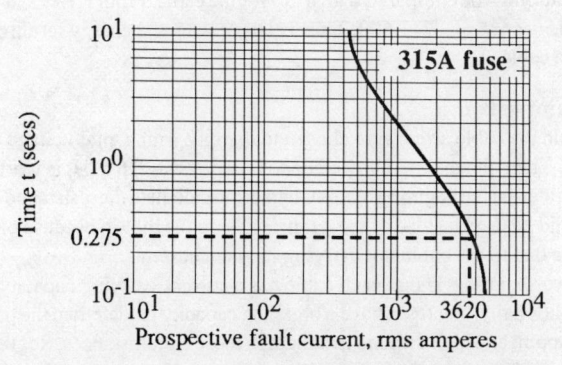

Figure 4 - Part 315 A fuse characteristic

For a short circuit current of 3620 A the disconnection time is 0.275 seconds. This has now to be checked by the maximum disconnection time allowed by the formula of 434-03-03. From K1A Part 4, the value of k for aluminium with a working temperature of 70 °C is 76.

$$t = \frac{k^2 S^2}{I^2} = \frac{76^2 \times 300^2}{3620^2} = 39.67 \text{ secs.}$$

This is greater than the actual disconnection time of .275 seconds, so the cable is protected.
Next the cable to the motor has to be sized and then checked.
From MC1 Part 4, 5.5 kW 415 V motor has a full load current of 11 A, and requires a 25 A GEC Type T 'Gg' fuse, or a GEC 20M25 type NIT fuse.
This is a three-phase circuit, so the minimum short-circuit current will occur when there is a short between two phases; this means that the Z_p is required from the source.
The resistance of one phase of the 4 mm2 cable from table RC1 Part 4 is 6.362 Ω / 1000 m at 115 °C, so the resistance of 50 m = 0.05×6.362 = 0.3181 Ω.
Working out the total impedance to the motor:

Value of Z_p at the source	0.0055219	+	j0.01786
Value of Z_p for cable 'C'	0.022218	+	j0.012236
Value of Z_p for cable 'F'	0.3181	+	j0.0000
Total	0.34584	+	j0.030096

Total Z_p of one phase = $\sqrt{0.34584^2 + 0.0301^2}$ = 0.3471Ω

$$I_{pp} = \frac{V_L}{2 Z_p} \quad \text{therefore } I_{pp} = \frac{415}{2 \times 0.3471} = 597.8 \text{ A}$$

Now, from the fuse characteristic in Part 4 obtain the disconnection time for the 25 A fuse, in a similar manner to that shown in Figure 4. The fault current is past the point at which the characteristic touches the current axis. The disconnection time is less than 0.1 seconds, so the manufacturer's $I^2 t$ let-through has to be compared with $k^2 S^2$, instead of using the formula.

From the $I^2 t$ chart in Part 4, the total $I^2 t$ for a type T 25 A fuse at 415 V is 1.85×1000 = 1,850 A^2sec. This must now be compared with $k^2 S^2$ for the cable. From K1A, k is 115, S is 4 mm^2, therefore $k^2 S^2 = 115^2 \times 4^2$ = 211,600. This is larger than the energy let-through the fuse, so the fuse protects the cable.

Comment on example two

In the above example the cable sizes were chosen to comply with a total voltage drop of 4%: i.e., 16.6 V three-phase. The voltage drop allowed up to board A was 0.6 V. It is worth noting that by using the formulae given iat the begining of this Chapter; to calculate the resistance of the conductor, a 2.5 mm^2 cable could be used instead of the 4 mm^2 cable to the motor, because of the reduction in conductor resistance due to the cable not carrying its rated current.

Another point worth noting is that each cable has a current-carrying capacity in excess of the size of the fuse. Additionally, the fuses have a breaking capacity greater than the prospective short-circuit current at the point at which they are installed. In these circumstances, Regulation 434-03-02 is applicable, even though the final circuit is to a motor; the calculations were therefore unnecessary.

SHORT-CIRCUIT CALCULATIONS - EXAMPLES

Example 3 - Maximum short-circuit current.
Calculate what short-circuit current the starter for the motor in example 2 must withstand, if the Z_p at board A at 20 °C is 0.0054 + j0.01786, and the cable length to the starter is 50 m.

Working
The calculated short-circuit currents in example 2 are only suitable for checking that the cables are protected. For checking that equipment, such as the starter is protected, the maximum prospective short-circuit current is required.

In a three-phase system, the maximum fault current is with a fault across all three-phases. This means that the impedance of one phase conductor only is required, but this must be taken at the appropriate temperature. Since the maximum current is needed, the impedance of the conductors should be taken at the lowest temperature at which the conductors could be when a fault occurs, as the damage could occur when the conductors were not carrying any load current. In the absence of any definite information, take the resistance of the conductors at 20 °C, and those of any transformer involved when the windings are cold.

Z_p at the L.V. board @ 20 °C	0.0054 + j0.01786
Z_p for the cable 'C' @ 20 °C = 0.161(0.1 + j 0.076)	0.0161 + j0.012236
Z_p for the cable 'F' @ 20 °C = 0.05 × 4.61	0.2305 + j0.000000
Total Z_p for the cable 'C' @ 20 °C	0.252 + j0.030096

$$\text{Total } Z_p = \sqrt{0.252^2 + 0.0301^2} = 0.2538 \, \Omega$$

$$I_p = \frac{240V}{0.2538} = 945.6A$$

It would now require checking that the starter would withstand a fault current of some 1000 amps. Note that the fault current is some 350 amps larger than that calculated for the protection of the cables.

Example 4 - Socket outlet ring circuit.
A ring circuit is to be installed from a consumer unit, the circuit being wired in 2.5/1.5 twin and cpc cable. The electricity supplier advises that the external earth loop impedance Z_E will be 0.35 ohms, and the system is TN-C-S. The cable is to be installed clipped direct, and will not be grouped with other cables, or installed in contact with thermal insulation. The circuit will be protected by a 30 A rewirable fuse. Determine (a) the length of the ring circuit and (b) whether the circuit is protected if a phase neutral fault develops 10 m from the fuse in one leg of the circuit.

Working
Strictly speaking, voltage drop cannot be worked out for a ring circuit at the design stage, since the designer has no control over what the user may plug into the circuit in the future. The length of the circuit is determined by the maximum floor area allowed for the circuit.

As an approximation, the average length of a ring circuit for voltage drop can be worked out by taking the worst and best current distribution for the circuit. This is how the total length is determined for this example.

First, calculate the pessimistic length. This will occur if all the load were at the mid point of the ring. A current equal to half the rating of the protective device will flow in each half of the ring circuit; therefore, using table CR7, mV/A/m =18.

SHORT-CIRCUIT CALCULATIONS - EXAMPLES

$$\text{Length of half ring circuit} = \frac{9.6V \times 1000}{15A \times 18\Omega} = 35.56 \text{ m}$$

Total length of ring circuit $= 35.56 \times 2 = 71.12$ m

Secondly, the optimistic length will be when each socket outlet carries an equal amount of current, and there are equal lengths of cable between each item. Assume 30 socket outlets. There will be 15 sockets in each leg of the ring circuit, and each socket will have a 1 A load; therefore, 9,600 mV = 18L(15A+ 14A + 13A + 12A + 11A + 10A + 9A + 8A + 7A + 6A + 5A + 4A + 3A + 2A + 1A)

$$9{,}600 \text{ mV} = 18 \times L \times 120: \text{therefore, } L = \frac{9600}{18 \times 120} = 4.44 \text{ m / cable section}$$

There are 31 equal sections of cable, so the total circuit length $= 4.44 \times 31 = 137.64$ m.

$$\text{Average ring circuit length} = \frac{71.12 + 137.64}{2} = 104.38 \text{ m.}$$

This indicates a maximum ring circuit length of 100 metres, providing the floor area is not exceeded, or the actual loading on the ring is not known. A 30 A protective device would give slightly different results.

The short-circuit current 10 m from the fuse with a phase to neutral fault can now be determined. The calculation is easier to follow if the conductors are labelled as in Figure 5.

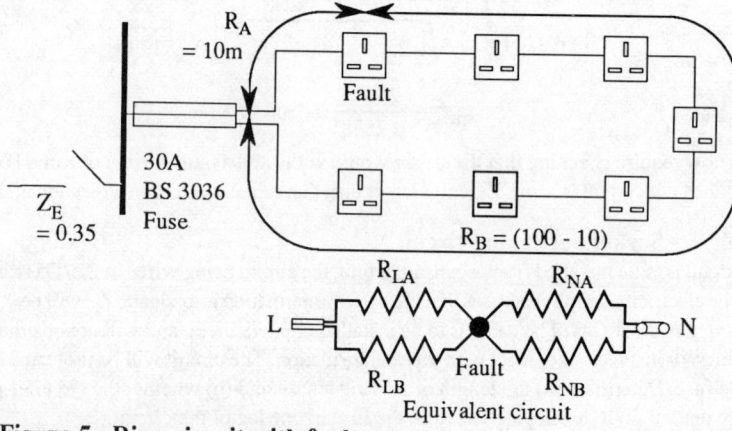

Figure 5 - Ring circuit with fault.

The resistance values of 2.5 mm^2 at 115 °C have been obtained from Table RC1 in Part 4: i.e. 10.226 Ω per 1000 m = 0.010226 Ω per metre.
The resistance R_{LA} = 10 × 0.010226 Ω = 0.10226 Ω.
The resistance R_{LB} = 90 × 0.010226 Ω = 0.92034 Ω.
These resistance are in parallel therefore :

$$\frac{R_{LA} \times R_{LB}}{R_{LA} + R_{LB}} = \frac{0.10226 \times 0.92034}{0.10226 + 0.92034} = 0.092\Omega$$

A similar calculation can be carried out for the neutral, but since this is the same size as the phase conductor, it will have the same resistance; i.e. $0.092\,\Omega$.

These two resistances are in series: therefore $Z_{pn} = 0.092 \times 2 = 0.184\,\Omega$.

As the system is TN-C-S, Z_E will give the phase/neutral impedance external to the installation.

$$\text{Total } Z_{pn} = Z_E + Z_{pn} = 0.35 + 0.184 = 0.534\,\Omega.$$

$$I_{pn} = \frac{240V}{0.534} = 449.44A.$$

The fuse will see, and react to the fault current of 449.4 amps. Looking at the BS 3036 fuse characteristics in Part 4, it will be seen that the disconnection time is 0.1 secs. The energy let-through the fuse will, therefore, be $450 \times 450 \times 0.1$ (See last paragraph.) $= 20{,}250\,A^2$ sec.

But as far as the ring circuit conductors are concerned, the fault current of 450 A will divide by the ratio of the conductor lengths of each leg of the ring circuit, since each leg is in parallel, i.e.,

$$I_{pn} \text{ in the 10m length } = \frac{90}{100} \times 449 = 404.1A.$$

Similarly, the short-circuit current in the other leg of the ring circuit will be:

$$\frac{10}{100} \times 449 = 44.9A.$$

It is now important to establish what current can be tolerated by each leg of the ring circuit. This can be done by using the equation $I^2 t = k^2 S^2$, and working out the value of I, for the actual disconnection time t found from the fuse characteristic.

$$I = \frac{k \times S}{\sqrt{t}} = \frac{115 \times 2.5}{\sqrt{0.1}} = 909.15\,A$$

This indicates that the conductors will tolerate a larger current than that to which they are subjected; i.e., 909.15 A is larger than 404.1 A. The fuse, therefore, protects the ring circuit against short-circuit current, and because of the lower fault current, the longer leg of the ring will also be protected.

By calculation, it can be proved that this circuit is also satisfactory as far as section 413 and 543 are concerned.

Energy let-through a rewirable fuse

When the disconnection time is 0.1 seconds or less, the manufacturer's $I^2 t$ energy let through the protective device has to be compared with $k^2 S^2$ for the conductor. Rewirable fuses are not energy limiting devices, since they only interrupt the fault current as it passes through zero. This means that the arcing time and energy are, in general, insignificant in relation to pre-arcing time and energy. For a rewirable fuse, a sufficiently accurate value of $I^2 t$ let-through energy can be obtained by squaring the current at 0.1 seconds, and multiplying this by 0.1. This would then be the value to compare with $k^2 S^2$.

14. Isolation and switching

Types of switching
As a result of international agreements, three types of switching have been introduced, namely:
- Isolation
- Switching off for mechanical maintenance
- Emergency switching

In addition to these three types of switching there is 'functional switching', this being related to the user of the installation.

Isolation
Isolation is used to prevent danger. Such a danger exists when work has to be carried out on parts that are live in normal use, or with mechanical equipment driven by electrical means.

When work has to be carried out on such equipment, it is essential to ensure that isolation from every source of electrical energy is complete. For example, in damp situations, motors are quite often provided with internal heaters to limit condensation within the motor. The circuits are designed so that the heater is switched on when the motor is switched off. In these circumstances the isolator at the motor must isolate both the power supply to the motor and the supply to the heating unit in the winding of the motor, if danger is to be avoided.

Another example of dual supplies going into equipment is with control circuits which derive their supply from a source of power different from that of the equipment. In this instance, isolation is not complete until the control circuits have also been isolated.

The requirements of the IEE Wiring Regulations for isolation and switching, are given in Chapter 46, Section 476 and 537. Compliance with these regulations is also likely to comply with the Electricity at Work Regulations. The IEE Regulations do not, however, lay down any rules for the management of safe systems of working.

Object and purpose
The object of isolation is to ensure that normally live parts are made dead before work commences. The purpose of isolation is to is to enable work to be carried out on parts that are normally live, in complete safety. One of the precautions to be taken, is to ensure the supply cannot be restored by a remote switch or control circuit. The electrical equipment to be worked on should not be accepted as safe until it has been proved dead by testing. Any test instrument used must have fusible leads. Testing to prove that a circuit or piece of equipment is dead is considered to be live working, so training is required before personnel are allowed to test.

Isolation by non-electrical personnel
Isolation can also include the isolation of equipment that is permanently being taken out of service, and can be used to enable non-electrical work to be carried out. However, where non-electrical personnel use isolators to enable mechanical work to be carried out, they should be taught that the load must be switched off first, since isolators are not designed to switch load on or off. It is better to have the isolator interlocked with a load breaking device, so that the load is switched off before the isolator's contacts open. This is best achieved by using isolators with additional contacts which

break first and make last; these are then wired into the stop circuit of the controlling contactor, which disconnects the load before the isolator contacts start to open. This arrangement also has the advantage that the equipment can only be restarted at the starter in the proper manner.

Where required
Isolation, in the form of a switch or linked circuit breaker, capable of switching the full load current of the installation, has to be provided as near as possible to the origin of every installation. With the exception of three-phase TN-S and TN-C-S systems where a solid neutral link is allowed, all live conductors must be disconnected.

An isolator or switch near the origin of the installation, can be used to switch a group of circuits, and can be the only isolator installed to comply with the regulations. An example is the consumer unit installed in a house.

Where there are more than one source of supply to an installation, a switch is required for each source, together with a notice warning that all switches must be operated to isolate the installation. Where one of the sources of electricity requires its earth to be independent, a switch may be inserted between the neutral point and earth, providing that the switch is linked so that it connects and disconnects the live conductors at the same time as the earth connection.

To comply with the regulations, isolators will be required at switchgear to enable maintenance to be carried out. The isolator must isolate all phase conductors in a TN-S or TN-C-S system, and all live conductors in a TT or IT system. The regulations also require every circuit and final circuit to have a means of switching the supply on load. They will also be required at every motor, and for high voltage discharge lighting.

Isolators can be placed remotely from the equipment or circuits they control, providing that the means of isolation can be secured in the open position. This means preventing inadvertent or accidental re-closure whilst being used to isolate the equipment or circuits they control. This requires their being secured against inadvertently or unintentionally being returned to the 'on' position by other personnel, or through mechanical shock or vibration.

Any locks or handles within the same installation used to make certain that the supply cannot be inadvertently switched on, must be unique to that installation. This is best achieved by each lock having a different type of key, one of which is kept by the manager, and the other by the skilled operative. An alternative arrangement when the primary isolator is remote, is to have another isolator adjacent to the equipment.

Warning notices required
Where an isolator is not capable of cutting off all the supplies to a piece of equipment, a notice must be displayed, warning that the remaining live parts must be isolated elsewhere, and it is more meaningful if the label specifies where. Alternatively, an interlocking arrangement may be used so that all circuits are isolated before access can be gained to the equipment. Except where previously mentioned, isolators must not be placed in protective conductors, nor in protective earthed neutrals.

Requirements for isolators
Since isolators are intended to stop an electrical supply getting to parts that are normally live whilst work is being carried out on those parts, they have to comply with certain requirements. They must have sufficient poles to disconnect all live conductors including neutrals, unless the system is TN-S or TN-C-S, in which case only the phase conductors need to be disconnected.

250 ISOLATION AND SWITCHING

They must also have a specified isolating distance between poles when open, and the position of the contacts shall be either externally visible, or clearly and reliably indicated only when the specified isolating distance has been achieved on each pole. This precludes the use of semi-conductor devices, or micro-switches being used as isolators.

Isolators must be selected and installed so as to prevent unintentional re-closure by mechanical shock or vibration, or by inadvertent or unauthorised operation by other personnel.

Where the neutral of an isolator or switch in a three-phase four wire TN-S and TN-C-S system is a solid link, it must either be only removed by using a tool, or accessible to skilled persons only.

What can be used as an isolator

Switches and circuit breakers can be used as isolators, providing they comply with the rules for isolators. Plugs and sockets can also be used providing they are suitable for breaking current whilst being disengaged.

Using lock off stop buttons that are mounted next to the equipment is not a safe means of isolation, since it would be possible for someone to tamper with the remote starter (pushing the contactor in would energise the cables), or a fault could energise the main circuit.

Solid links or fuse links can be used for isolation, but will require a management procedure to ensure that the load is switched off first to ensure that no danger will arise.

The difference between an isolator and a switch, or an isolating switch, is often misunderstood. Isolators are not designed to make or break load or fault currents; the load should therefore be switched off before the isolator is operated. On the other hand, a switch or an isolating switch is designed to make and break load current.

Where an isolator is used in conjunction with a circuit breaker for maintenance, it must either be interlocked with it, or accessible only to skilled persons.

Fundamental requirements for safety

There are three fundamental requirements in Chapter 13 of the Wiring Regulations that have to be taken into account. Regulations 130-06-01 and 130-06-02 are generally taken into account in Chapter 46, Sections 476 and 537, but account should also be taken of 130-07-01; this calls for every piece of equipment which requires operation or attention by a person to be installed so that adequate and safe means of access and working space are afforded for such operation or attention. The E W Regulations also have a similar requirement in Regulation 15.

15. Mechanical maintenance

What it means
Mechanical maintenance includes the replacement, refurbishment or cleaning of lamps and non-electrical parts of equipment, plant and machinery. Confusion arises between 'isolation' and 'switching off for mechanical maintenance', because 'isolation' is a general term which has normally been used when maintenance has to be carried out on equipment. As a result of the international discussions, however, a distinction has been drawn between 'isolation', and 'switching off for mechanical maintenance'.

The difference is, isolation enables work to be carried out on normally live parts in safety, whereas switching off for mechanical maintenance allows work to be carried out on non-electrical parts, without the risk of those parts being moved by electrical means. Thus isolation is for the use of electrically skilled or instructed persons, to enable them to be able to work in safety.

General requirements
Where there is a risk of physical injury, some means of switching off for mechanical maintenance has to be provided. The devices used for switching off for mechanical maintenance must be suitably placed, readily identifiable, and convenient for their intended use.

Suitably placed means that the switch should be operable safely without personnel having to stretch over moving equipment to reach the switch.

Labelling
In process plants, it is quite common for the switches to be grouped adjacent to each other. In this case, each switch should be labelled so that no confusion arises as to which switch to use. 'Convenient for their intended use,' implies having the switch local to the equipment it controls, otherwise personnel will not be bothered to walk the distance to switch off whilst they make some small adjustment to the machinery.

Where required
A device for switching off for mechanical maintenance has to be provided for every motor, electrically heated surfaces which can be touched, and electromagnetic equipment, the operation of which could cause a mechanical accident. The device should be inserted in the main supply circuit.

The regulations do allow a device to be inserted in the control circuit, providing precautions are taken to ensure the same degree of safety as that if the device were inserted in the main supply circuit. This, however, can be dangerous, and it can be very difficult to ensure such safety. This method should only be used where specified in a British Standard, and only then as a last resort.

A fundamental requirement is that the switch must be under the control of the person carrying out the mechanical maintenance. This means that precautions have to be taken to prevent the equipment from being unintentionally or inadvertently reactivated. Padlocking the switch in the OFF or OPEN position is one method of achieving this requirement.

The regulations allow devices such as lock off type stop buttons, to be inserted into control circuits, but this practice can be dangerous, since a fault could start up the equipment; therefore this mode of switching off is not recommended.

The replacing and cleaning of lamps is considered to be mechanical maintenance, and the local lighting switches can be used for switching off; this naturally precludes two way switches, since one of the switches can never be under the control of the operator. It is also going to be very difficult to comply with this requirement in industrial and commercial premises when all the lighting switches are grouped in one position, since the regulations call for precautions to be taken against the unintentional or inadvertent re-activation of the switch. Another factor which has to be taken into account is whether the lighting switches comply with the requirements laid down for isolating devices given in the paragraph of type of device below.

Type of device

The type of device used for switching off for mechanical maintenance can be:
- a switch,
- plug and socket outlet,
- isolator, providing there is a means of switching off the load before the isolator is operated,
- circuit breakers, providing the manufacturer confirms that the circuit breaker is suitable for this duty,
- control switches operating contactors (not recommended).

Each of the devices listed above must comply with certain requirements. They should be inserted in the main supply circuit, should be manually operated, have an externally visible contact gap, or clearly and reliably indicate the OFF or OPEN position only when each pole is open, and be capable of cutting off the full load current of the circuit they control.

Switches for mechanical maintenance are provided for the use of non-electrical personnel. In general, it is far safer to provide isolating switches (instead of isolators) for mechanical equipment, to enable maintenance to be carried out.

The regulations allow one type of device to be used for different functions, providing it complies with the requirements for each of the functions for which it is used. This means that a switch can be used to perform the functions of: isolation, switching off for mechanical maintenance, or emergency switching; the switch must, however, comply with the requirements for each function.

16. Emergency switching

Definition
Emergency switching is defined as the rapid cutting off of electrical energy to remove any hazard to persons, livestock or property which may occur unexpectedly and is provided for use by any person. People should therefore be advised of the location of emergency switching devices and the devices should be sited so that they will always be available for use.

Where required
Every fixed or stationary appliance that can give rise to danger and is not connected to the supply by a plug and socket, shall be provided with a means of interrupting the supply on load. The means of interrupting the supply shall be so placed that no danger to the operator of the equipment shall occur. Emergency switching devices are operated when danger or a hazard is occurring, and as such they can be operated by anyone; they should therefore be installed at every point at which it may be necessary to disconnect the supply rapidly to prevent or remove the hazard that is occurring.

Where a machine driven by electrical means may give rise to danger, an emergency switch shall be provided; it shall be readily accessible and easily operated by the person in charge of the machine. Where more than one means of manually stopping the machine is provided, and a danger could arise by the unexpected restarting of the machine, arrangements have to be made to prevent the restarting of the machine.

In process plants, or where conveyors are used, additional emergency switching will be required at other hazardous points. In the case of open conveyors, these will be required throughout the conveyor's length. This is best achieved by the use of a trip wire running the full length of the conveyor, and in the case of long conveyors, with intermediate emergency switching points along the length of the conveyor. When installing trip wires, it is important to ensure that the switch is operated by a pull on the wire in either direction. A trip wire with a single switch at one end would not be acceptable.

Operation of emergency switches
Emergency switches should preferably be manually operated, and should be of the latching type of device, or be capable of being restrained in the off position; the handles or buttons must be coloured red.

The means of interrupting the supply must be capable of cutting off the full load current of the installation, acting directly on the appropriate supply conductors, so that a single action will cut off the supply. The means of interrupting the supply shall, with the exception of a three-phase TN-S and TN-C-S system, disconnect all live conductors. Where a three-phase TN-S or TN-C-S system is used, the neutral need not be disconnected unless there is a risk of electric shock.

Where devices such as circuit breakers and contactors are operated by an emergency switch or stop button which is remote, they shall be arranged to open on the de-energisation of the coils.

Where emergency switching could result in a greater danger arising, for instance, by the disconnection of safety services, then it is permissible for the emergency switching to be carried out only by skilled or instructed persons.

As stated earlier, emergency switches can be operated by anyone. Although not a requirement of the regulations, staff should be instructed as to the location and purpose of emergency switching devices.

Emergency stopping

Where mechanical movements caused by electrical means can give rise to danger, a means of emergency stopping has to be provided. Emergency stopping is only concerned with stopping mechanical movement, and can be achieved by energising a circuit to stop the mechanical movement, such as energising a circuit to feed d.c., into the motor stator at the same time as the a.c., mains supply to the motor is cut off. This is known as 'd.c., injection breaking'. It is important to note that the de-energising of the main circuit and the energising of the braking circuit is achieved by one initiation of the emergency stopping device.

Colour of emergency switches

Emergency switches should preferably be coloured red.

Fireman's switch

A fireman's emergency switch has to be provided in the low voltage circuit for exterior discharge lighting and for interior discharge lighting operating at a voltage above 1000 volts. A fireman's switch is not required for a portable discharge lighting unit or a sign which does not exceed a rating of 100 Watts, if it is fed from a readily accessible socket outlet.

Exterior and interior lighting

Covered markets, arcades and shopping malls are all considered to be exterior installations, but temporary installations in permanent exhibition halls are considered to be interior installations.

In general, switches for discharge lighting should be mounted adjacent to the discharge lighting fitting, at a height not exceeding 2.75 metres from the ground, unless the local Fire Authority agree the switch position. The switch can be placed in another position, but a nameplate, which is clearly readable from the ground, has to be placed next to the discharge lighting, advising where the switch is located, and the switch must be clearly labelled to indicate which lighting it controls.

For exterior installations the switch should be mounted outside the building, whilst for interior installations the switch should be mounted in the main entrance to the building. In both cases the switch position can be agreed with the local Fire Authority. Switches should be installed in a conspicuous position, and where more than one switch is installed, each switch must be marked to indicate the installation it controls.

Colour of fireman's switch

The switches should be coloured red and be labelled 'FIREMAN'S SWITCH'. They should have the ON and OFF positions clearly marked, so that they can easily be read by a person standing on the ground. The OFF position of the switch must be at the top.

17. Other means of switching

Difference between switching and interrupting the supply
A main switch or circuit breaker has to be provided for every installation and every circuit and final circuit has to be provided with a means of *interrupting* the supply on load, or in any foreseen fault conditions. There is a difference between switching and *interrupting* the supply; *interrupting* does not necessarily mean using a mechanical switching device; the *interruption* could be brought about electronically.

Circuits and appliances
A group of circuits may be switched by a common device, such as the isolator on a consumer unit.

Every circuit and final circuit shall be provided with a means of switching for *interrupting* the supply on load. A plug and socket-outlet with a rating not exceeding 16 A may be used as a switching device. Only with an a.c., supply may a socket exceeding 16 A be used as a switching device, and only then, if it has a breaking capacity appropriate to its intended use.

Functional switching
Functional switching is the device used to control a circuit or group of circuits in normal use. The device used may not comply with the requirements given previously for switches or isolators. Thus micro-gap switches and semi-conductor devices are functional switches.

Cooking appliances
Every fixed or stationary cooking appliance has to be controlled by a switch which is separate from the appliance. The switch must be in a position where it can be operated without putting the operator of the switch in danger. A danger can occur should a fire start due to cooking oil or fat overflowing on the hot plates of the cooker. The switch should, therefore, be in a position where it will be clear of any flames emanating from the cooker and where it can be operated by persons without them being placed in danger from the fire.

One switch can be used to control more than one appliance in the same room, but consideration has to be given to the total load on the switch, and, if danger is to be avoided, the location of the switch relative to the equipment.

This arrangement is usually adopted where split level cookers are installed in domestic kitchens.

18. Swimming pools

The International Electrotechnical Commission produced an International Standard for swimming pools in 1983. This standard has now been incorporated into the 16th Edition Wiring Regulations.

Swimming pools can be large or small: they can be indoors and out: they can be private, or open to the public. Most of the swimming pools installed in the gardens of domestic properties are in the open, and sunk into the ground. There are a few domestic pool installations which are installed in a building, (so as to make the pool usable all the year round) and a few that are in the open, but built above ground level. The regulations are to be applied to all types of swimming pool installations.

A swimming pool looks harmless enough, but in practice, it is more dangerous than a bathroom, as far as electrical services are concerned. In the first place, a bathroom is usually carpeted and the floor itself has a certain amount of insulation; secondly, most of the equipment in the modern bathroom is made up of insulating material, such as plastic baths, wash hand basins etc. Additionally, over the years, stringent rules have been applied to the installation of electrical equipment in the bathroom. The swimming pool area, on the other hand, is in direct contact with the ground: the area round it is usually wet, and people are walking round the pool with bare feet, so more care is needed with the electrical services.

Classification of Zones

To make it easier to understand the requirements of the Standards or Regulations, the area round the pool is divided into Zones. The IEC classify these areas as Zone 0, Zone 1, and Zone 2. The IEE classify them as Zone A, Zone B and Zone C as shown in Figures 1 and 2. Zone A, covers the pool area, Zone B, is an area extending from the pool edge for a distance of 2 metres and Zone C extends from Zone A for a further 1.5 metres, anything outside these zones can be considered to be Zone D or Zone 3. The zones extend upwards for 2.5 metres (this being the limit of arm's reach); where diving boards are installed, the zone is extended by the height of the board above the datum, as shown in Figure 1.

Zone A (0)

Looking first at Zone A (0), which is the pool itself. The only electrical services which are allowed to be taken to the pool are for those appliances which are directly associated with the pool: these are usually pool lights. The degree of protection for the equipment should be IPX8. This means that they should be suitable for total immersion in water under a specified pressure. They must also be designed to resist the chemicals in the water. The electrical supply to each pool light must be from its own transformer, or an individual winding on a multi-secondary winding transformer; in each case, the secondary's open circuit voltage must not exceed 18 V. The electrical supply to any other type of equipment must be from a SELV source situated outside Zones A, B and C; it must comply with the SELV regulations, and be on its own circuit and must not exceed 12 volts. The wiring should be of Class II construction (i.e. double insulated) without any metallic covering. No junction boxes or other switchgear or accessories are permitted in the Zone.

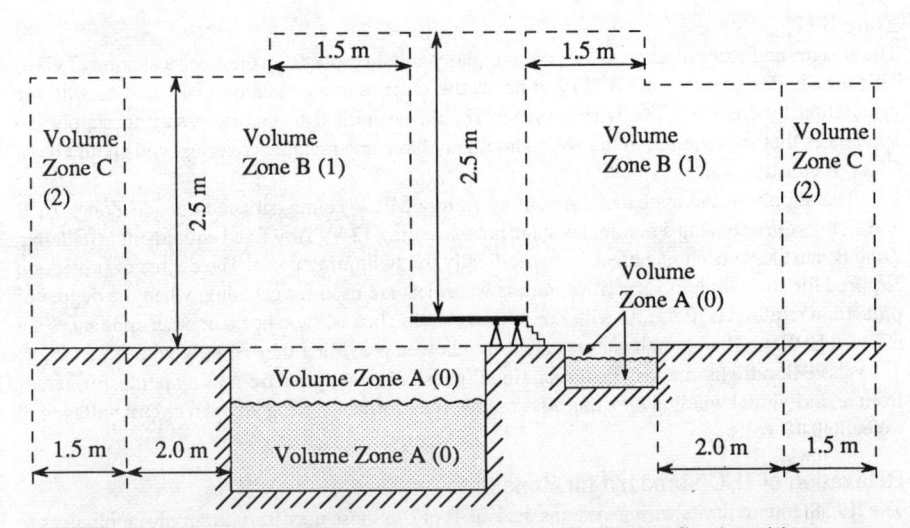

NOTES: The dimensions are measured taking account of walls and fixed partitions.
IEC 364-7-702 Zones are 0, 1 & 2, IEE Regulations Zones are A, B & C.

Figure 1 - Zone dimensions of swimming pools and paddling pools

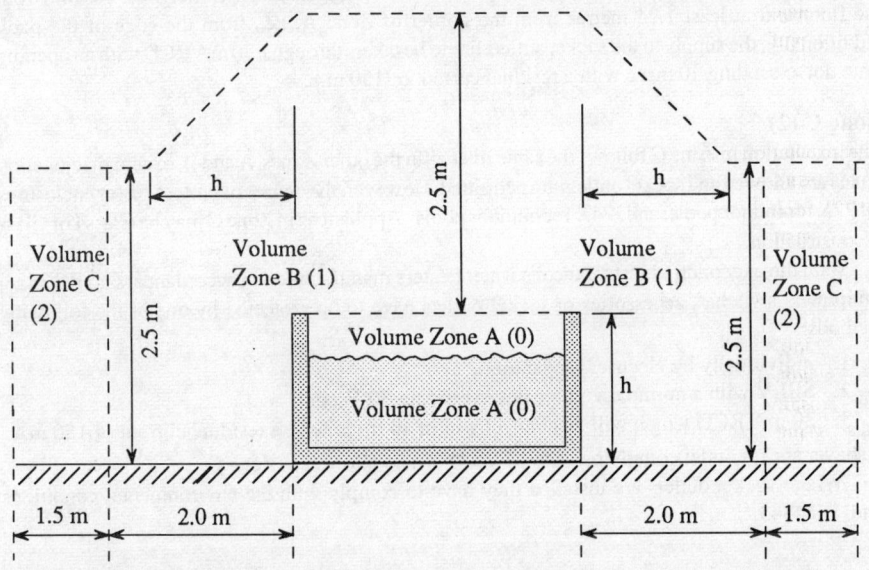

NOTES: The dimensions are measured taking account of walls and fixed partitions.
IEC 364-7-702 Zones are 0, 1 & 2, IEE Regulations Zones are A, B & C.

Figure 2 - Zone dimensions for basins above ground

Zone B (1)

The area immediately adjacent to the pool is equally as dangerous from the electrical point of view. This area is designated Zone B (1) and needs the same amount of care to be taken as with the installation to Zone A. The wiring system should be limited to that necessary to supply the appliances that are required in the zone, should not have any metallic covering, and should be of Class II construction.

The supply to the appliances should be from a SELV source situated outside Zones A, B and C; the source having a nominal voltage not exceeding 12 V. Any fixed equipment installed in Zone B must have been manufactured specifically for swimming pools. The degree of protection required for such equipment is IPX4, unless water jets are used for cleaning, when the degree of protection required is IPX5. As with Zone A, no junction boxes, switchgear or other appliances are allowed in Zone B, although water heaters are allowed providing they comply with BS 3456.

Where floodlights are installed, each floodlight shall be supplied from its own transformer, or from an individual winding on a multi-secondary transformer, having an open circuit voltage not exceeding 18 volts.

Relaxation of IEC Standard for Zone B

The IEC Standard limits wiring systems in Zone B (1), to those necessary to supply appliances in Zone B; no switchgear or accessories being allowed. However, there is a relaxation in the IEE Wiring Regulations for swimming pools concerning socket outlets installed in Zone B. Socket outlets are allowed in Zone B where it is impossible to install them outside the Zone. Any socket outlets installed in Zone B must be manufactured to BS 4343, be installed at a height of 300 mm from the floor and at least 1.25 metres from the border of Zone A, i.e., from the edge of the pool. Additionally, the supply to the socket outlets has to be taken through a 30 mA RCD with an opening time not exceeding 40 msec with a residual current of 150 mA.

Zone C (2)

The installation in Zone C follows the same lines as in the other Zones A and B, except that junction boxes are allowed and socket outlets are permitted. However, the degree of protection for enclosures is IP2X for indoor pools, and IP4X for outdoor pools. Appliances in Zone C may be Class I or Class II construction.

With the exception of instantaneous water heaters manufactured in accordance with BS 3456, equipment, switches, accessories or socket outlets have to be protected by one of the following methods:

1. individually by electrical separation,
2. SELV with a nominal voltage not exceeding 50 V, or
3. 30 mA RCD which will disconnect within 40 msec with a residual current of 150 mA.

A shaver socket outlet complying with BS 3535 is also allowed in Zone C.

Where socket outlets are installed they have to comply with the environmental conditions and BS 4343.

Electric heating

One of the biggest dangers with indoor private pools, is where central heating radiators have been installed in the pool room to keep the air temperature above that of the pool in order to stop condensation; these should be included in the local supplementary bonding.

Heating units embedded in the floor are allowed in Zones B and C provided that the units have an earthed metallic sheath connected to the local supplementary bonding conductors. The units must also be covered by a metallic grid connected to the same local supplementary bonding conductors.

Only water heaters complying with BS 3456 are allowed in Zone B, unless the equipment is supplied by a 12 volt SELV circuit; this being part of the relaxation mentioned earlier.

Equipment outside Zones A, B & C (Zone D)

Boilers and pumps for swimming pools are usually placed in their own room, or in the garage for outdoor pools on private property. The pump should be one manufactured for swimming pools, and made of plastic, so that there is no contact between the water in the pump and the metal parts of the motor; for safety the supply to the pump is taken through a 30 mA RCD.

In domestic installations, the boiler generates its own millivolts for the control of the boiler. The only precaution needed is when time switches, to ensure the boiler switches off before the pump, are installed, there must be electrical separation between the supply mains to the timer and the contacts connected to the boiler wiring; this separation should be the same as that for a safety isolating transformer manufactured to BS 3535.

Additional requirements

All other requirements of the 16th Edition Wiring Regulations should be complied with, such as the requirements for overloads, fault currents, voltage drop, and installation design.

Where SELV circuits are used, whatever the nominal voltage, protection against direct contact has to be provided by barriers or enclosures affording at least IP2X protection, or insulation capable of withstanding a test voltage of 500 V for 1 minute. The measures of protection by non-conducting location, placing out of reach, obstacles and earth-free equipotential bonding, are not allowed.

Local supplementary bonding

Local supplementary bonding is required to connect all extraneous conductive parts and exposed conductive parts together. In the case of swimming pools, this means anything in Zones A, B and C. Additionally, the floor round the pool (Zones B & C) is an extraneous conductive part and must be included in the supplementary bonding. This is best achieved, if the floor is not reinforced concrete, by the installation of an earth rod connected to the supplementary bonding conductors. Where solid floors are installed, an equipotential bonded grid has to be provided in the floor, and this used as part of the equipotential bonding; in practice, the reinforcing mesh in the floor could be used. Where socket outlets are installed, the earth terminal of the socket outlet must be included in the supplementary bonding.

Supplementary bonding must not, however, be made to those appliances which have been energised by a SELV source.

Size of supplementary bonds

Where there are no exposed conductive parts in Zones A, B and C, the minimum size of supplementary bonding conductor is 2.5 mm² if it is sheathed or otherwise mechanically protected, or 4 mm² if is not mechanically protected.

Where exposed conductive parts are in Zones A, B and C, then two sets of rules have to be complied with. One of the rules determines the actual size of the supplementary bonding conductor needed, and the other specifies the minimum size allowed.

$$\text{The allowed resistance of the supplementary bonding conductor} = \frac{50}{I_a}$$

where I_a is the current which will disconnect an overcurrent protective device within 5 seconds, or is the residual current $I_{\Delta n}$ of an RCD.

The minimum size is determined from 547-03, and is illustrated in Figure 3.

Exposed conductive part to exposed conductive part

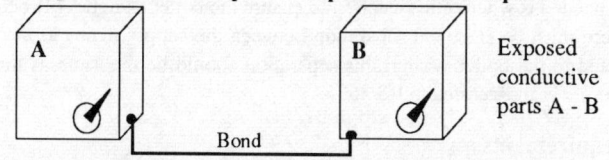

Exposed conductive parts A - B

Same conductance as the smallest cpc feeding A or B providing that it is mechanically protected: otherwise minimum size is 4 mm²

Exposed conductive part C to extraneous conductive part

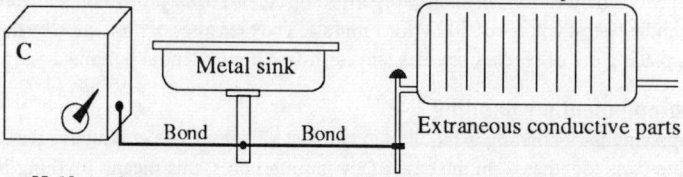

Extraneous conductive parts

Half conductance of the cpc in 'C' if sheathed or mechanically protected: otherwise minimum size required is 4 mm²

Extraneous conductive part to extraneous conductive part

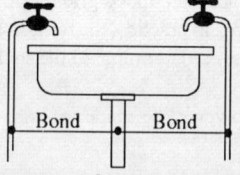

Minimum size 2.5 mm² if sheathed or otherwise mechanically protected: 4 mm² if not mechanically protected

Figure 3 - Minimum size of supplementary bonding conductors

19. Inspection and testing

The fundamental requirements for safety in the regulations call for an inspection and test to be carried out. The test has to be carried out on completion of an installation, on completion of an alteration, or on completion of an addition to an installation. Moreover, inspection and testing has to be carried out before the installation is put into use. The regulations do not specify any size of alteration or addition, so no matter how small an alteration or addition is, an inspection and test is required.

The regulations are concerned with providing safety, especially from fire, shock, burns and mechanical movement actuated by electrical means. The object is to ensure that all the fundamental requirements for safety are complied with. To assist in compliance with these requirements for safety, parts three to six in the regulations give methods and practices which will satisfy these safety requirements.

The regulations specify that the tests carried out shall present no danger to persons, property, or equipment, even if the circuit tested is defective. This means that the installation has to be proved before the application of any mains voltage necessary to complete the tests, and that the necessary steps have been taken to ensure that any equipment that can be damaged by insulation testers is disconnected before the tests are carried out.

Obviously the only person or persons who should carry out an inspection and test are those who are competent to do so. This also means that they must have a thorough understanding of the IEE Wiring Regulations and their application, as well as the instruments they have to use.

When to start inspection

On the larger contract, the inspection and testing should start at the same time as the contract, and continue as the contract proceeds. Quite often the contract can be broken down into areas. As each area is completed, the inspection and testing can be completed; for example, each floor in a multistorey building. This may be impractical for the small contract, so the inspection and test will have to be carried out on completion of the work.

Care is needed when inspecting and approving areas of a contract, that the installation which joins all the areas together, or branch circuits feeding separate areas, do not get omitted from the inspection and testing schedule.

Information to be provided

The person carrying out the inspection and test has to be provided with the details of the assessment of general characteristics. These details will comprise:

1. the maximum demand of the installation,
2. the number and type of live conductors, i.e., whether the supply is single-phase, three-phase three-wire, or three-phase four wire etc.,
3. the type of earthing arrangement: i.e. of which system is the installation part, is it, TN-S, TN-C-S, or TT. The system is unlikely to be TN-C, but where the supply is obtained from the customer's own transformer, it is possible to find an IT system,
4. the nominal voltage of the supply and its frequency,
5. the prospective short-circuit current at the origin of the installation. In a three-phase four wire supply, both the single-phase and three-phase value will be required,

6. the phase earth loop impedance external to the installation,
7. the type and rating of the overcurrent protective device installed at the origin of the installation by the supplier,
8. whether any supplies for safety services are provided, and whether they are automatic or non-automatic. The characteristics of the source of the safety service will also be required in the same way that the characteristics of the main supply is provided,
9. a legible diagram, chart or table showing in particular: the type and composition of each circuit. This means an indication of the outlet points or equipment served by the circuit, and the number and size of conductors, and the type of wiring used (pvc single core cable in conduit, for instance),
10. the information to enable the identification of the devices that perform the functions of protection, isolation, and switching, together with their locations,
11. a description of how the installation's exposed and extraneous conductive parts are protected against the persistence and magnitude of earth fault voltages appearing on the metalwork.

Checking design

The information provided either by the person who designed the installation, or who carried out the installation has to be checked as listed above. In checking the above, it is important to check the following items, since these are the foundation stones on which the installation was built.
- Prospective short circuit current at the origin.
- Size and type of protective device at the origin.
- Phase earth loop impedance external to the installation.
- Type of earthing arrangement (i.e. system type).
- Is the supply suitable for the installation's maximum demand?

As far as the prospective short-circuit current is concerned, where the supply is three-phase, it is important to obtain the single-phase as well as the three-phase short-circuit current. Both values are required, since the single-phase value is needed to check that conductors are protected, and the three-phase value is required to check that the switchgear, starters etc., have the correct withstand capacity.

Inspection: visual and physical

Visual inspection has to precede testing to ensure danger cannot arise when tests are carried out. The visual inspection is made with the supply disconnected. This is a good safety measure, since equipment will be opened and live conductors touched during the visual inspection.

A visual inspection of the installation has to be made to establish that the equipment complies with British Standards. On large contracts this is best done by inspecting equipment, cables, conduit etc., for British Standard labels before the installation is carried out (see Figure 1).

British Standard Kite Mark

British Standard Safety Mark

Figure 1 - British Standard Kite and Safety Mark

If the designer (installer) has stated the equipment complies with a Harmonization Document or IEC code, it should be verified that the designer certifies that it complies with the safety requirements of the regulations.

Visual inspection also means physically checking the installation to make certain that the electrical equipment (which includes cables etc.) has been correctly selected and erected, and not damaged during erection.

Items that should be checked whilst carrying out a visual inspection include:
1. the connection of conductors: are they properly made and tight, and include all the strands of the conductor in the lug or termination, i.e., have strands been removed to enable the connection to be made?
2. that conductors are properly identified, and are properly protected against mechanical damage, and that single pole switches have only been installed in the phase conductors,
3. that the number of cables enclosed in conduit or trunking specified in the design has not been exceeded, and that burrs have been removed from conduit etc.,
4. that conductors have the correct current-carrying capacity, and where they have been derated for grouping that the correct formula from Appendix 4 has been used. For instance, the formula used when simultaneous overload cannot occur has not been used for socket-outlet circuits,
5. that the correct connections have been made at socket-outlets and lampholders, such as the phase conductor being connected to the centre contact of an E.S. lampholder,
6. that where equipment passes through a fire barrier, the holes round the equipment have been filled to the same degree of fire resistance as the wall, floor or ceiling through which the equipment passes, and that internal fire barriers have been installed in the equipment,
7. that protection against direct contact has been made, and where RCDs are used for supplementary protection, they will disconnect within 40 msec with an earth fault of 150 mA,
8. a check should be made to ensure that all barriers have been installed, and that enclosures comply with the regulations. Holes in equipment should not exceed IP2X generally, or IP4X in top surfaces,
9. that the methods of protection against indirect contact comply with the regulations. In particular, that main equipotential bonding conductors have been installed, and are the correct size, so that where disconnection cannot be made within the times specified in the regulations either supplementary bonding conductors have been installed, or RCDs, or the impedance of the protective conductor has been limited to a value given in Table 41C, Table PCZ 8 Part 3,
10. where distribution boards contain circuits that have to be disconnected within 5 secs or 0.4 secs, that either the protective conductor impedance back to the main earth bar (to which the equipotential bonding conductors are connected) is limited in accordance with Table 41C (PCZ 8), or equipotential bonding is carried out at the distribution board to the same extraneous conductive parts as the main equipotential bonding,
11. a check has to be made to ensure that equipment which requires labelling has been properly labelled: for example: labels attached to the earth clamps of bonding conductors,
12. that distribution boards, circuits, switches and terminations are correctly identified,
13. that there is sufficient access and working space at equipment, and that danger or warning notices are installed at the appropriate places. Posting of notices also includes those required for the periodic testing of RCDs, and notice of when the installation should be inspected again.

MEGGER AHEAD OF THE REG'S

Europe's leading electrical test equipment available from AVO International.

A range of high quality instruments to meet the 16th Edition IEE Wiring Regulations.

Each one manufactured to national and international standards of accuracy, reliability and performance.

BM100/3 Insulation & Continuity Tester
- Complies with the 16th Edition Regulations
- 500 V d.c. test voltage at 0,5 MΩ and above
- Also available BM101/3 with an extra resistance range replacing the voltage range.

LT5 Digital Loop Tester
- Complies with 16th Edition Regulations
- Automatic test – no push button
- Resistance 20 Ω TN: 200 Ω TT

CBT 3 Digital RCD Tester
- Automatic testing of conventional and 'S' & General type RCD's
- 12 RCD trip current ratings 6 mA to 500 mA
- Timing (maximum time duration) up to 2000 milliseconds

BM206 Insulation and Continuity Tester
- Complies with 16th Edition Regulations
- Test Voltages 250 V, 500 V & 1000 V d.c.
- Analogue/Digital l.c.d. display

AVO INTERNATIONAL

ARCHCLIFFE ROAD, DOVER, KENT CT17 9EN, ENGLAND
TEL: 0303 202620. FAX: 0304 207342. TX: 96283.

Testing

Before any tests are carried out it is essential that the operator is fully familiar with and competent to use the instruments required. It is also essential that the instruments used comply with British Standards, and have their calibration checked regularly. Since some of the tests are to be made with the mains voltage on, it is important to ensure that those particular instruments are equipped with fusible leads.

Sequence of tests

Once the visual inspection is complete, tests have to be carried out. These can be summarised in a list.

The following tests have to be carried out.
1. Continuity of protective conductors. This involves every protective conductor including main and supplementary bonding conductors.
2. Continuity of final ring circuit conductors.
3. Insulation resistance.
4. Insulation applied on site.
5. Protection by separation of circuits.
6. Protection against direct contact by barriers or enclosures.
7. Insulation of non-conducting floors and walls.
8. Earth electrode resistance.
9. Polarity.
10. Earth fault loop impedance.
11. Operation of residual current operated devices.
12. Prospective short circuit current test.

Tests 1 to 8 do not need the supply connected. Test 9 can be carried out before the installation is energised, but it is usually done at the same time as the earth loop impedance or the RCD tests are made, the instrument indicating whether or not the test should proceed.

Protective conductors

The regulations call for every protective conductor to be separately tested to ensure that it is electrically sound and correctly connected. This requirement means that every circuit protective conductor, main equipotential bonding conductors, and supplementary bonding conductors are tested.

Non-ferrous protective conductors

Providing the protective conductor is not a steel enclosure or a conduit, then the tests can be made with an ohmmeter. The resistances can be checked against the installation design to make certain that there are no high resistance joints.

Providing the phase conductor is the same length and follows the same route as the protective conductor, the ratio of the area of the phase conductor to the protective conductor can be used to determine the resistance of the protective conductor; the phase conductor can be used as the return conductor to the ohmmeter, and by temporarily connecting the phase conductor to the earth terminal at the distribution board, readings can be taken at the end of the circuit.

Where a radial circuit has several outlet points, this test will combine some of the polarity tests which are required later in the test schedule.

The resistance of the protective conductor obtained by this method is given by the formula:

$$R_{pc} = R_o \times \frac{A_p}{A_p + A_{pc}}$$

Where:
- A_{pc} = Area of protective conductor.
- A_p = Area of the phase conductor.
- R_o = The ohmmeter reading.
- R_{pc} = Resistance of protective conductor.

To obtain the phase earth loop impedance by measuring the circuit's resistance with an ohmmeter, connect the phase conductor to earth at the consumer unit, then measure the resistance at each accessory or outlet point between phase and earth.

There is an alternative to the above methods. This has to be used when checking the continuity of bonding conductors. In this case it involves using long test leads. One lead of the continuity tester is connected to the main earthing terminal, whilst the other test lead is connected to various exposed conductive parts of the circuit, or accessory earth terminals. This resistance obtained includes the resistance of the test leads, so this must be measured and deducted from the readings obtained.

When measuring the resistance of bonding conductors, the test leads will be placed at each end of the bonding lead.

This test will also be checking in part the polarity of the conductors and connections.

Ferrous protective conductors

Where a protective conductor is a steel enclosure or a conduit, the continuity test has to be made by passing a current 1.5 times the design (full load) current of the circuit. The test current need not exceed 25 amps, and the test voltage must not exceed 50 volts a.c., or d.c., Where an a.c., test is applied, the frequency has to be that of the supply.

If d.c., is used, the circuit has first of all to be checked to make certain that there are no inductors incorporated in it. It is also important that the test currents are obtained from a safety source of supply; in the case of a.c.,, this would be through a double insulated isolating transformer, the test being shown in Figure 2.

The resistance of the test leads T_A and T_B are also required, the resistance of the protective conductor being: $R_{pc} = \frac{V}{I} - (R_{TA} + R_{TB})$

Where a supply is not available, then a battery can be used, providing there are no inductors in the circuit. Again the neutral can form a convenient return to the test supply, but its resistance will need to be deducted from the calculated value of resistance.

Test instruments have been developed by the instrument manufacturers to carry out this test.

Ring circuit continuity

The continuity of ring circuit conductors has to be proved. This will include the protective conductor where it is installed in the form of a ring, for instance: when twin & cpc cable is used.

Test Method 1

It is first important to identify the outward leg conductors and the inward leg conductors of the ring. Connect the phase conductor of the outward leg to the neutral of the inward leg at the distribution board. Now take the resistance reading between the inward phase conductor and the outward neutral conductor (i.e. the phase and neutral conductors that have not been connected together).

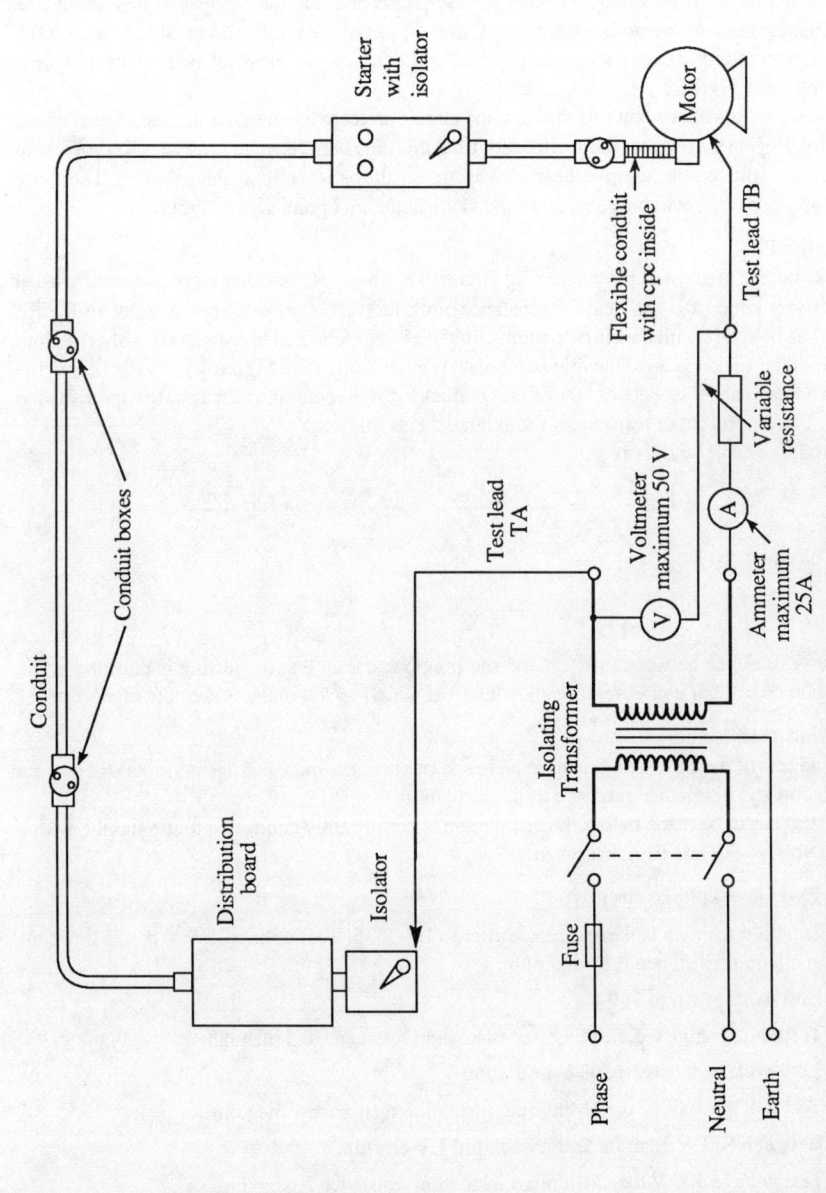

Figure 2 – Continuity and resistance of ferrous protective conductor

The next stage is to connect these remaining phase and neutral conductors together. Now measure the resistance at each socket outlet between the phase and neutral conductor; this should be approximately the same at each socket outlet. If the connections at the distribution board are wrong, the resistance reading at each socket will get smaller the further you move away from the mid point of the ring. (See Figure 3.)

Now repeat the above, but this time use the circuit protective conductor instead of the neutral conductor. Repeat the tests at each socket outlet, but this time between phase and cpc. Keep a record of these test results, since they give the total resistance of the phase and cpc round the ring. The phase earth loop impedance will be a quarter of the value at the mid point socket outlet.

Test method 2

There is another method of proving a ring circuit. The first test is made before the ends of each conductor are joined together, and the resistance of the loop is obtained for each conductor R_p, R_n and R_E The next test is made with the conductors of each loop joined together, and a shorting-out plug put in the socket nearest the mid point of the ring circuit. (See Figure 4.)

The resistance is obtained between the phase conductor and the neutral, and the phase conductor and the cpc. The second set of resistances should then be as follows:

For the phase and neutral loop:

$$R = \frac{R_p}{4} + \frac{R_n}{4} = \frac{R_p}{2} \text{ or } \frac{R_n}{2}$$

For the phase to cpc loop:

$$R = \frac{R_p}{4} + \frac{R_E}{4}$$

The main difference between this test and the previous one is that the testing is all done at the distribution board, but it does require knowledge of which socket outlet is near the mid point.

Insulation resistance

The soundness of the conductor's insulation has to be checked, since leakage currents will cause a deterioration and ultimate breakdown of the insulation.

The test has to be made before the installation is permanently connected to the supply, with a d.c., test voltage not less than that given below.

Extra-low voltage circuits

Supplied from an isolating transformer to BS 3535. Test voltage 250 V d.c., Minimum insulation resistance 0.25 megohm.

Low-voltage up to 500 V

Test voltage 500 V d.c., Minimum insulation resistance 0.5 megohm

Low-voltage between 500 V and 1000 V

Test voltage 1000 V d.c., Minimum insulation resistance 1 megohm.

Between SELV circuits and associated LV circuits

Test voltage 500 V d.c., Minimum insulation resistance 5 megohm.

Withstand test between SELV circuits and LV circuits

Test voltage 3750 r.m.s., a.c., for 1 minute.

INSPECTION AND TESTING   269

CONTINUITY TEST ON FINAL RING MAIN CIRCUIT METHOD 1
OPEN CIRCUIT TEST

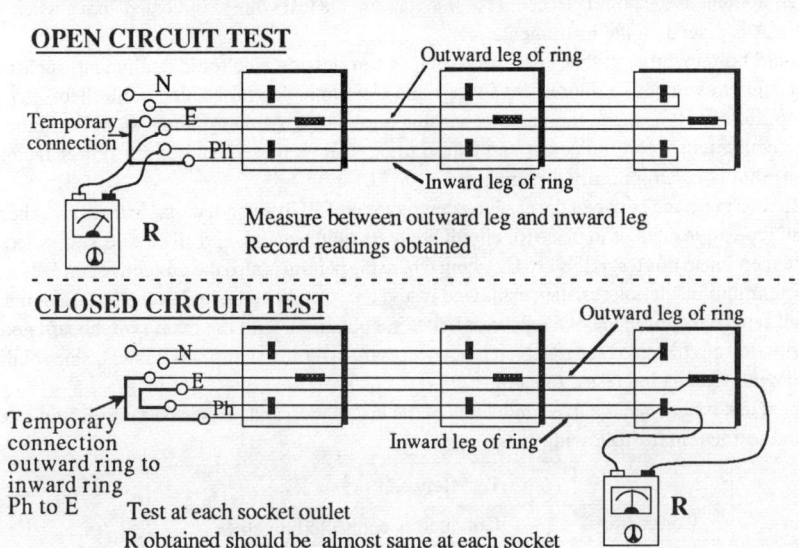

Temporary connection

Outward leg of ring

Inward leg of ring

Measure between outward leg and inward leg
Record readings obtained

CLOSED CIRCUIT TEST

Temporary connection
outward ring to
inward ring
Ph to E

Outward leg of ring

Inward leg of ring

Test at each socket outlet
R obtained should be almost same at each socket

Figure 3 - Testing ring circuit at each socket outlet

CONTINUITY TEST ON FINAL RING MAIN CIRCUIT METHOD 2
OPEN CIRCUIT TEST

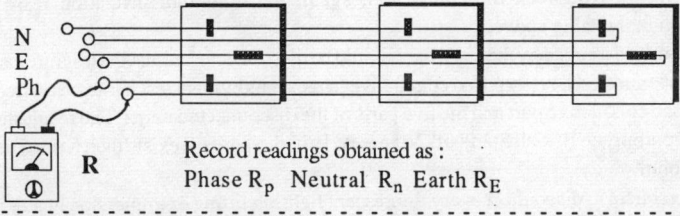

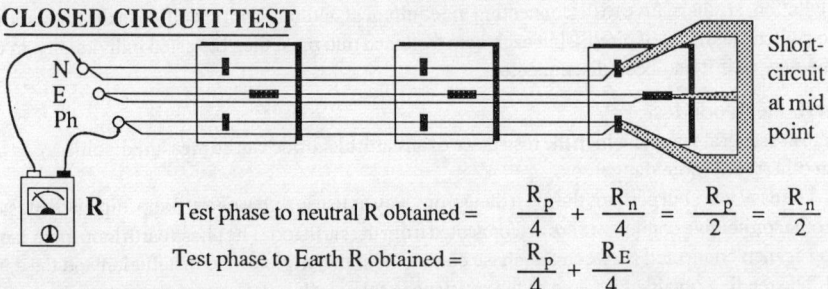

Record readings obtained as :
Phase R_p Neutral R_n Earth R_E

CLOSED CIRCUIT TEST

Short-circuit at mid point

Test phase to neutral R obtained = $\dfrac{R_p}{4} + \dfrac{R_n}{4} = \dfrac{R_p}{2} = \dfrac{R_n}{2}$

Test phase to Earth R obtained = $\dfrac{R_p}{4} + \dfrac{R_E}{4}$

Figure 4 - Testing ring circuit from distribution board

The test equipment used should be capable of maintaining the test voltage indicated above when a load of 1 mA is placed on the instrument.

It should be remembered that these test voltages can destroy electronic equipment, such as electronic starter switches in fluorescent fittings and electronic central heating controllers, electronic amplifiers in RCDs, and temperature sensing equipment. All these items as well as other electronic equipment, neon indicators and capacitors (such as those installed for power factor correction) should be temporarily disconnected.

The first test is made between the earthing conductor and all live conductors, with all switches closed, all protective devices in place (or circuit breakers switched on), and all circuits connected. The only exception to this test is the TN-C system where the neutral is also the protective conductor.

The minimum insulation resistance allowed is given in the above table, but the installation is considered satisfactory if the above values or better are obtained, with the main switchboard and each distribution circuit tested separately. It is not necessary to remove lamps for this test, since both phase and neutral are being tested to earth.

The next test is between the live conductors of the installation and, this is best carried out to a pattern as illustrated in the following chart.

Conductor	Test Between
	Conductors connected together
Red phase	Yellow & Blue phase and Neutral
Yellow phase	Red & Blue phase & Neutral
Blue phase	Red & Yellow phase & Neutral
Neutral	Red, Yellow & Blue phases

As this test is carried out between live conductors, it is necessary to disconnect equipment and remove lamps connected between the conductors; again, the minimum insulation resistance required is as given in the table above.

Where items can be damaged by the tests, they should be disconnected whilst the tests are carried out. If the items disconnected have exposed conductive parts, an individual test has to be carried out between the exposed conductive part and the live parts of the disconnected items. The result should comply with the appropriate British Standard. Where no British Standard exists then it must not be less than 0.5 megohm.

Since it is impractical to disconnect every fluorescent lighting fitting or immersion heater in an installation, an alternative to disconnecting the equipment, although not specified in the regulations, is to switch them out of circuit, but each item switched out, must then be tested individually in the same way as if it had been disconnected.

Earth electrode test

There are several ways in which the resistance of an earth electrode can be measured; some are easier than others, but more dangerous.

The easy way, but perhaps the most dangerous way, is to use a phase earth loop impedance tester. All other protective conductors are disconnected from the earth rod. The phase earth loop impedance tester is then connected between the phase conductor at the origin of the installation and the earth rod. The reading obtained is taken as the resistance of the earth rod. Extreme care is required when using this method of testing, since the test is being carried out at mains voltage. The connection from

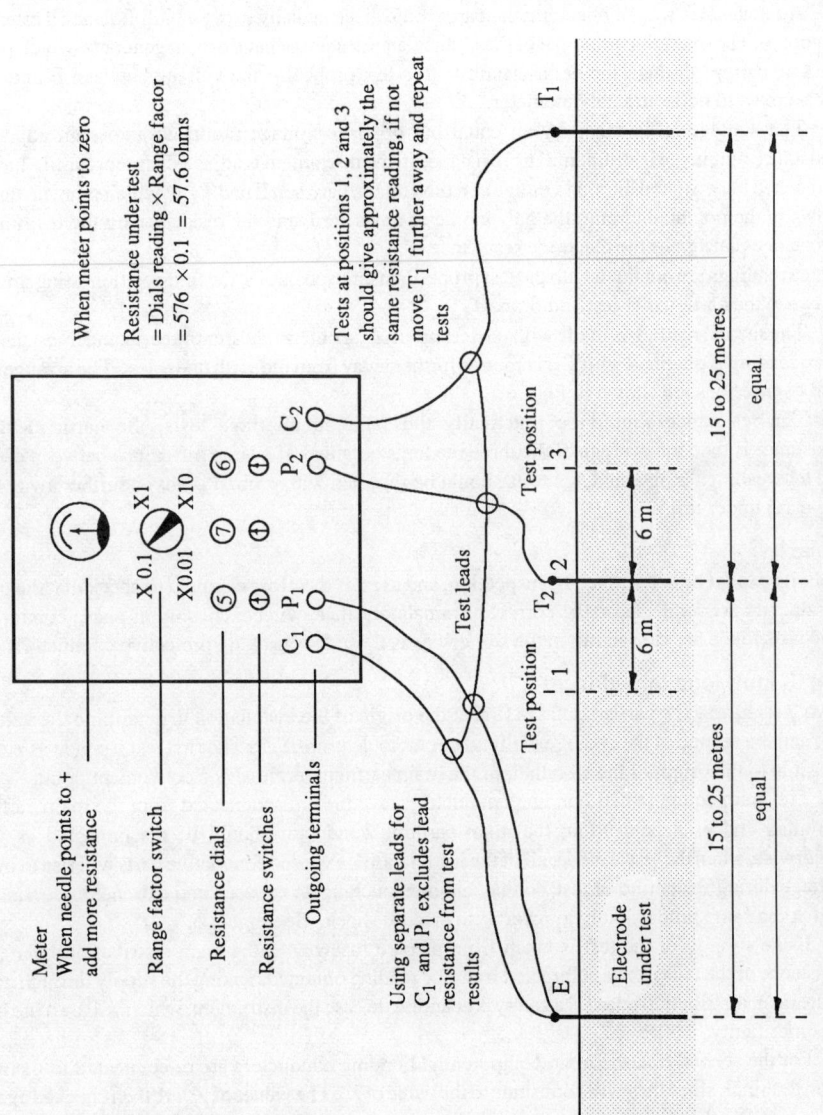

Figure 5 - Null Balance Tester - used for measuring the resistance of an earth rod.

the tester should be made to the earth rod first, and everyone kept away from it. Connection is made to the mains last, and the instrument operated immediately, then disconnected from the mains.

A much safer way of checking the earth rod resistance is by using a Null Balance Tester (see Figure 5). The operating principal of the Null balance tester is a hand driven generator which passes an a.c., current through the earth electrode to a test probe T_1; the voltage between E and T_2 is measured and deflects a galvanometer.

This voltage is balanced by an equal but opposite voltage produced across an adjustable resistance within the instrument. The resistance in the instrument is adjusted in steps, until a balance is reached between the internal voltage and the voltage between E and T_2, at which point no current flows in the potential circuit, the galvanometer reads zero, and the resistance on the instruments resistance dials gives the electrode resistance.

Three readings are obtained with the test probe T_2 in three positions, the first position being midway between the earth under test and probe T_1

The second reading is made with the test probe T_2 six metres nearer the earth under test, and the third reading is obtained with T_2 six metres further away from the earth under test. The arrangement and connections are shown in Figure 5.

The resistance should be practically the same in all three tests; the earth electrode resistance is then the average of the three readings obtained. If a large difference between each of the test readings is obtained, the tests should be repeated with probe T_1 moved further away from the earth under test.

Polarity

Polarity cannot be determined by inspection, and a test is therefore essential to check that the phase conductors are connected to the correct terminals, and that switches are only in phase conductors. The tests made are carried out in the same way as those for checking protective conductors.

Earth fault loop impedance

Two tests have to be carried out: the first at the origin of the installation to determine the value of Z_E, and the second at the end of each final circuit to determine Z_S. The first test is made as near as possible to the origin of the installation: the main earthing terminal is a convenient point.

The main equipotential bonding conductors have to be disconnected from the main earthing terminal, and as a precaution, the main earthing conductor should be disconnected as well. Otherwise, when the test is made, all of the installation's exposed conductive parts will rise to mains voltage during the period of test, so that anyone touching an exposed and extraneous conductive part at the same time is likely to receive an electric shock. See Figure 6.

If the supply for the test is taken from a spare fuseway in the main distribution board, the resistance of the tails can be subtracted from the reading obtained. Taking the supply through a fuse in the main distribution board is a safety precaution, in case the instrument, which will be hand held, becomes faulty.

For the second test, the main equipotential bonding conductors are re-connected to the main earth terminal, since they will contribute to the value of Z_S. The values of Z_S are then checked against the tables for shock protection, to verify that they comply with the limits laid down for shock protection. It must be remembered that the design calculations for shock protection will have been based on a conductor temperature which allows for the temperature rise caused by the phase earth fault current I_f.

INSPECTION AND TESTING 273

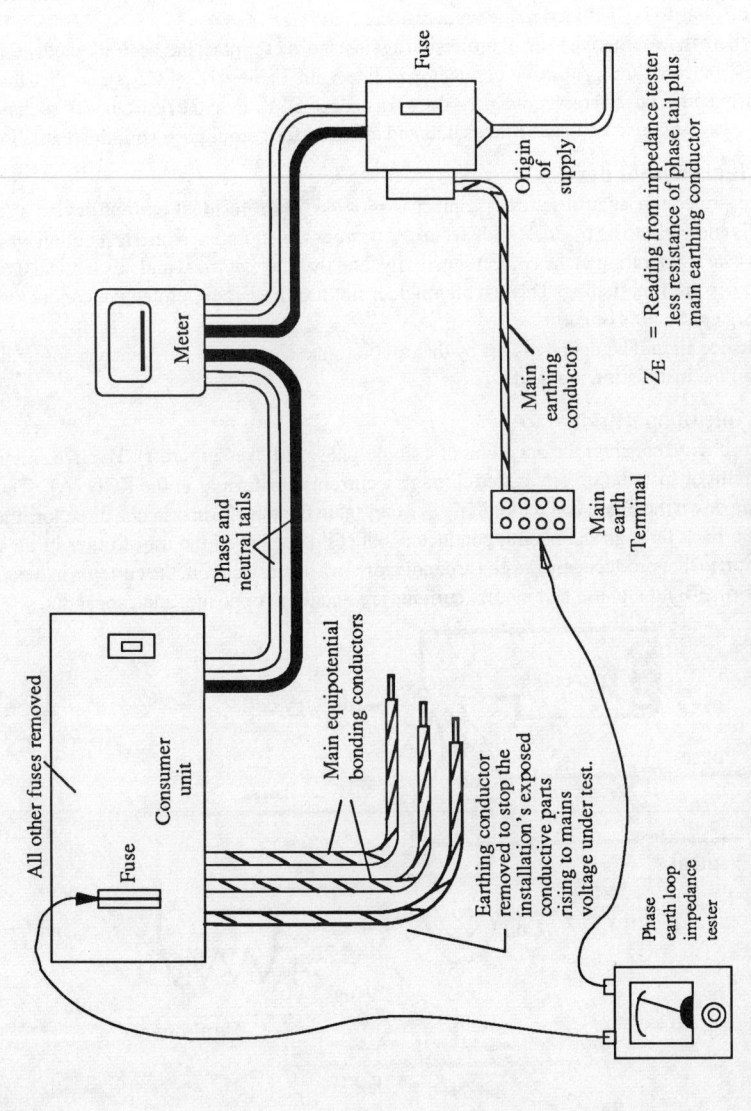

Figure 6 - Determination of Z_E at the origin of the installation

Z_E = Reading from impedance tester less resistance of phase tail plus main earthing conductor

This temperature will be higher than those prevailing at the time of test. The test results should therefore be reduced by an amount determined by the temperature at the time of test, and the temperature allowed in the design calculations.

Tables ZS 1 to 7 in Part 3 of the handbook give the Z_s values at various temperatures when copper conductors are used as cpcs, such as twin & cpc cable. On TT installations, the resistance of the earth electrode has to be included in the values of Z_s

Where the alternative form of protection against indirect contact has been adopted, i.e., limiting the resistance of the protective conductor in line with Table 41C (PCZ 8), so that the voltage appearing on exposed conductive parts does not exceed 50 V, then the resistance of such protective conductors will also have to be measured and checked for compliance with the regulations.

Residual current devices

Where protection against indirect contact is provided by a residual current device a test of its effectiveness has to be made. Residual current devices are equipped with a test button which tests the device is operating at the correct sensitivity, and that the electrical and mechanical elements of the device are functioning. The test button does not test the circuit, protective conductors, or any earthing conductors or earth .

The device should be operated first by the test button to check that the device is not faulty before tests on the installation are made.

Principle of operation

For simplicity, consider the operation of a single-phase unit (see Figure 7). The live conductors of the circuit or installation are passed through a current transformer in the RCD (A). The current flowing down the phase conductor induces a current in the transformer in one direction, the current flowing back through the neutral conductor induces a current in the transformer in the opposite direction to that produced by the phase conductor. In a healthy circuit, the currents induced into the transformer by the phase and neutral currents are equal but opposite, and cancel out.

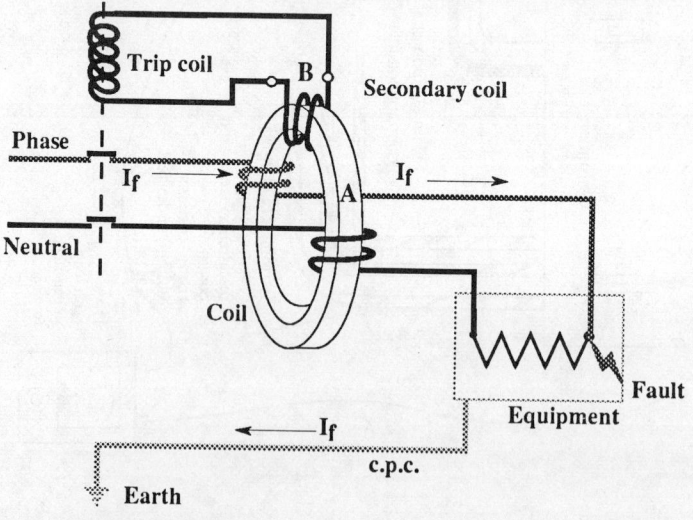

Figure 7 - RCD operation

When an earth leakage occurs, some of the current flowing through the phase conductor flows back to the source of energy through the protective conductor. The current flowing back through the neutral is therefore less than that flowing through the phase conductor, causing an imbalance in the current transformer (B), which generates a voltage across the trip coil of the RCD. When the earth leakage current is between 50% and 100% of the rated residual operating current of the RCD it will trip, disconnecting the circuit or the installation. The RCD does not provide either overload or short-circuit current protection, and does not limit the magnitude of the fault current.

RCD testing

The loads normally supplied by the RCD are disconnected during the test. The test is then made on the load side of the RCD, between the phase conductor and the circuit protective conductor, so that a suitable test current flows.

The first test is made with tripping current set at 50% of the rated tripping current of the RCD for a period of two seconds; during this period the RCD should not trip.

With a test current equal to the residual operating current of the RCD, a general purpose RCD should open within 0.3 secs; the test should not be applied for more than 0.5 secs. For an RCD manufactured to BS 4293, the RCD should open in less than 0.2 sec.

If an RCD has been installed to reduce the risk of shock from direct contact, a residual test current of 150 mA should cause the RCDs breaker to open in less than 40 msecs. The test current must not be applied for a period exceeding 40 msecs.

Having completed the test, the effectiveness of the test button on the RCD should again be checked to ensure that everything is satisfactory.

Where an S type RCD with a time delay is used, it should not open in less than 0.13 secs, but should open within 0.5 secs. The maximum time permitted for the test current is 0.5 secs.

Where an RCD manufactured to BS 4293 incorporating a time delay is used, it should trip within 50% and 100% of the rated time delay plus 0.2 secs.

Verification should also be made that the RCD has the correct withstand capacity for possible short-circuit or earth fault currents, and that discrimination between RCDs or between RCDs and other devices will be achieved where this is required to prevent danger.

Completion Certificates

When an installation is complete, and has been tested, a completion certificate has to be filled in, together with an inspection certificate, examples of which will be found in Appendix 6 of the regulations.

The completion certificate requires the designer of the installation, the installer and the person who carries out the inspection and testing, to certify that the installation complies with the IEE Wiring Regulations. This completion certificate is required even for alterations or additions. Additionally an inspection notice of the form outlined in Regulation 514-12-01 has also to be fixed at the origin of the installation.

IMPORTANT

This installation should periodically be inspected and tested, and a report on its condition obtained, as prescribed in the Regulations for Electrical Installations issued by the Institution of Electrical Engineers.

Date of last inspection ....

Recommended date of next inspection ....

Periodic inspection and testing

When an inspection is carried out, only those parts requiring more careful examination need to be partially dismantled. The first requirement is to check the installation to see whether any alterations or additions have been carried out. If they have, then the next step is to ask to see the completion and inspection certificates. If these are not available, then the inspection of that part of the installation is as though the installation had only just been completed.

As far as the rest of the installation is concerned, careful inspection and testing is required to see if any deterioration due to environmental conditions, or to mechanical damage has occurred. The installation should then be checked to ensure connections are still tight, barriers are still in place, and protective devices are of the correct size etc. The inspection and testing must be carried out so that no danger to persons, livestock or property will occur.

A report on the condition of the installation shall be given by the person carrying out the inspection or his authorised representative, to the person ordering the inspection and test.

This report shall also include any dangerous conditions arising from non-compliance with the regulations, and should detail the limits within which the inspection and test were carried out.

Needless to say, the inspection and testing must be carried out by a competent person.

Periods of inspection

The frequency of the periodic inspection and testing has to be determined by the type of installation, the environmental conditions and the quality and frequency of maintenance that the installation is going to receive.

For installations having a normal environment, where maintenance is regularly carried out, the following periods between inspections are offered for guidance.

General installations	5 years
Temporary installations on construction sites	3 months
Caravan sites	1 year
Agricultural premises	3 years

Inspection certificates

The results of tests should be carefully recorded, so that if a problem occurs in the future, a properly completed test certificate will help to prove that the installation was in good condition when handed over to the customer. Where tests are carried out by maintenance staff in a factory or large commercial premises, recording the results enables a comparison with previous tests to be made; this can be helpful in preventing a fault developing.

The larger installation requires test schedules for each type of test; this is best achieved by producing a test schedule which also details the circuits installed from a distribution board. If these test schedules are numbered, it is only necessary to record the number on the Inspection Certificate, and attach the test schedules to it.

Samples of an inspection certificate and distribution board test schedule are illustrated on the following pages.

Alterations and additions

A test and inspection have to be carried out when either an alteration or addition to an installation has been made. Any defects found in the electrical installation related to the alteration or addition, have to be notified to the person ordering the work. It must also be verified that the alteration or addition will not impair the safety of the existing installation.

INSPECTION CERTIFICATE   Number: _____

Address of installation tested _____

Type of installation; New / alteration / addition	Supply Voltage		Frequency		
System of which installation is part Answer Yes or No to System type	TN-C	TN-S	TN-C-S	T T	I T
Origin of the installation (Tick appropriate box)	Transformer L.V.box		Supliers fuse (cut out)		
Three phase prospective short-circuit current at origin					
Single phase prospective short-circuit current at origin					
Phase earth loop impedance at origin					
Maximum demand per phase	Red phase		Yellow phase	Blue phase	
Type of protective devices used throughout the installation & Size of main incoming switch or CB	BS 3036	BS 1361	BS 88	MCB'S	MCCB'S	RCD'S

Test Results recorded on the attached test sheet numbers	TEST SHEET NUMBERS				
Resistance of supplementary bonding conductors					
Continuity of protective conductors					
Continuity of ring final circuit conductors					
Insulation resistance of the fixed installation					
Insulation resistance of equipment tested separately					
Earth electrode resistance in Ω and location of earth electrode					
Resistance of equipotential bonding conductors and size					
Polarity					
Earth fault loop impedance					
Operation of residual current devices and operating current					
Checks of I_b, I_n, I_z with respect to G.A.T.S & I_n					
Condition of switches, plugs, sockets, flexible cable and cords					
Segregation of Category 1, 2 & 3 circuits checked					

Special measures taken for protection against direct and indirect contact and recorded on separate test sheets	TEST SHEET NUMBERS				
Protection against direct contact, by insulation applied on site					
Protection against direct contact, by barriers or enclosures					
Protection against indirect contact by non-conducting floors and walls					
Protection by separation of circuits					

Deviations from the IEE Wiring Regulations (120-02, 120-05)

Comments on the existing installation if applicable (743-01-01)

I CERTIFY that I have inspected and tested the installation at the above premises and the results above along with the test schedules represent a true indication of the state of the installation.

Name (Block capitals) _____ Signed _____ Date _____

For and on behalf of _____

Address _____

_____

This installation should be further inspected and tested on or before :- _____

The tests outlined on this sheet are based on Part 7 of the 16th Edition IEE Wiring Regulations   © TEM 1987

278 INSPECTION AND TESTING

DISTRIBUTION BOARD OR SWITCHGEAR REF:																							TEST CERTIFICATE NUMBER :								
Insulation resistance at busbars			Maximum demand	R	Y	B																	D.B. chart fitted		Date of test			Signatures		Weather conditions	Voltage
R-Y-B & N	R to Y-B-N	Y to R-B-N	B to R-Y-N	Design value																			Z_S at busbars		Person making test						Calculated Volt drop
				Actual value																			I_p at busbars		Witness						
Instrument serial numbers																																
Circuit		Protection		Cable						Equipment				Continuity of ring circuit						Continuity				Phase earth loop impedance Z_S							
														Open circuit test			Closed circuit test			Protective conductors				Design Z_s	Test		Z_s allowed @ C	Position of test			
Nº	I_b	Checked	Type	I_n	Checked	Method of installation	Applicable derating factors	I_{ab}	Checked	Phase	Insulation resistance Phase to Earth		Polarity checked	Single pole switches in phase conductors	RCD's		P Ω	N Ω	CPC Ω	P - N Ω	P - CPC Ω						Z_s @ °C				
1	R																															
	Y																															
	B																															
2	R																															
	Y																															
	B																															
3	R																															
	Y																															
	B																															
4	R																															
	Y																															
	B																															
5	R																															
	Y																															
	B																															
8	R																															
	Y																															
	B																															
9	R																															
	Y																															
	B																															
10	R																															
	Y																															
	B																															
11	R																															
	Y																															
	B																															
12	R																															
	Y																															
	B																															

© TEM 1987

PART THREE

INSTALLATION TABLES

SPACE FOR NOTES AND AMENDMENTS

Contents

1. Introduction 284
2. How to use the tables, with examples. 289

Installation tables giving maximum lengths of circuits

INST 1	Maximum lengths for a disconnection time of 0.4 seconds individual circuits	297
INST 2	Maximum length for a disconnection time of 5 seconds individual circuits	298
INST 3	Voltage drop - Determination of voltage drop at a lighting point	299
INST 4	Maximum lengths for 13 A socket outlet circuits	300

Current ratings for single core cables

CR 1	PVC insulated cables enclosed in conduit, trunking, or open and clipped direct	301
CR 1a	PVC insulated cables enclosed in conduit, trunking, or open and clipped direct	302
CR 2	PVC insulated cable enclosed in conduit in contact with thermal insulation	303
CR 2	PVC insulated and sheathed cables installed on cable tray	303
CR 3	XLPE insulated cables enclosed in conduit, trunking, or open and clipped direct	304
CR 3a	XLPE insulated cables enclosed in conduit, trunking, or open and clipped direct	305
CR 4	XLPE insulated cables enclosed in conduit in contact with thermal insulation	306
CR 4	XLPE insulated and sheathed cables installed on cable tray	306
CR 5	85 °C rubber insulated cable installed in conduit, trunking, or open and clipped direct	307
CR 6	Aluminium cables open and clipped direct, or installed on cable tray	308
CR 13	Micc insulated and pvc sheathed cables, light, and heavy duty	319
CR 14	Micc insulated bare copper cables exposed to touch, light, and heavy duty	320
CR 15	Micc insulated bare copper cables exposed to touch installed on cable tray	321
CR 16	Micc insulated bare cables not exposed to touch, open and clipped direct	322

Current ratings for multicore cables.

CR 7	Installed in conduit or trunking, or open and clipped direct	309
CR 7a	Installed in conduit or trunking, or open and clipped direct	310
CR 8	Installed in contact with thermal insulation, or on cable tray	311
CR 8a	Installed in contact with thermal insulation, or on cable tray	312
CR 17	Flexible cords to BS 6500	323

Current ratings for single core and multicore micc. cables.

CR 13	PVC sheathed, open and clipped direct, or lying on a non-metallic surface	319
CR 14	Bare copper cables, exposed to touch, open and clipped direct	320
CR 15	PVC sheathed cables installed on cable tray	321
CR 16	Bare copper cables, not exposed to touch, open and clipped direct	322

Current ratings for multicore armoured cables copper conductors
CR 9	PVC insulated cables, open and clipped direct, or on cable tray	313
CR 9a	PVC insulated cables, open and clipped direct, or on cable tray	314
CR 10	XLPE insulated & pvc sheathed cables, open and clipped direct, or on tray	315
CR 10a	XLPE insulated & pvc sheathed cables, open and clipped direct, or on tray	316

Current ratings for multicore armoured cables aluminium conductors.
CR 11	PVC insulated and sheathed cables, open and clipped direct, or on tray	317
CR 12	XLPE insulated & pvc sheathed cables, open and clipped direct, or on tray	318

Maximum values of Z_S for protection against indirect contact.
ZS 1	Z_S for design and testing for BS 88 HRC fuses, disconnection time 0.4 seconds	324
ZS 1a	Z_S for design and testing for BS 88 HRC fuses, disconnection time 5 seconds	325
ZS 2	Z_S for design and testing for BS 1361 fuses, disconnection time 0.4 & 5s	326
ZS 3	Z_S for design and testing for BS 3036 (rewirable) fuses, disconnection time 0.4 & 5s	327
ZS 4	Z_S for design and testing for Type 1 mcbs, disconnection time 0.1 to 5s	328
ZS 5	Z_S for design and testing for Type 2 mcbs, disconnection time 0.1 to 5s	329
ZS 6	Z_S for design and testing for Type B mcbs, disconnection time 0.1 to 5s	330
ZS 7	Z_S for design and testing for Type 3 & type C mcbs, disconnection time 0.1 to 5s	331
ZS 8	Z_S for design of reduced voltage supplies with different protective devices	332
ZS 9	Z_S for design of special sites with different protective devices, disconnection time 0.2s	333

Values of Z_S for thermal protection of protective conductors
PCZ 1	Maximum value of Z_S for BS 88 HRC fuses	334
PCZ 2	Maximum value of Z_S for BS 1361 cartridge fuses	335
PCZ 3	Maximum value of Z_S for BS 3036 (rewirable) fuses	336
PCZ 4	Maximum value of Z_S for Type 1 miniature circuit breakers	337
PCZ 5	Maximum value of Z_S for Type 2 miniature circuit breakers	338
PCZ 6	Maximum value of Z_S for Type B miniature circuit breakers	339
PCZ 7	Maximum value of Z_S for Type 3 & type C miniature circuit breakers	340
PCZ 8	Maximum value of cpc impedance for protection of portable equipment disconnection 5s	341

Maximum design impedance to give short-circuit protection.
SCZ 1	Maximum value of Z_{pn} for Type 1 miniature circuit breakers	342
SCZ 2	Maximum value of Z_{pn} for Type 2 miniature circuit breakers	343
SCZ 3	Maximum value of Z_{pn} for Type B miniature circuit breakers	344
SCZ 4	Maximum value of Z_{pn} for Type 3 & type C miniature circuit breakers	345
SCZ 5	Maximum value of Z_{pn} for BS 88 HRC fuses	346
SCZ 6	Maximum value of Z_{pn} for BS 1361 cartridge fuses	347
SCZ 7	Maximum value of Z_{pn} for BS 3036 (rewirable) fuses	348

Cable sizes for grouped circuits enclosed in conduit or trunking
CSG 1	Single core cables not in contact with thermal insulation	349
CSG 3	Single core cables in conduit in contact with thermal insulation	351
CSG 4	Multicore cable in conduit or trunking not in contact with thermal insulation	352

Cable sizes for grouped circuits bunched and clipped direct
CSG 2	Single core cables not in contact with thermal insulation	350
CSG 5	Multicore, or twin & cpc cable not in contact with thermal insulation	353
CSG 6	Multicore, or twin & cpc cable in contact with thermal insulation	354
CSG 9	PVC insulated & pvc sheathed steel wire armoured cables	357
CSG 12	XLPE insulated & pvc sheathed steel wire armoured cables	360

Cable sizes for grouped circuits clipped direct in single layer
CSG 7	PVC multicore, or twin & cpc cable not in contact with thermal insulation	355
CSG 10	PVC insulated & pvc sheathed steel wire armoured cables	358
CSG 13	XLPE insulated & pvc sheathed steel wire armoured cables	361
CSG 15	Micc pvc sheathed cables	363
CSG 17	Aluminium pvc insulated and armoured multicore cables	365
CSG 19	Aluminium XLPE insulated and armoured multicore cables	367

Cable sizes for grouped multicore cables installed on metal cable tray, sheaths touching
CSG 8	PVC multicore or twin & cpc cable not in contact with thermal insulation	356
CSG 11	PVC insulated steel wire armoured cable with sheaths touching	359
CSG 14	XLPE insulated & pvc sheathed steel wire armoured cable with sheaths touching	362
CSG 16	Micc pvc sheathed cable with sheaths touching	364
CSG 18	Aluminium pvc insulated and armoured multicore cables with sheaths touching	366
CSG 20	Aluminium XLPE insulated and armoured multicore cables with sheaths touching	368

Cable sizes for pvc insulated cables when grouped and protected by a BS 3036 rewirable fuse used for overload protection
CSG 21	Multicore or twin & cpc cables bunched and clipped direct	369
CSG 22	Multicore or twin & cpc cables clipped direct in a single layer, sheaths touching	370
CSG 23	Multicore or twin & cpc cables single layer on cable tray, sheaths touching	371
CSG 24	Multicore or twin & cpc cables enclosed in conduit or trunking	372

Wiring capacity of conduit and trunking
CCC 1	Cable capacity of conduit up to 50 mm^2 cable, and 38 mm conduit	373
CCT 1	Cable capacity of trunking up to 150 mm^2 cable, and 150 x 150 mm trunking	374

Note 1: If the method of installation, or the protective device rating required is not included in the CSG tables, refer to the CRG tables in Part 4, which give the current rating required for all methods of installation for different types of protective devices. Then look up the cable size required from the CR Tables.

Note 2: When a BS 3036 (semi-enclosed/rewirable) fuse only is used for short-circuit protection, i.e., when the load is not likely to cause an overload, as is the case with immersion heaters, electric radiators etc., tables CSG 1 to CSG 20 can be used for determining the size of conductor. When the BS 3036 fuse is used for overload protection, then tables CSG 21 to 24 should be used.

1. Introduction to Parts 3 & 4

Part 3 of the handbook is produced to assist the contractor, electrician, and engineer to size cables for installations, without the need for complicated calculations. It can be used by technicians and engineers to make a quick check that cable sizes are satisfactory on larger projects, especially where computer programmes are used for sizing cables. To enable a check with the Wiring Regulations to be made, the reference number of the applicable table number in the Regulations is placed in brackets in the bottom left hand corner of the page. Thus readers can carry out spot checks and gain confidence in using the Tables.

The choice of table references used in Part 3 and Part 4 gives some idea of the table contents. For example, CR stands for current rating or cable rating and CSG stands for cable size grouped. With familiarity of use, this should enable the table required to be found quickly, obviating reference to the contents for the page number.

Part 4 includes tables that give information necessary when designing an installation, and will therefore be useful for the design engineer.

Installation tables

Four tables have been provided at the beginning of Part 3 to aid the small contractor, particularly with houses or small commercial premises. The first two tables give the maximum length of cable of sizes 1mm^2 to 16 mm^2 that can be installed, from each type of protective device for different sized loads. The third table enables the voltage drop at a lighting point to be determined easily. The fourth table gives the maximum length of socket outlet circuits for pvc cable, pvc cable installed in conduit, and micc cable.

Sizing conductors

The order of checks that should be made to determine the size of conductor are listed as follows
      Overload rating: I_z greater than or equal to I_n.
      Derating factors,
      a) Grouping,
      b) Ambient,
      c) Thermal insulation,
      d) Type of protective device,
      Voltage drop,
      Shock protection,
      Short-circuit protection,
      Thermal protection of the protective conductor.

Derating factors

The most prevalent derating factor will be the one for grouping. To make the sizing of cables easier and quicker, for those instances where the only derating factor is for grouping, tables in Part 3 are provided. These show the size of cable required to be used for both single-phase and three-phase circuits. These are covered by Tables CSG 1 to 24.

The Tables CSG 1 to 20 can be used for BS 3036 rewirable fuses, when the fuse is not being used for overload protection, such as when the fuse is being used for a heating load.

INTRODUCTION

Tables CSG 21 to 24 are provided for those occasions when BS 3036 rewirable fuses are used for overload protection, and take into account the derating for the rewirable fuse. Tables CSG 21 to 24 are not required when the fuse is being used for short-circuit protection, such as where the load is an immersion heater, or fixed heating load.

Where circuits are subject to an ambient temperature higher than 30 °C, and are grouped, the CRG Tables in Part 4 can be used. The current obtained from the appropriate CRG table is divided by the derating factor for the ambient temperature obtained from the bottom of the appropriate CR table. This revised current value can then be used to size the cable, using cable ratings from the CR Tables.

Tables CRG 1 to 11 give the current-carrying capacity required for single core and multicore cables, for each type of protective device with each type of cable grouping. The Tables for BS 3036 rewirable fuses take into account the derating factor for the fuse.

As mentioned earlier, the tables for BS 3036 fuses are only required where the fuse is being used for overload protection. Part 4 contains Table GF1, which gives the derating factors for grouping, to enable those who would like to check Tables CRG1 -11 to do so.

Voltage drop

To make voltage drop calculations simpler, Tables CR1 to 17 are provided in Part 3. These tables give the current-carrying capacity of conductors and a factor. The factor is based on length multiplied by amps for 4% voltage drop. This factor enables a quick assessment of whether a chosen cable size is suitable for 4% voltage drop (it is easier to work with whole numbers than with 10^{-3}).

All that is necessary is to multiply the length of the circuit by the circuit's full load current (design current), then choose a factor from the appropriate table, equal to, or larger than, the length × amps value calculated. This gives the minimum size of conductor required to satisfy voltage drop. A check must then be made to ensure that the conductor selected has sufficient current-carrying capacity for the load to be placed on it.

The factor (length × amps) in the tables can be used in several ways: this is more fully explained in the chapter 'How to use the Tables'.

Where the cable length required for a circuit is limited by the voltage drop to a few metres less than the length required for the circuit, and where the circuit is feeding a load with a power factor, then the r and x values from the CR tables can be used to calculate the voltage drop in the circuit, by using the following formula:

$$\text{Voltage drop} = \frac{(\cos\emptyset\, r + \sin\emptyset\, x) \times \text{cable length} \times \text{full load current}}{1000}$$

where $\cos\emptyset$ is the power factor, and $\sin\emptyset$ is the sine of the power factor angle. For convenience, a table is provided in Part 4 giving the sine of the power factor and the corresponding angle in degrees and radians. See Chapter on 'How to use the Tables'.

Shock protection

Having determined the length of circuit, it is now necessary to check protection against indirect contact (shock protection). The first requirement is to find the value of Z_S up to the distribution board from which the circuit is taken. Where the circuit is connected to a distribution board at the incoming supply point, then the value of Z_E at the origin of the installation is required.

INTRODUCTION

The next step is to work out the value of the impedance of the phase conductor and circuit protective conductor for the circuit. The impedance used should be that which the conductor will attain due to the fault current increasing the temperature of the conductor, before the protective device disconnects the circuit.

Since it is difficult to determine the final temperature of the conductor with the fault current flowing, an average is taken between the working temperature of the conductor and the limit temperature of the conductor insulation.

Resistance per metre is given at various temperatures in Table RC1 for copper conductors, Table RA1 for aluminium conductors with pvc and xlpe insulation, and RXLPE 1 for copper conductors with thermosetting insulation.

The most common temperature for pvc insulation is 115 °C: i.e., 160 °C + 70 °C divided by 2, where 160 °C is the limit temperature for pvc, and 70 °C is the working temperature of the conductor.

The impedances taken from the Tables for the phase and cpc are then multiplied by the circuit length, divided by 1000, and the total is then added to Z_s up to the start of the circuit.

The total Z_S obtained is then checked against the design value for the type of protective device being used, and the disconnection time required either of 0.4 or 5 seconds, in Tables ZS 1 to ZS 7.

Short-circuit protection

Where it cannot be assumed that the protective device is protecting the live conductors against short-circuit current, such as motor circuits, or with certain types of circuit breaker, the thermal suitability of the live conductors has to be checked.

The impedance of the phase and neutral supply conductors up to the origin of the installation is required for single-phase, and three-phase and neutral circuits. With a three-phase three-wire supply, the impedance of one phase conductor only is needed. If the supply is PME (TN-C-S system) then Z_E is the phase and neutral impedance up to the origin of the installation.

Single-phase & TP&N circuits

Where resistance (impedance) is referred to throughout these notes, it is the average between the limit temperature for the conductor insulation and the working temperature of the conductor.

Step 1 Obtain the resistance (impedance) of phase and neutral conductor up to the origin of the installation.

Step 2 Determine the resistance (impedance) of the phase and neutral conductor from the origin to the circuit distribution board.

Step 3 Determine the resistance (impedance) of the phase and neutral conductor of the circuit: i.e., the value for the phase conductor worked out for shock protection multiplied by two.

Step 4 Add resistances (impedances) obtained together: i.e., Step 1 + Step 2 + Step 3; this equals Z_{pn}.

Step 5 Now check the value of Z_{pn} worked out against the value obtained from Tables SCZ 1 to 7 for the type and size of protective device being used. If the calculated value does not exceed the value given in the table, the conductors are protected against short circuit current.

INTRODUCTION 287

Three-phase 3 wire circuits
Step 1 Obtain the resistance (impedance) of one phase conductor up to the origin of the installation.
Step 2 Determine the resistance (impedance) of one phase conductor from the origin to the circuit distribution board.
Step 3 Determine the resistance (impedance) of one phase conductor in the circuit (note: already worked out for shock protection).
Step 4 Add the resistances (impedances) obtained together: i.e., Step 1 + Step 2 + Step 3, this gives Zp. Multiply this figure by 2, and divide by $\sqrt{3}$.
Step 5 Check the value of $2Zp \div \sqrt{3}$ worked out against the value from Tables SCZ 1 to 7 for the size and type of circuit protective device used. If the calculated value of $2Zp \div \sqrt{3}$ does not exceed the table value, the phase conductors are protected against short circuit current.

Circuit protective conductor
The suitability of the protective conductor can be checked, either by using the table in Chapter 54 of the Regulations, or the modified Table M54G given in part 4, or by calculation.

The formula $S = \dfrac{\sqrt{I^2 t}}{k}$ is used for checking the suitability of the protective conductor to withstand the earth fault current flowing through it. Where the earth fault current is less than the short-circuit current, it will also be necessary to re-check that the phase conductor is still protected against damage by the fault current. To obviate the need to carry out calculations for the protective conductor, Tables PCZ 1 to 7 can be used.

Since the phase earth loop impedance Z_S has already been worked out for shock protection, it is only necessary to check that the Z_S value for the circuit does not exceed the Z_S value given in the tables, for the type and size of circuit protective device being used.

Since most of the circuit protective conductors will either be contained in the cable, or in conduit, or trunking, the values given in the PCZ tables are based on the conductor temperature at the start of the fault being 70 °C, and the final conductor temperature being 160 °C: i.e., a 'k' factor of 115 has been used to obtain the values given. The Tables ZS 1 to 7 also give the phase earth loop impedance values at different temperatures, for testing purposes.

Additional Tables
For design engineers who want to do the calculations, additional tables, provided in Part 4, contain information that will be required when designing an installation. Part 4 contains tables such as the earth loop impedance of the various types of armoured cable, including the reactance of the cable armouring; the earth loop impedance of micc cable, together with the resistance, reactance, and impedance of cables up to 300 mm^2. A table giving the 'k' factor for different types of conductor and conductor insulation along with the area of conduit and trunking is provided.

Since steel conduit and trunking will be used as protective conductors, a table giving the impedance of conduit and trunking has been included in Table ZCT 1. To obtain the fault current required for the conduit table, divide the phase to earth voltage by the Z_S worked out for the circuit.

Two values of impedance for conduit, depending on the earth fault current, are given in Table ZCT 1. It is easier to start by using the table giving the maximum impedance of the conduit, and if the circuit satisfies Z_s with this value, it will also satisfy Z_s when the fault current is more than 100A. In the extreme case, it may be necessary to work out two values of Z_s, using both conduit impedances from ZCT 1.

Tables for the current-carrying capacity of flexible cords, and the cable capacity of trunking and conduit are also included, together with a table giving the full load current for various sizes of motor.

Basis of Tables

Since most installations will be operating at 240V between phase and neutral, and 240V phase to earth, the tables have been based on this voltage, so they will require modifying if a different voltage from 240V is being used.

Modification of the values of Z_p, Z_{pn}, or Z_s from the ZS, PCZ or SCZ tables for other phase to earth, or phase to neutral voltage, can be carried out. Multiply the values from the tables by the actual phase to earth, or phase to neutral voltage, and then divide the result by 240 volts.

2. How to use the tables

Installation tables

These have been provided for the small contractor and electrician. The loads specified in the tables are those which can be expected in domestic premises, or in small commercial premises. The loads cover Infra-red heaters, showers, cookers etc. Two tables are provided: INST 1 gives maximum cable lengths for 0.4 second disconnection time, for areas such as bathrooms, outside equipment, or cooker units containing a socket outlet; INST 2 gives cable lengths for 5 seconds' disconnection time, and covers equipment inside the equipotential zone together with cooker units without a socket outlet.

Several sizes of cable have been given for each load. It should be noted that where the load is not likely to cause an overload, the conductor's current-carrying capacity does not need to be as large as the protective device rating. The conductor does, however, need to be capable of carrying the circuit's full load current. Although twin and cpc cable sizes have been given in tables INST 1 and 2, they can be used for conduit and micc cable, providing the live conductors are of the same size as those given in the table.

Where circuits are grouped, in contact with thermal insulation, or where the ambient temperature exceeds 30 °C, the circuit conductors need to be sized accordingly, but the lengths for each size of cable in the tables are still applicable.

Table INST 3 enables the voltage drop at a given lighting point to be worked out quickly, by adding the necessary values from each of the three tables contained in INST 3. Adjustments can be made for different sizes of cable. The load can also be adjusted, providing each lighting point has the same load.

Again, INST 3, can be used for single circuits in conduit, or micc cable, providing the 'loop-in system' of wiring is used: otherwise it is only approximate.

For socket outlet circuits, table INST 4 is provided. The figures in the table cover protection against indirect contact, thermal protection of the cpc, and the thermal protection of the live conductors. Voltage drop has been taken into consideration in the table. Voltage drop for ring circuits has been taken as the average of the minimum and maximum voltage drop, the maximum length being where the socket outlets are equally distributed round the ring, and each socket is equally loaded. The minimum length of the ring circuit is based on all the load being placed at the mid point. For both ring and radial circuits the load has been taken to be equal to the size of the protective device size. This arrangement has been chosen since voltage drop cannot be calculated for socket outlet circuits, because the load that can be connected by the user at any time is unknown. Because of this, the formula for circuits not subject to simultaneous overload cannot be used, neither can the circuit be considered lightly loaded. The length of the circuit is determined by the floor area allowed for the circuit. Where the circuits are subjected to grouping, ambient temperature higher than 30 °C, or thermal insulation, they must be derated. However, derating for grouping is only required when more than two ring circuits are grouped together, or are grouped with other circuits. See notes at foot of table.

Current rating tables

The tables give the maximum current rating for single circuits based on an ambient temperature of 30 °C, when the circuits are carrying the tabulated current (referred to as I_{tab}) given in the table.

The operating temperature of the conductors will be that shown at the top of each table.

Ambient temperature correction factors for the most common form of conductor insulation are given at the bottom of each table, both for BS 3036 fuses, and for other types of protective device.

As explained in the introduction, a factor comprising length × amps is given to enable a quick assessment of the cable size required for a voltage drop of 4%. In practice, with the exception of socket outlet circuits, the equipment the circuit is to supply determines the length of the circuit, and the full load current that the conductors have to carry. Therefore, to determine the cable size required, just multiply the length of the circuit by the full load current of the circuit, then use the factor column in the CR table, select a factor equal to, or larger than the value calculated. This gives the cable size required. The current rating for the cable size chosen should now be checked to ensure that the cable is capable of carrying the circuit's full load current. The columns headed 'r' and 'x' will be dealt with later.

Using the factor (length × amps)

To simplify the examples the following notation is used.
1. 'Factor' from the table equals length × amps
2. 'Allowed Vd' is 4% voltage drop, i.e. 9.6 V for single-phase, or 16.6 V for three-phase.
3. 'Available Vd' is the amount of voltage drop available for the circuit under consideration. It would be 9.6 V single-phase, or 16.6 V three-phase, if there were no voltage drop up to the origin of the circuit.
4. I_b is the design, or full load current taken by the circuit.

This means that only two numbers have to be remembered: 9.6 V for single-phase and 16.6 V for three-phase.

Determination of maximum circuit length when conductors are not carrying I_{tab}.

(1) Where the full voltage drop allowance either for single-phase or three-phase is available,

$$\frac{\text{Factor}}{I_b} = \text{actual length allowed.}$$

Circuit length when 'Available Vd' is less than 'Allowed Vd'.

(2) Where the length, the full load current I_b, and the Vd available are all known, and where the cable size is required to suit the available voltage drop, then

$$\text{Factor} = \frac{\text{Length} \times I_b \times \text{Allowed Vd}}{\text{Available Vd}}$$

Now look in a suitable table, and choose a Factor equal to, or larger than the Factor calculated.

Circuit length when f.l.c. is less than I_{tab} and Available Vd. is less than the Allowed Vd.

(3) Where both conditions (1) and (2) are applicable, the formula can be combined.

$$\frac{\text{Factor}}{I_b} \times \frac{\text{Available Vd}}{\text{Allowed Vd}} = \text{actual length allowed}$$

Finding the actual volt drop in a circuit for a given size and length of cable.

(4) Quite often the length of circuit is known, along with the full load current. What is required is the voltage drop for a particular conductor size. This can be calculated using the formula

$$\text{Voltage drop in cable} = \frac{\text{Actual length} \times I_b \times \text{Allowed Vd}}{\text{Factor}}$$

Factor for any other value of voltage drop
Where the percentage voltage drop allowed is different from 4%, the factor can be adjusted to the different percentage voltage drop. This situation may occur where the equipment requires a lower voltage drop than that allowed by the Wiring Regulations. The following formula may be used.

$$\frac{\text{Factor from table} \times \text{Desired volt drop \%}}{4\%} = \text{Revised Factor}$$

Alternatively, a ratio of the two voltage drop percentages can be worked out so that this ratio can be applied to any factor in the tables.

$$\frac{\text{Desired voltage drop percentage}}{4\%} = \text{Ratio to be applied to cable factor}$$

Determination of impedance mV/A/m
There may be occasions when an engineer requires to know the mV/A/m for a given cable; this can be obtained as follows:

Single-phase circuits

$$\frac{9.6V}{\text{Factor for cable}} = \text{mV/A/m impedance}$$

Three-phase circuits

$$\frac{16.6V}{\text{Factor for cable}} = \text{mV/A/m impedance}$$

Power factor
If having worked out the cable length by using the cable factor, the cable length falls short of that required, and if the circuit is supplying a load with a power factor, allowance can be made for the power factor when calculating the voltage drop by using the 'r' and 'x' values from the tables. Table MISC 2 Part 4 gives the sine of power factor.

$$\text{Voltage drop} = \frac{(\cos\varnothing r + \sin\varnothing x) \times \text{Length} \times \text{Load}}{1000}$$

Alternatively

$$\text{Cable length allowed} = \frac{\text{Available Vd} \times 1000}{(\cos\varnothing r + \sin\varnothing x)I_b}$$

Allowance for conductor temperature
Where the ambient temperature is not less than 30 °C, where the protective device is not a BS 3036 fuse, where the conductor is not carrying its full rated current, and where the conductors are larger than 16 mm^2, allowance can be made for the reduction in voltage drop due to the resistance of the conductor being lower than its stated operating temperature.

$$\text{Cable length allowed} = \frac{\text{Available Vd} \times 1000}{(C_t \cos\varnothing r + \sin\varnothing x)I_b}$$

For cable up to 16 mm^2, the increase in length can be determined by increasing the Factor in the CR tables, as given in the following formula.

$$\frac{\text{Factor}}{C_t} = \text{New factor to use in formulae.}$$

The calculation of C_t is given in Part 2, in the Chapter on Voltage drop.

Earth loop impedance tables

Seven tables comprising ZS 1 to ZS 7 are provided for checking that a circuit will comply with the regulations for protection against indirect contact.

When designing a circuit, the Z_S values in the row headed "Design' should be used. The average of the conductor operating temperature plus the limit temperature of the conductor insulation should be used when selecting the resistance (or impedance) of the phase and protective conductors. Tables are provided in Part 4 giving the earth loop impedance of various types of cable. The value Z_S is obtained from

$$Z_S = Z_E + Z_{inst}$$

where Z_E is the earth loop impedance up to the origin of the circuit, and Z_{inst} is the phase earth loop impedance of the circuit up to the point being checked. When testing an installation, the conductors will not be operating at the temperature used when designing the circuit: i.e., 115 °C in the case of pvc, so the values used from the tables are those listed under testing for the ambient temperature ruling at the time of the test.

The values in the tables are based on $U_o = 240$ V, for any other nominal voltage divide the figure in the table by 240 and multiply by the actual voltage.

Protective conductor tables

Having worked out the Z_S for the protection against indirect contact, that value can now be used to check whether the circuit protective conductor is suitable for the thermal stresses caused by the fault current, by using Tables PCZ 1 to 7

Look in the column for the size of protective device protecting the circuit, and then in the row for the protective conductor size. If the Z_S is less than the figure in the table, the protective conductor is satisfactory. If the Z_S is larger than the figure in the table, re-calculate Z_S using a larger protective conductor until the circuit Z_S is less than that given in the table.

When mcbs are used as the protective device, the circuit Z_S must not be less than the minimum value given in the last column.

Although the PCZ tables are based on the thermal capacity of copper conductors, they can be used when conduit or trunking is used as a protective conductor, since the thermal capacity of the conduit is larger than that of the copper conductor which can be installed through it, in accordance with the regulations. The suitability of conduit can easily be checked by using Table M54G in part 4. Where micc is used, the area of the sheath is given in Table ELI 5.

Short-circuit protection

The tables SCZ 1 to 7 are used to check that the live conductors are suitable for the thermal stresses caused by a short-circuit.

The table works in a similar manner to that of the PCZ tables, but this time, the impedance of the phase and neutral conductor is worked out at the design temperature, and then checked with the value given in the table under the protective device size, and opposite the conductor size.

Where three-phase three-wire circuits are concerned, an adjustment has to be made to the calculated impedance. The impedance should be divided by 1.732 before comparing it with the value in the table. The reason for this adjustment is that the table is based on the single-phase voltage of 240 volts.

In TN-C-S systems, the declared Z_E at the origin of the installation can be used to give the external phase/neutral impedance. This value can then be used for calculating the minimum short-circuit current. In order to check the thermal capability of the conductors, the Z_E is added to the phase and neutral impedance up to the point of fault, to give the total Z_{pn}.

When checking breaking capacity (maximum fault current) in a system, the external Z_{pn} will be equal to 240 V divided by the prospective short-circuit current at the origin of the installation. Thus, for domestic premises, the external $Z_{pn} \Omega = 240\text{ V} \div 16,000\text{ A} = 0.015\ \Omega$ This figure is then used in the calculations to give the total phase/neutral impedance, for determining the maximum short-circuit current, to check whether equipment has the correct breaking capacity.

Cable sizes when grouped together

Twenty four tables, CSG 1 to CSG 24 are provided; these give the cable size required for different types of grouping, and different methods of installation. All that is required, is to look in the column under the size of the protective device for the number of circuits or cables grouped together, to obtain the cable size.

Tables 21 to 24 deal specifically with BS 3036 fuses, and take into account the derating factor 0.725 when a rewirable fuse is used for overload protection.

Where the load is not likely to cause overload current (resistive load), the design current is used instead of the rating of the protective device, when derating for grouping. In these circumstances, use the full load current of the circuit, and not the fuse size in the CSG tables.

If any particular grouping has not been covered by one of these tables, it is only necessary to look at the CRG (cable rating grouped) tables in Part 4, which give the current-carrying capacity required for the conductor, when grouped.

Cable capacity of conduit and trunking

The details of how to use the tables CCC 1 and CCT 1 are given with the tables.

Worked examples

To illustrate the use of the tables, the following worked examples are given, firstly by using the quick reference tables from Parts 3 and 4, and secondly by working out the problem the long way, again using the tables in the handbook.

Example 1 (Heating load)

A single-phase circuit is to supply a fixed heating load of 6 kW; the cable is to be twin &cpc cable clipped direct to a surface along with three other cables, with their sheaths touching. If the circuit length is 25 metres and is to be supplied from a BS 1361 fuse, what size of fuse and cable is required, if the system is TN-C-S, and Z_E is 0.35? The ambient temperature is 30 °C, and the cables will not be in contact with thermal insulation.

Working

Fixed load: so use INST 2, with a disconnection time of 5 seconds.
From INST 2, the BS 1361 fuse size required is 30 A, but the load is a heating load, so I_b can be used for derating. Full load current is 25 A. From CSG 5, the cable size required for 4 circuits grouped, and an I_b of 25 A is 6 mm^2.
From INST 2, the maximum length for 6 mm^2 cable for a TN-C-S system is 52.6 metres; this is twice the length required for the circuit, so the cable is suitable.

Example 2
Same example as 1 but worked out the long way.

Working
Full load current = 6000 W ÷ 240 V = 25 A. Use 30 A BS 1361 fuse. Number of cables grouped together = 3 + 1 = 4. From GF1 derating factor = 0.65. Since load is a heating load, I_b can be used instead of I_n. Therefore, calculated current-carrying capacity I_t required for conductors is:

$$I_t = \frac{25}{0.65} = 38.46 \text{ A}$$

From Table CR7, minimum cable size required is 6 mm², with an I_{tab} of 46 A, and Factor 1315.
Voltage drop

$$\text{Actual length allowed for 4\% voltage drop} = \frac{\text{Factor}}{I_b} = \frac{1315}{25} = 52.6\text{m}$$

Voltage drop requirements satisfied.
Protection against indirect contact
From MISC 1, cpc size in 6 mm² cable is 2.5 mm².
From table RC1, resistance at 115 °C for 6 mm² is 4.25 Ω per 1000 m (R_1)
From table RC1, resistance at 115 °C for 2.5 mm² is 10.226 Ω per 1000 m (R_2)
$Z_{inst} = R_1 + R_2 = 4.25 \, \Omega + 10.226 \, \Omega = 14.476 \, \Omega / 1000 \text{ m}$
Actual $Z_{inst} = 25 \text{ m} \times 0.014476 \, \Omega = 0.3619 \, \Omega$
$Z_S = Z_E + Z_{inst} \therefore Z_S = 0.35 \, \Omega + 0.3619 \, \Omega = 0.7119 \, \Omega$
Maximum Z_S allowed from table ZS 2 for 30 A fuse = 1.92 Ω, so the circuit is satisfactory.
Protective conductor
From PCZ 2, maximum Z_S allowed for 2.5 mm² cpc with 30 A fuse is 1.92 Ω This is greater than the actual Z_S for the circuit (already worked out), so cpc is protected.
Short circuit protection
Resistance of both the phase conductor and the neutral conductor are required from the source. Since the system is TN-C-S, the value of Z_E is also the phase/neutral impedance.
Total $Z_{pn} = (2 \times 25 \text{ m} \times 0.00425 \Omega) + 0.35 \, \Omega = 0.5625 \, \Omega$
From table SCZ 6, maximum impedance Z_{pn} allowed for 30 A fuse and 6mm² conductor is 2.43 Ω. This is greater than the actual Z_{pn} for the circuit, so 6 mm² cable is protected.

Example 3 (Voltage drop)
A 75 kW DOL 415 V motor with a power factor of 0.7 is to be supplied from an HRC fuse board 100 metres from the motor. The installation is to be carried out with an XLPE armoured cable installed on a cable tray, with one other cable. The starter will be adjacent to the motor; the overload settings will be 130 A minimum 160 A maximum. If the length of the cable is 100 metres, and the three-phase voltage drop up to the distribution board is 4 volts, will the cable length be suitable for an overall voltage drop of 4%? Also, determine the size of the HRC fuse required for the circuit.

Working
From Table MC1, 75 kW @ 415 V gives f.l.c. of 136 A, HRC fuse size required is 200M250
Maximum overload setting is 160 A. This must be used since overloads can be turned up.
From CSG 14, two circuits with an overload setting of 160 A, so cable size required is 50 mm²

From CR10, 50 mm² cable, with a current-carrying capacity of 197 A, with a Factor of 19080.

$$\text{Maximum length allowed for cable} = \frac{19080}{160} \times \frac{16.6 - 4}{16.6} = 90.15\text{m}$$

The cable length appears to be too long, so see if allowing for power factor will help.
From CR10 $R = 0.86$ and $x = 0.135$
From MISC 2 with a power factor of 0.7 sine of the angle $= 0.7141$

$$\text{Maximum length of cable allowed} = \frac{(16.6 - 4) \times 1000}{(0.7 \times 0.86 + 0.7141 \times 0.135) \, 160\text{A}}$$

$$= \frac{12600}{111.744} = 112.758 \text{ metres}$$

The length of the cable will be suitable.

$$\text{Actual voltage drop in cable} = \frac{(0.602 + 0.0964) \times 100\text{m} \times 160\text{A}}{1000} = 11.17\text{V}$$

Add to this the 4 volts drop up to the distribution board = 15.17 V total.
Note: If this circuit were checked for protection against indirect contact, the 50 mm² cable would not be large enough. Neither would the armour be suitable as the protective conductor.

Example 4

The supply to a 415 V three-phase 28 A welding load is to be taken from an HRC fuse in a sub-distribution board. The cables which are to be pvc insulated with copper conductors, are to be installed in conduit, along with the cables for a single-phase circuit using 1.5 mm² cable, and a three-phase circuit using 2.5 mm² cable; it is anticipated that there will be two bends between draw- in boxes, which will be spaced 3 m apart. The three-phase voltage drop up to the distribution board is 4 V, and the voltage drop in the final circuit is to be limited to 4%. Z_s at the distribution board bus bars is 0.5 Ω, and Z_p at the bus bars is 0.3 Ω. What size cable will be required if the circuit length is 80 metres?

Working

Fuse size required for 28 A load is 32 A, assuming there is no inrush current.

Grouping - cable size
Number of circuits = 3. From CSG 1 - three-phase, conductor size required = 10 mm²
From Table CR 1, $I_{tab} = 50$ A, Factor = 4368.

Voltage drop
Using formula previously given:

$$\text{Length allowed} = \frac{4368}{28} \times \frac{(16.6 - 4)}{16.6} = 118.4 \text{ m}$$

Conduit size
From table CCC 1 Table 1   factor for 1.5 mm² cable = 22 × 2 = 44
                             factor for 2.5 mm² cable = 30 × 3 = 90
                             factor for 10 mm² cable   = 105 × 3 = 315
                                                         Total   = 449

From CCC 1 Table 2, two bends 3 m between boxes so conduit size required = 32 mm

Protection against indirect contact

From table RC 1 Part 4 resistance of 10 mm² cable at 115 °C = 0.002525 Ω/m
From table ZCT 1 Part 4 impedance of 32 mm conduit with joints = 0.0025 Ω/m
Total 0.005025 Ω/m

Z_{inst} = 80 × 0.005025 Ω = 0.402 Ω.
Now Z_S at the distribution board busbar is the Z_E for the following circuit.
Therefore Z_S = 0.5 + 0.402 Ω = 0.902 Ω.
From ZS 1 A, for 5 seconds and 32 A fuse, Z_S allowed = 1.92 Ω - This is satisfactory.

Protective conductor

Checking with Table M54G, MISC 1 in Part 4.
10 mm² phase conductor, protective conductor area = 10 × 2.45 = 24.5 mm².
From MISC 1 Part 4, area of 32 mm heavy gauge conduit = 167 mm².
So conduit is satisfactory as a circuit protective conductor.

Short-circuit protection

Z_p up to the distribution board is 0.3 Ω
Resistance of 10 mm² phase conductor = 0.002525 Ω × 80 = 0.202 Ω
Total Z_{pp} = 2(0.3 + 0.2)Ω = 1.0 Ω

To compare the value (1.0 Ω) with readings in SCZ1 ÷ √3 = $\frac{1.0}{1.732}$ = 0.577 Ω

From SCZ 5, maximum impedance allowed = 1.7 Ω - This is satisfactory.

General

Short cuts could obviously be made in the examples given, but all the steps have been included to illustrate how the various tables are used. Additionally, the answers have been given to three or four decimal places; this has been done so that those who want to work through the examples can compare answers. In practice answers will be rounded up or down to whole numbers.

Conduit

As far as conduit is concerned, there will be no problem using it in a normal installation as a protective conductor, because its thermal capacity will always withstand the fault current flowing through it.

Installing a protective conductor within the conduit is really a waste of time and money. In the past, heavy gauge galvanised solid drawn conduit has been installed in hazardous areas without any additional protective conductors, and without any problems arising from the installation.

The point is often made that a protective conductor is installed in the conduit in case it becomes broken, corroded, or otherwise damaged. The answer to this point is that the conduit has not been installed in accordance with the regulations in the first place, as can be seen if the regulations given under 'Conduit' in Part 1 are studied.

Conduit, like all other parts of the installation, relies very heavily on good workmanship. (Regulation 13.01.01.) Since conduit and trunking are often specified for the wrong environmental conditions, and rely heavily on good workmanship, the view could be taken that installations using conduit and trunking as a protective conductor should be subject to regular inspection and testing. The need for such inspection and testing is, however, the fault of the designer, not of the conduit or trunking.

DISCONNECTION TIME 0.4 SECONDS — INST 1 — MAXIMUM CABLE LENGTHS

Maximum lengths in metres of single-phase circuits with pvc cable with various loads in areas requiring a DISCONNECTION TIME OF 0.4 seconds

Maximum length in metres for voltage drop, protection against indirect contact, thermal protection of cpc and short-circuit current

Resistive load in KW	Full load current in amperes	Protective device type	Protective device rating I_n	1.0 mm² twin & cpc TN-C-S $Z_E=0.35$	1.0 mm² twin & cpc TN-S $Z_E=0.8$	1.5 mm² twin & cpc TN-C-S $Z_E=0.35$	1.5 mm² twin & cpc TN-S $Z_E=0.8$	2.5 mm² twin & cpc TN-C-S $Z_E=0.35$	2.5 mm² twin & cpc TN-S $Z_E=0.8$	4 mm² twin & cpc TN-C-S $Z_E=0.35$	4 mm² twin & cpc TN-S $Z_E=0.8$	6 mm² twin & cpc TN-C-S $Z_E=0.35$	6 mm² twin & cpc TN-S $Z_E=0.8$	10 mm² twin & cpc TN-C-S $Z_E=0.35$	10 mm² twin & cpc TN-S $Z_E=0.8$	16 mm² twin & cpc TN-C-S $Z_E=0.35$	16 mm² twin & cpc TN-S $Z_E=0.8$
1	4.2	BS 88	6	52.3	52.3	79.4	79.4	128.0	128.0	209.0	209.0	315.6	315.6	523.6	523.6	822.8	822.8
3	12.5	"	16	17.4	17.4	26.4	26.4	42.6	42.6	69.8	69.8	105.2	105.2	174.5	174.5	274.2	274.2
6	25.0	"	25					21.3	21.3	34.9	30.3	52.6	48.3	87.2	78.7	137.0	119.9
7	29.2	"	32					18.2	10.7	29.9	12.5	45.0	20.0	74.8	32.6	117.5	49.6
8.3	34.6	"	40							19.9		35.2	4.0	57.3	6.7	87.3	10.2
10	41.7	"	50									19.9		31.5		47.9	
1	4.2	BS 1361	5	52.3	52.3	79.4	79.4	128.0	128.0	209.0	209.0	315.6	315.6	523.6	523.6	822.8	822.8
3	12.5	"	15	17.4	17.4	26.4	26.4	42.6	42.6	69.8	69.8	105.2	105.2	174.5	174.5	274.2	274.2
6	25.0	"	30					21.3	14.8	34.9	17.3	52.6	27.6	87.2	45.0	137.0	68.5
7	29.2	"	30					18.2	14.8	29.9	17.3	45.0	27.6	74.8	45.0	117.5	68.5
8.3	34.6	"	45							4.8		17.2		28.1		42.8	
10	41.7	"	45									17.2		28.1		42.8	
1	4.2	BS 3871 Type 1	5	52.3	52.3	79.4	79.4	128.0	128.0	209.0	209.0	315.6	315.6	523.6	523.6	822.8	822.8
3	12.5	"	15	17.4	17.4	26.4	26.4	42.6	42.6	69.8	69.8	105.2	105.2	174.5	174.5	274.2	274.2
3	12.5	"	16	17.4	17.4	26.4	26.4	42.6	42.6	69.8	34.9	105.2	52.6	174.5	87.2	274.2	137.0
6	25.0	"	25					21.3	21.3	34.9	29.9	52.6	45.0	87.2	74.8	137.0	117.5
7	29.2	"	30					18.2	18.2	29.9	29.9	45.0	45.0	74.8	74.8	117.5	117.5
7	29.2	"	32					18.2	18.2	29.9	29.9	45.0	45.0	74.8	74.8	117.5	117.5
8.3	34.6	"	40							25.2		38.0		63.0		99.1	99.1
10	41.7	"	45									31.5		52.3		82.2	82.2
1	4.2	BS 3871 Type 2	5	52.3	52.3	79.4	79.4	128.0	128.0	209.0	209.0	315.6	315.6	523.6	523.6	822.8	822.8
3	12.5	"	15	17.4	17.4	26.4	26.4	42.6	42.6	69.8	64.6	105.2	102.9	174.5	167.6	274.2	255.2
3	12.5	"	16	17.4	17.4	26.4	26.4	42.6	42.6	69.8	58.1	105.2	92.5	174.5	150.7	274.2	229.5
6	25.0	"	25					21.3	21.0	34.9	24.7	52.6	39.3	87.2	64.1	137.0	97.6
7	29.2	"	30					18.2	12.6	29.9	14.7	45.0	23.4	74.8	38.2	117.5	58.2
7	29.2	"	32					18.2	10.0	29.9	11.7	45.0	18.6	74.8	30.8	117.5	46.2
8.3	34.6	"	40							22.0	2.6	35.2	4.1	57.3	6.7	87.3	10.2
10	41.7	"	45									28.3		46.0		70.2	
1	4.2	BS 3036	5	52.3	52.3	79.4	79.4	128.0	128.0	209.0	209.0	315.6	315.6	523.6	523.6	822.8	822.8
3	12.5	"	15	17.4	17.4	26.4	26.4	42.6	42.6	69.8	69.8	105.2	105.2	174.5	174.5	274.2	274.2
6	25.0	"	30					21.3	12.6	34.2	14.7	52.6	23.4	87.2	38.2	135.3	58.2
7	29.2	"	30					18.2	12.6	29.9	14.7	45.0	23.4	74.8	38.2	117.5	58.2

Notes: Table based on 240V and voltage drop of 4%. Cables to be installed open and clipped direct, circuits need derating when grouped, in contact with thermal insulation, or when ambient temperature exceeds 30 °C, although the lengths in the table are still applicable. Where the load is not likely to cause overload, e.g. heating load, immersion heater etc, the cable's current-carrying capacity does not have to equal the rating of the protective device. The table gives lengths for different sizes of cable, for each load for TN-C-S and TN-S systems, and can be used in any area where disconnection time has to be 0.4 seconds, e.g. bathroom, or fixed equipment outdoors. The table can also be used for micc cable, or pvc cable single circuits in conduit and trunking. If installed with other circuits, the cable size would need increasing in accordance with the CSG Tables.

DISCONNECTION TIME 5 SECONDS INST 2 MAXIMUM CABLE LENGTHS

Maximum lengths in metres of single-phase circuits with pvc cable with various loads in areas requiring a
DISCONNECTION TIME OF 5 seconds

Maximum length in metres for voltage drop of 4%, protection against indirect contact, thermal protection of cpc and short-circuit current

Resistive load in KW	Full load current in amperes	Protective device type	Protective device rating I_n	1.0 mm² twin & cpc			1.5 mm² twin & cpc			2.5 mm² twin & cpc			4 mm² twin & cpc			6 mm² twin & cpc			10 mm² twin & cpc			16 mm² twin & cpc		
				TN-C-S $Z_E=0.35$	TN-S $Z_E=0.8$		TN-C-S $Z_E=0.35$	TN-S $Z_E=0.8$		TN-C-S $Z_E=0.35$	TN-S $Z_E=0.8$		TN-C-S $Z_E=0.35$	TN-S $Z_E=0.8$		TN-C-S $Z_E=0.35$	TN-S $Z_E=0.8$		TN-C-S $Z_E=0.35$	TN-S $Z_E=0.8$		TN-C-S $Z_E=0.35$	TN-S $Z_E=0.8$	
1	4.2	BS 88	6	52.3	52.3	17.4	79.4	79.4	26.4	128.0	128.0	42.6	209.4	209.4	69.8	315.6	315.6	105.2	523.6	523.6	174.5	822.8	822.8	274.2
3	12.5	"	16	17.4	17.4		26.4	26.4		42.6	42.6	21.3	69.8	69.8	34.9	105.2	105.2	52.6	174.5	174.5	87.2	274.2	274.2	137.1
6	25.0	"	25							21.3	21.3	18.2	34.9	34.9	21.6	52.6	52.6	45.0	87.2	87.2	74.8	137.1	137.1	117.5
7	29.2	"	32							18.2	18.2		29.9	29.9		45.0	45.0	24.8	74.8	74.8	63.0	117.5	117.5	99.1
8.3	34.6	"	40										19.5			38.0	38.0		63.0	63.0	32.6	99.1	99.1	49.6
10	41.7	"	50													26.2			52.3	52.3				
1	4.2	BS 1361	5	52.3	52.3	17.4	79.4	79.4	26.4	128.0	128.0	42.6	209.4	209.4	69.8	315.6	315.6	105.2	523.6	523.6	174.5	822.8	822.8	274.2
3	12.5	"	15	17.4	17.4		26.4	26.4		42.6	42.6	21.3	69.8	69.8	34.9	105.2	105.2	52.6	174.5	174.5	87.2	274.2	274.2	137.1
6	25.0	"	30							21.3	21.3	18.2	34.9	34.9	29.4	52.6	52.6	45.0	87.2	87.2	74.8	137.1	137.1	117.5
7	29.2	"	32							18.2	18.2		29.9	29.9		45.0	45.0		74.8	74.8	63.0	117.5	117.5	99.1
8.3	34.6	"	45										4.8			23.2	23.2		22.5	22.5		99.1	99.1	34.2
10	41.7	"	45													23.2								34.2
1	4.2	BS 3871 Type 1	5	52.3	52.3	17.4	79.4	79.4	26.4	128.0	128.0	42.6	209.4	209.4	69.8	315.6	315.6	105.2	523.6	523.6	174.5	822.8	822.8	274.2
3	12.5	"	15	17.4	17.4		26.4	26.4		42.6	42.6	21.3	69.8	69.8	34.9	105.2	105.2	52.6	174.5	174.5	87.2	274.2	274.2	137.1
6	25.0	"	16							42.6	42.6	21.3	69.8	69.8	34.9	105.2	105.2	52.6	174.5	174.5	87.2	274.2	274.2	137.1
7	29.2	"	30							18.2	18.2	18.2	29.9	29.9	29.9	45.0	45.0	45.0	74.8	74.8	74.8	117.5	117.5	117.5
7	29.2	"	32							18.2	18.2		29.9	29.9	29.9	45.0	45.0	45.0	74.8	74.8	74.8	117.5	117.5	117.5
8.3	34.6	"	40										25.2	25.2		38.0	38.0		63.0	63.0		99.1	99.1	
10	41.7	"	45													31.5	31.5		52.3	52.3		82.2	82.2	
1	4.2	BS 3871 Type 2	5	52.3	52.3	17.4	79.4	79.4	26.4	128.0	128.0	42.6	209.4	209.4	69.8	315.6	315.6	105.2	523.6	523.6	174.5	822.8	822.8	274.2
3	12.5	"	15	17.4	17.4		26.4	26.4		42.6	42.6	21.1	69.8	69.8	34.9	105.2	105.2	52.6	174.5	174.5	87.2	274.2	274.2	137.1
6	25.0	"	16							42.6	42.6	12.6	69.8	69.8	24.7	105.2	105.2	58.1	167.6	167.6	64.1	255.2	255.2	137.1
7	29.2	"	25							12.6	12.6	10.0	34.9	34.9	14.7	52.6	52.6		150.7	150.7	38.2	229.5	229.5	117.5
7	29.2	"	30										29.9	29.9	11.7	45.0	45.0		92.9	92.9	30.3	97.6	97.6	87.3
8.3	34.6	"	32										29.9	29.9	2.6	45.0	45.0		39.3	39.3	6.7	58.2	58.2	
10	41.7	"	45										22.1	22.1		35.2	35.2		23.4	23.4		46.2	46.2	
																28.3	28.3		18.6	18.6		70.2	70.2	10.2
																			4.1	4.1				
																			46.1	46.1				
1	4.2	BS 3036	5	52.3	52.3	17.4	79.4	79.4	26.4	128.0	128.0	42.6	209.4	209.4	69.8	315.6	315.6	105.2	523.6	523.6	174.5	822.8	822.8	274.2
3	12.5	"	15	17.4	17.4		26.4	26.4		42.6	42.6	21.3	69.8	69.8	34.9	105.2	105.2	52.6	174.5	174.5	87.2	274.2	274.2	137.1
6	25.0	"	30							21.3	21.3	18.2	34.9	34.9	29.9	52.6	52.6	45.0	87.2	87.2	74.8	137.1	137.1	117.5
7	29.2	"	30							18.2	18.2		29.9	29.9		45.0	45.0		74.8	74.8		117.5	117.5	

Notes: Table based on 240V and voltage drop of 4%. Cables to be installed open and clipped direct, circuits need derating when grouped, in contact with thermal insulation, or when ambient temperature exceeds 30 °C, although the lengths in the table are still applicable. Where the load is not likely to cause overload, e.g. heating load, immersion heater etc., the cable's current-carrying capacity does not have to equal the rating of the protective device. The table gives lengths for different sizes of cable, for each load for TN-C-S and TN-S systems, and can be used in any area where disconnection time has to be 0.4 seconds, e.g. bathroom, or fixed equipment outdoors. The table can also be used for mice cable, or pvc cable single circuits in conduit and trunking. If installed with other circuits, the cable size would need increasing in accordance with the CSG Tables.

VOLTAGE DROP — INST 3 — LIGHTING

Voltage drop in volts, in a lighting circuit with 240V single-phase supply, using the loop-in method of wiring (see notes).

TABLE 1 - Distance from consumer unit to first lighting point in metres

Number of lighting points	3	4	5	6	7	8	9	10	11	12	13	14	15	16	17	18	19	20	21
										Voltage drop									
1	0.06	0.07	0.09	0.11	0.13	0.15	0.17	0.18	0.20	0.22	0.24	0.26	0.28	0.29	0.31	0.33	0.35	0.37	0.39
2	0.11	0.15	0.18	0.22	0.26	0.29	0.33	0.37	0.40	0.44	0.48	0.51	0.55	0.59	0.62	0.66	0.70	0.73	0.77
3	0.17	0.22	0.28	0.33	0.39	0.44	0.50	0.55	0.61	0.66	0.72	0.77	0.83	0.88	0.94	0.99	1.05	1.10	1.16
4	0.22	0.29	0.37	0.44	0.51	0.59	0.66	0.73	0.81	0.88	0.95	1.03	1.10	1.17	1.25	1.32	1.39	1.47	1.54
5	0.28	0.37	0.46	0.55	0.64	0.73	0.83	0.92	1.01	1.10	1.19	1.28	1.38	1.47	1.56	1.65	1.74	1.83	1.93
6	0.33	0.44	0.55	0.66	0.77	0.88	0.99	1.10	1.21	1.32	1.43	1.54	1.65	1.76	1.87	1.98	2.09	2.20	2.31
7	0.39	0.51	0.64	0.77	0.90	1.03	1.16	1.28	1.41	1.54	1.67	1.80	1.93	2.05	2.18	2.31	2.44	2.57	2.70
8	0.44	0.59	0.73	0.88	1.03	1.17	1.32	1.47	1.61	1.76	1.91	2.05	2.20	2.35	2.49	2.64	2.79	2.93	3.08
9	0.50	0.66	0.83	0.99	1.16	1.32	1.49	1.65	1.81	1.98	2.15	2.31	2.48	2.64	2.81	2.97	3.14	3.30	3.46
10	0.55	0.73	0.92	1.10	1.28	1.47	1.65	1.83	2.02	2.20	2.38	2.57	2.75	2.93	3.12	3.30	3.48	3.67	3.85

TABLE 2 - Voltage drop between lighting fittings

Length between lights in metres	1st & 2nd	2nd & 3rd	3rd & 4th	4th & 5th	5th & 6th	6th & 7th	7th & 8th	8th & 9th	9th & 10th
					Voltage drop				
2	0.33	0.29	0.26	0.22	0.18	0.15	0.11	0.07	0.04
3	0.50	0.44	0.39	0.33	0.28	0.22	0.17	0.11	0.06
4	0.66	0.59	0.51	0.44	0.37	0.29	0.22	0.15	0.07
5	0.83	0.73	0.64	0.55	0.46	0.37	0.28	0.18	0.09
6	0.99	0.88	0.77	0.66	0.55	0.44	0.33	0.22	0.11
7	1.16	1.03	0.90	0.77	0.64	0.51	0.39	0.26	0.13
8	1.32	1.17	1.03	0.88	0.73	0.59	0.44	0.29	0.15
9	1.49	1.32	1.16	0.99	0.83	0.66	0.50	0.33	0.17
10	1.65	1.47	1.28	1.10	0.92	0.73	0.55	0.37	0.18
11	1.81	1.61	1.41	1.21	1.01	0.81	0.61	0.40	0.20
12	1.98	1.76	1.54	1.32	1.10	0.88	0.66	0.44	0.22
13	2.15	1.91	1.67	1.43	1.19	0.95	0.72	0.48	0.24
14	2.31	2.05	1.80	1.54	1.28	1.03	0.77	0.51	0.26
15	2.48	2.20	1.93	1.65	1.38	1.10	0.83	0.55	0.28

TABLE 3 - Voltage drop between lighting point and switch, where switch cables are connected at lighting point

Distance - lighting point to switch in metres							
	1	2	3	4	5	6	7
Volt drop	0.02	0.04	0.06	0.07	0.09	0.11	0.13

Distance - lighting point to switch in metres							
	8	9	10	11	12	13	14
Volt drop	0.15	0.17	0.18	0.20	0.22	0.24	0.26

Distance - lighting point to switch in metres							
	15	16	17	18	19	20	21
Volt drop	0.28	0.29	0.31	0.33	0.35	0.37	0.39

HOW TO USE: The table is based on 1mm² pvc cable, 240 V supply, with 100 watt load on each lighting point, and the loop-in system of wiring. Add the voltage drop from the consumer unit to the first lighting point (Table 1), then add the voltage drop between each lighting point (Table 2) and then add the voltage drop for the length of cable from the lighting point to switch (Table 3). This will give the voltage drop at that lighting point. To obtain the voltage drop for a larger cable, multiply the total voltage drop obtained by 44, and divide by R volt drop from the CR Table for the larger cable. For larger, but equal loads on each lighting point, multiply the voltage drop by the increase in load, e.g. for 200 watt at each lighting point, multiply the voltage drop worked out by 2.

DISCONNECTION TIME 0.4 SECONDS — INST 4 — MAXIMUM CABLE LENGTHS

240V SINGLE-PHASE BS 1363 SOCKET OUTLET CIRCUITS (See notes)

Maximum length in metres for protection against indirect contact, thermal protection of cpc and short-circuit current

Circuit type	Protective device type	Protective device rating I_n	Maximum floor area allowed in sq. metres	2.5 mm² twin & cpc				4 mm² twin & cpc				2.5 mm² single core p.v.c. in steel conduit				4 mm² single core p.v.c. in steel conduit				2 core 1.5 mm² mi cc cable				2 core 2.5 mm² mi cc cable			
				TN-C-S $Z_E=0.35$		TN-S $Z_E=0.8$		TN-C-S $Z_E=0.35$		TN-S $Z_E=0.8$		TN-C-S $Z_E=0.35$		TN-S $Z_E=0.8$		TN-C-S $Z_E=0.35$		TN-S $Z_E=0.8$		TN-C-S $Z_E=0.35$		TN-S $Z_E=0.8$		TN-C-S $Z_E=0.35$		TN-S $Z_E=0.8$	
1	2	3	4	5		6		7		8		9		10		11		12		13		14		15		16	
Ring	BS 88	32	100	95		43		128		50		95		72		155		94		60		54		80		80	
Radial	"	32	50	-		-		27		12		-		-		27		23		-		-		17		17	
Radial	"	20	20	27		27		44		44		27		27		44		44		17		17		28		28	
Ring	BS 1361	30	100	95		59		147		69		95		95		155		130		60		60		80		80	
Radial	"	30	50	-		-		29		17		-		-		29		29		-		-		19		19	
Radial	"	20	20	27		27		44		42		27		27		44		44		17		17		28		28	
Ring	BS 3871 Type 1	32	100	95		95		155		155		95		95		155		155		60		60		80		80	
Ring	"	30	100	104		105		170		170		104		105		170		170		65		65		110		110	
Radial	"	32	50	-		-		27		27		-		-		27		27		-		-		17		17	
Radial	"	30	50	-		-		29		29		-		-		29		29		-		-		19		19	
Radial	"	20	20	27		27		44		44		27		27		44		44		17		17		28		28	
Ring	BS 3871 Type 2	32	100	95		40		124		46		95		67		155		88		60		31		80		52	
Ring	"	30	100	104		50		137		58		104		84		170		111		60		35		110		66	
Radial	"	32	50	-		-		27		11		-		-		27		22		-		-		17		13	
Radial	"	30	50	-		-		29		14		-		-		29		27		-		-		19		16	
Radial	"	20	20	27		27		44		39		27		27		44		44		17		17		28		28	
Ring	BS 3036	30	100	104		50		137		58		104		84		170		111		67		64		110		102	
Radial	"	20	20	27		27		44		44		27		27		44		44		17		17		28		28	

Notes:
(1) Twin & cpc cable to be installed open and clipped direct. Where circuits are grouped or in contact with thermal insulation, or the ambient temperature exceeds 30 °C, larger cables must be installed; the lengths given in the Table for each cable size will still be applicable, but see note 3. The table is equally applicable to single circuits in plastic conduit if a minimum size of 1.5mm² cpc. is also installed.

(2) The Table gives the maximum length for protection against indirect contact, for the thermal suitability of the protective conductor and for live conductors with a fault. Voltage drop has been taken into account by assuming that radial circuits are fully loaded. For ring circuits, the average of the minimum and maximum voltage drop that could occur with the ring fully loaded has been used (i.e. assume all load at mid point, and load evenly distributed round the ring). In practice, the voltage drop for socket circuits cannot be calculated, since the load can be changed at any time. Voltage drop is therefore limited by the floor area allowed for each type of circuit.

(3) Grouping: Two circuits can be installed without derating for grouping. For ring circuits only, up to five circuits can be bunched together when 4mm² twin & cpc cable is used. For derating radial circuits, use I_n divided by derating factor G (Table GF1). For ring circuits use I_n x 0.67 divided by G, then select the cable to carry the calculated current. No further diversity is allowed. Any other formula should not be used, because simultaneous overload can occur with socket circuits, and the conductors cannot be assumed to be lightly loaded.

I_tab SINGLE CORE CABLES — CR 1 COPPER CONDUCTORS

Non-armoured single core p.v.c. insulated copper cables with or without sheath to BS6004 or BS 6346

Full thermal ratings — Ambient temperature 30 °C — Conductor operating temperature 70 °C

Conductor cross-sectional area in sq.mm.	Enclosed in conduit, trunking, underground conduit (with sheath). Not in contact with thermal insulation.								Open, clipped direct, or lying on a non-metallic surface, or enclosed in non-thermal insulating plaster. Sheaths touching.								
	Two cables single-phase a.c.				Three or four cables three-phase a.c.				Two cables single-phase a.c.				Three or four cables three-phase a.c.				
	Current rating amperes	Voltage drop mV/A/m		Factor length × amperes	Current rating amperes	Voltage drop mV/A/m		Factor length × amperes	Current rating amperes	Voltage drop mV/A/m		Factor length × amperes	Current rating amperes	Voltage drop mV/A/m		Factor length × amperes
		r	x			r	x			r	x			r	x	
1	2	3	4	5	6	7	8	9	10	11	12	13	14	15	16	17
1.0	13.5	44.0	-	218	12	38.0	-	437	15.5	44.0	-	218	14	38.0	-	437
1.5	17.5	29.0	-	331	15.5	25.0	-	664	20	29.0	-	331	18	25.0	-	664
2.5	24	18.0	-	533	21	15.0	-	1107	27	18.0	-	533	25	15.0	-	1107
4.0	32	11.0	-	873	28	9.50	-	1747	37	11.0	-	873	33	9.50	-	1747
6.0	41	7.30	-	1315	36	6.40	-	2594	47	7.30	-	1315	43	6.40	-	2594
10	57	4.40	-	2182	50	3.80	-	4368	65	4.40	-	2182	59	3.80	-	4368
16	76	2.80	-	3429	68	2.40	-	6917	87	2.80	-	3429	79	2.40	-	6917
25	101	1.80	0.33	5333	89	1.50	0.29	10710	114	1.75	0.200	5486	104	1.50	0.25	10710
35	125	1.30	0.31	7385	110	1.10	0.27	15091	141	1.25	0.195	7680	129	1.10	0.24	15091

Ambient temperature correction factors for p.v.c insulated cable

	For rewirable fuses only								For all other protective devices							
Ambient temperature °C	25	30	35	40	45	50	55	60	65	25	30	35	40	45	50	55	60	65
Correction factor	1.03	1.0	0.97	0.94	0.91	0.87	0.84	0.69	0.48	1.03	1.0	0.94	0.87	0.79	0.71	0.61	0.50	0.35

Note: Factor gives cable size for 4% voltage drop. Multiply load current by cable length, then select a factor equal to, or larger than the calculated value of load × length. For loads with a power factor where length is critical, use the values of r and x in the formulas given in the introduction, or in Part 2 Voltage drop. Cables need derating when BS 3036 fuses are used for overload protection.

(4D1)

Itab SINGLE CORE CABLES   CR 1(a)   COPPER CONDUCTORS

Non-armoured single core p.v.c. insulated copper cables with or without sheath to BS6004 or BS 6346

Full thermal ratings   Ambient temperature 30 °C   Conductor operating temperature 70 °C

Conductor cross-sectional area in sq.mm.	Enclosed in conduit, trunking, underground conduit (with sheath). Not in contact with thermal insulation							Open, clipped direct, or lying on a non-metallic surface, or enclosed in non-thermal insulating plaster. Sheaths touching						
	Two cables single-phase a.c.				Three or four cables three-phase a.c.			Two cables single-phase a.c.				Three or four cables three-phase a.c.		
	Current rating amperes	Voltage drop mV/A/m		Factor length × amperes	Current rating amperes	Voltage drop mV/A/m	Factor length × amperes	Current rating amperes	Voltage drop mV/A/m		Factor length × amperes	Current rating amperes	Voltage drop mV/A/m		Factor length × amperes
		r	x			mV/A/m			r	x			r	x	
1	2	3	4	5	6	7	8	9	10	11	12	13	14	15	16	17
50	151	0.950	0.30	9600	134	0.81	0.26	19529	182	0.930	0.190	10105	167	0.800	0.24	19762
70	192	0.650	0.29	13333	171	0.56	0.25	27213	234	0.630	0.185	14545	214	0.550	0.24	27667
95	232	0.490	0.28	17143	207	0.42	0.24	34583	284	0.470	0.180	19200	261	0.410	0.23	35319
120	269	0.390	0.27	20426	239	0.33	0.23	40488	330	0.370	0.175	23415	303	0.320	0.23	41500
150	300	0.310	0.27	23415	262	0.27	0.23	46111	381	0.300	0.175	28235	349	0.260	0.23	48824
185	341	0.250	0.27	25946	296	0.22	0.23	51875	436	0.240	0.170	33103	400	0.210	0.22	53548
240	400	0.195	0.26	29091	346	0.17	0.23	57241	515	0.185	0.165	38400	472	0.160	0.22	61481
300	458	0.160	0.26	30968	394	0.14	0.23	61481	594	0.150	0.165	43636	545	0.130	0.22	66400
400	546	0.130	0.26	33103	467	0.12	0.22	66400	694	0.120	0.160	48000	634	0.105	0.21	69167

Ambient temperature correction factors for p.v.c insulated cable

	For rewirable fuses only						For all other protective devices					
Ambient temperature °C	25	30	35	40	45	50	55	60	65	25	30	35	40	45	50	55	60	65
Correction factor	1.03	1.0	0.97	0.94	0.91	0.87	0.84	0.69	0.48	1.03	1.0	0.94	0.87	0.79	0.71	0.61	0.50	0.35

Note: Factor gives cable size for 4% voltage drop. Multiply load current by cable length, then select a factor equal to, or larger than the calculated value of load × length. For loads with a power factor where length is critical, use the values of r and x in the formulas given in the introduction, or in Part 2 Voltage drop. Cables need derating when BS 3036 fuses are used for overload protection.

(4D1)

I_tab SINGLE CORE CABLES   CR 2   COPPER CONDUCTORS

Non-armoured single core p.v.c. insulated copper cables with or without sheath to BS6004 or BS 6346

Full thermal ratings   Ambient temperature 30 °C   Conductor operating temperature 70 °C

Conductor cross-sectional area in sq.mm.	Enclosed in conduit in thermally insulating wall or ceiling, conduit in contact with thermally conductive surface								Conductor cross-sectional area in sq.mm.	Sheathed cables on perforated cable tray, bunched and unenclosed. Holes in tray occupy 30% of tray area							
	Two cables single-phase a.c.				Three or four cables three-phase a.c.					Two cables single-phase a.c.				Three or four cables three-phase a.c.			
	Current rating amperes	Voltage drop mV/A/m		Factor length × amperes	Current rating amperes	Voltage drop mV/A/m		Factor length × amperes		Current rating amperes	Voltage drop mV/A/m		Factor length × amperes	Current rating amperes	Voltage drop mV/A/m		Factor length × amperes
		r	x			r	x				r	x			r	x	
1	2	3	4	5	6	7	8	9	10	11	12	13	14	15	16	17	18
1.0	11.0	44	-	218	10.5	38.0	-	437	25	126	1.750	0.200	5486	112	1.5	0.25	10710
1.5	14.5	29	-	331	13.5	25.0	-	664	35	156	1.250	0.195	7680	141	1.1	0.24	15091
2.5	19.5	18	-	533	18	15.0	-	1107	50	191	0.930	0.190	10105	172	0.8	0.24	19762
4.0	26.0	11	-	873	24	9.50	-	1747	70	246	0.630	0.185	14545	223	0.55	0.24	27667
6.0	34.0	7.3	-	1315	31	6.40	-	2594	95	300	0.470	0.180	19200	273	0.41	0.23	35319
10	46.0	4.4	-	2182	42	3.80	-	4368	120	349	0.370	0.175	23415	318	0.32	0.23	41500
16	61.0	2.8	-	3429	56	2.40	-	6917	150	404	0.300	0.175	28235	369	0.26	0.23	48824
25	80.0	1.8	0.33	5333	73	1.50	0.29	10710	185	463	0.240	0.170	33103	424	0.21	0.22	53548
35	99.0	1.3	0.31	7385	89	1.10	0.27	15091	240	549	0.185	0.165	38400	504	0.16	0.22	61481
50	119.0	0.95	0.30	9600	108	0.81	0.26	19529	300	635	0.150	0.165	43636	584	0.13	0.22	66400

Ambient temperature correction factors for p.v.c insulated cable

For rewirable fuses only

Ambient temperature °C	25	30	35	40	45	50	55	60	65
Correction factor	1.03	1.0	0.97	0.94	0.91	0.87	0.84	0.69	0.48

For all other protective devices

Ambient temperature °C	25	30	35	40	45	50	55	60	65
Correction factor	1.03	1.0	0.94	0.87	0.79	0.71	0.61	0.50	0.35

Note: Factor gives cable size for 4% voltage drop. Multiply load current by cable length, then select a factor equal to, or larger than the calculated value of load × length. For loads with a power factor where length is critical, use the values of r and x in the formulas given in the introduction, or in Part 2 Voltage drop. Cables need derating when BS 3036 fuses are used for overload protection.

(4D1)

I tab SINGLE CORE CABLES CR 3
COPPER CONDUCTORS

Non-armoured single core thermosetting (XLPE) insulated copper cables with or without sheath to BS 7211 or BS 5467

Full thermal ratings

Ambient temperature 30 °C Conductor operating temperature 90 °C

Conductor cross-sectional area in sq.mm.	Enclosed in conduit, trunking, underground conduit (with sheath). Not in contact with thermal insulation.							Open, clipped direct, or lying on a non-metallic surface, or enclosed in non-thermal insulating plaster.							
	Two cables single-phase a.c.				Three or four cables three-phase a.c.				Two cables single-phase a.c.				Three or four cables three-phase a.c.			
	Current rating	Voltage drop mV/A/m		Factor length ×	Current rating	Voltage drop mV/A/m		Factor length ×	Current rating	Voltage drop mV/A/m		Factor length ×	Current rating	Voltage drop mV/A/m		Factor length ×
	amperes	r	x	amperes	amperes	r	x	amperes	amperes	r	x	amperes	amperes	r	x	amperes
1	2	3	4	5	6	7	8	9	10	11	12	13	14	15	16	17
1.0	17	46	-	209	15	40	-	415	19	46	-	209	17.5	40.0	-	415
1.5	22	31	-	310	19	27	-	615	25	31	-	310	23	27.0	-	615
2.5	30	19	-	505	26	16	-	1038	34	19	-	505	31	16.0	-	1038
4.0	40	12	-	800	35	10	-	1660	46	12	-	800	41	10.0	-	1660
6.0	51	7.90	-	1215	45	6.80	-	2441	59	7.90	-	1215	54	6.80	-	2441
10	71	4.70	-	2043	63	4.00	-	4150	81	4.70	-	2043	74	4.00	-	4150
16	95	2.90	-	3310	85	2.50	-	6640	109	2.90	-	3310	99	2.50	-	6640
25	126	1.85	0.31	5053	111	1.60	0.27	10061	143	1.85	0.19	5189	130	1.60	0.19	10375
35	156	1.35	0.29	7111	138	1.15	0.25	14435	176	1.35	0.18	7111	161	1.15	0.18	14435
50	189	1.00	0.29	9143	168	0.87	0.25	18444	228	0.99	0.18	9600	209	0.86	0.18	19080

Ambient temperature correction factors for thermosetting insulation

	For rewirable fuses only								For all other protective devices							
Ambient temperature °C	25	30	35	40	45	50	55	60	65	25	30	35	40	45	50	55	60	65
Correction factor	1.02	1.0	0.98	0.95	0.93	0.91	0.89	0.87	0.85	1.02	1.0	0.96	0.91	0.87	0.82	0.76	0.71	0.65

Note: Factor gives cable size for 4% voltage drop. Multiply load current by cable length, then select a factor equal to, or larger than the calculated value of load × length. For loads with a power factor where length is critical, use the values of r and x in the formulas given in the introduction, or in Part 2 Voltage drop. Cables need derating when BS 3036 fuses are used for overload protection.

(4E1)

Itab SINGLE CORE CABLES **CR 3 (a)** **COPPER CONDUCTORS**

Non-armoured single core thermosetting (XLPE) insulated copper cables with or without sheath to BS 7211 or BS 5467

Full thermal ratings   Ambient temperature 30 °C   Conductor operating temperature 90 °C

Conductor cross-sectional area in sq.mm.	Enclosed in conduit, trunking, underground conduit (with sheath). Not in contact with thermal insulation.					Open, clipped direct, or lying on a non-metallic surface, or enclosed in non-thermal insulating plaster.								
	Two cables single-phase a.c.					Three or four cables three-phase a.c.				Two cables single-phase a.c.			Three or four cables three-phase a.c.	
	Current rating amperes	Voltage drop mV/A/m		Factor length × amperes		Current rating amperes	Voltage drop mV/A/m		Factor length × amperes	Current rating amperes	Voltage drop mV/A/m			Factor length × amperes
		r	x				r	x			r	x	Voltage drop mV/A/m r	x	
1	2	3	4	5		6	7	8	9	10	11	12	13	14	15	16	17
70	240	0.700	0.28	12800		214	0.600	0.24	25538	293	0.680	0.175	13521	268	0.590	0.175	26774
95	290	0.510	0.27	16552		259	0.440	0.23	33200	355	0.490	0.170	18462	326	0.430	0.170	36087
120	336	0.410	0.26	20000		299	0.350	0.23	39524	413	0.390	0.165	22326	379	0.340	0.165	43684
150	375	0.330	0.26	22326		328	0.290	0.23	44865	476	0.320	0.165	26667	436	0.280	0.165	51875
185	426	0.270	0.26	25946		370	0.230	0.23	51875	545	0.260	0.165	32000	500	0.220	0.165	59286
240	500	0.210	0.26	29091		433	0.185	0.22	57241	644	0.200	0.160	38400	590	0.170	0.165	69167
300	513	0.175	0.25	30968		493	0.150	0.22	61481	743	0.160	0.160	43636	681	0.135	0.160	79048
400	683	0.140	0.25	33103		584	0.125	0.22	66400	868	0.130	0.155	48000	793	0.110	0.160	85128
500	783	0.120	0.25	34286		666	0.100	0.22	69167	990	0.105	0.155	51892	904	0.088	0.160	92222
630	900	0.100	0.25	35556		764	0.088	0.21	72174	1130	0.086	0.155	54857	1033	0.071	0.160	97647

Ambient temperature correction factors for thermosetting insulation

	For rewirable fuses only						For all other protective devices						
Ambient temperature °C	25	30	35	40	45	50	55	60	65	25	30	35	40	45	50	55	60	65
Correction factor	1.02	1.0	0.98	0.95	0.93	0.91	0.89	0.87	0.85	1.02	1.0	0.96	0.91	0.87	0.82	0.76	0.71	0.65

Note: Factor gives cable size for 4% voltage drop. Multiply load current by cable length, then select a factor equal to, or larger than the calculated value of load × length. For loads with a power factor where length is critical, use the values of r and x in the formulas given in the introduction, or in Part 2 Voltage drop. Cables need derating when BS 3036 fuses are used for overload protection.

(4EI)

I tab SINGLE CORE CABLES — CR 4 — COPPER CONDUCTORS

Non-armoured single core thermosetting (XLPE) insulated copper cables with or without sheath to BS 7211 or BS 5467

Full thermal ratings

Ambient temperature 30 °C — Conductor operating temperature 90 °C

Conductor cross-sectional area in sq.mm.	Enclosed in conduit in thermally insulating wall or ceiling, conduit in contact with thermally conductive surface.							Conductor cross-sectional area in sq.mm.	Sheathed cables on perforated cable tray, bunched and unenclosed. Holes in tray occupy 30% of tray area.						
	Two cables single-phase a.c.				Three or four cables three-phase a.c.				Two cables single-phase a.c.				Three or four cables three-phase a.c.		
	Current rating amperes	Voltage drop mV/A/m		Factor length × amperes	Current rating amperes	Voltage drop mV/A/m		Factor length × amperes		Current rating amperes	Voltage drop mV/A/m		Factor length × amperes	Current rating amperes	Voltage drop mV/A/m		Factor length × amperes
		r	x			r	x				r	x			r	x	
1	2	3	4	5	6	7	8	9	10	11	12	13	14	15	16	17	18
1.0	14	46	-	209	13	40	-	415	25	158	1.85	0.190	5189	140	1.6	0.19	10375
1.5	18	31	-	310	17	27	-	615	35	195	1.35	0.180	7111	176	1.15	0.18	14435
2.5	24	19	-	505	23	16	-	1038	50	293	0.99	0.180	9600	215	0.86	0.18	19080
4.0	33	12	-	800	30	10	-	1660	70	308	0.68	0.175	13521	279	0.59	0.175	26774
6.0	43	7.9	-	1215	39	6.8	-	2441	95	375	0.49	0.170	18462	341	0.43	0.170	36087
10	58	4.7	-	2043	53	4.0	-	4150	120	436	0.39	0.165	22326	398	0.34	0.165	43684
16	76	2.9	-	3310	70	2.5	-	6640	150	505	0.32	0.165	26667	461	0.28	0.165	51875
25	100	1.85	0.31	5053	91	1.6	0.27	10061	185	579	0.26	0.165	32000	530	0.22	0.165	59286
35	124	1.35	0.29	7111	111	1.15	0.25	14435	240	686	0.20	0.160	38400	630	0.17	0.165	69167
50	149	1.00	0.29	9143	135	0.87	0.25	18444	300	794	0.16	0.160	43636	730	0.135	0.160	79048
70	189	0.70	0.28	12800	170	0.60	0.24	25538	400	915	0.13	0.155	48000	849	0.110	0.160	85128

Ambient temperature correction factors for thermosetting insulation

For rewirable fuses only

Ambient temperature °C	25	30	35	40	45	50	55	60	65
Correction factor	1.02	1.0	0.98	0.95	0.93	0.91	0.89	0.87	0.85

For all other protective devices

Ambient temperature °C	25	30	35	40	45	50	55	60	65
Correction factor	1.02	1.0	0.96	0.91	0.87	0.82	0.76	0.71	0.65

Note: Factor gives cable size for 4% voltage drop. Multiply load current by cable length. Select a factor equal to, or larger than the calculated value of load × length. For loads with a power factor where length is critical, use the values of r and x in the formulas given in the introduction, or in Part 2 Voltage drop. Cables need derating when BS 3036 fuses are used for overload protection.

(4E1)

Itab SINGLE CORE CABLES   CR 5   COPPER CONDUCTORS

Non-armoured single core 85 °C rubber insulated copper cables with or without sheath to BS6007 or BS 6883

Full thermal ratings   Ambient temperature 30 °C   Conductor operating temperature 85 °C

Conductor cross-sectional area in sq.mm.	Enclosed in conduit, trunking, underground conduit (with sheath). Not in contact with thermal insulation								Open, clipped direct, or lying on a non-metallic surface, or enclosed in non-thermal insulating plaster. Sheaths touching							
	Current rating amperes	Two cables single-phase a.c.		Factor length × amperes	Three or four cables three-phase a.c.			Factor length × amperes	Current rating amperes	Two cables single-phase a.c.		Factor length × amperes	Three or four cables three-phase a.c.			Factor length × amperes
		Voltage drop mV/A/m			Voltage drop mV/A/m					Voltage drop mV/A/m			Voltage drop mV/A/m			
		r	x		r	x				r	x		r	x		
1	2	3	4	5	6	7	8	9	10	11	12	13	14	15	16	17
1.0	17	46	-	209	15	40	-	415	19	46	-	209	17.5	40	-	415
1.5	22	31	-	310	19.5	26	-	638	25	31	-	310	23	26	-	638
2.5	30	18	-	533	27	16	-	1038	34	18	-	533	31	16	-	1038
4.0	40	12	-	800	36	10	-	1660	45	12	-	800	42	10	-	1660
6.0	52	7.7	-	1247	46	6.7	-	2478	59	7.7	-	1247	54	6.7	-	2478
10	72	4.6	-	2087	63	4.0	-	4150	81	4.6	-	2087	75	4.0	-	4150
16	96	2.9	-	3310	85	2.5	-	6640	108	2.9	-	3310	100	2.5	-	6640
25	127	1.85	0.32	5053	112	1.60	0.28	10061	143	1.85	0.200	5189	133	1.60	0.25	10375
35	157	1.35	0.31	6857	138	1.15	0.27	13833	177	1.30	0.195	7111	164	1.15	0.24	14435
50	190	1.00	0.30	9143	167	0.87	0.26	18242	215	0.97	0.19	9697	199	0.84	0.24	18864

Ambient temperature correction factors for 85 °C rubber insulation

	For rewirable fuses only							For all other protective devices						
Ambient temperature °C	25	30	35	40	45	50	55	60	65	25	30	35	40	45	50	55	60	65
Correction factor	1.02	1.0	0.97	0.95	0.93	0.91	0.88	0.86	0.83	1.02	1.0	0.95	0.90	0.85	0.80	0.74	0.67	0.60

Note: Factor gives cable size for 4% voltage drop. Multiply load current by cable length, then select a factor equal to, or larger than the calculated value of load × length. For loads with a power factor where length is critical, use the values of r and x in the formulas given in the introduction, or in Part 2 Voltage drop. Cables need derating when BS 3036 fuses are used for overload protection.

(4F1)

Itab SINGLE CORE CABLES   CR 6   ALUMINIUM CONDUCTORS

Non-armoured single core p.v.c. insulated aluminium cables with or without sheath to BS6004 or BS 6346

Full thermal ratings   Ambient temperature 30 °C   Conductor operating temperature 70 °C

Conductor cross-sectional area in sq.mm.	Open, clipped direct, or lying on a non-metallic surface, or enclosed in non-thermal insulating plaster. Sheaths touching.							Installed in free air on perforated cable tray, where the perforations occupy 30% of the tray area. Laid flat and touching							
	Two cables single-phase a.c.				Three or four cables three-phase a.c.				Two cables single-phase a.c.				Three or four cables three-phase a.c.			
	Current rating amperes	Voltage drop mV/A/m		Factor length × amperes	Current rating amperes	Voltage drop mV/A/m		Factor length × amperes	Current rating amperes	Voltage drop mV/A/m		Factor length × amperes	Current rating amperes	Voltage drop mV/A/m		Factor length × amperes
		r	x			r	x			r	x			r	x	
1	2	3	4	5	6	7	8	9	10	11	12	13	14	15	16	17
50	134	1.55	0.190	6194	123	1.35	0.24	12296	144	1.55	0.190	6194	132	1.35	0.24	12296
70	172	1.05	0.185	9143	159	0.91	0.24	17660	185	1.05	0.185	9143	169	0.91	0.24	17660
95	210	0.77	0.185	12152	194	0.67	0.23	23380	225	0.77	0.185	12152	206	0.67	0.23	23380
120	245	0.61	0.180	15000	226	0.53	0.23	28621	261	0.61	0.180	15000	240	0.53	0.23	28621
150	283	0.49	0.175	18462	261	0.42	0.23	34583	301	0.49	0.175	18462	277	0.42	0.23	34583
185	324	0.40	0.175	22326	299	0.34	0.23	40488	344	0.40	0.175	22326	317	0.34	0.23	40488
240	384	0.30	0.170	27429	354	0.26	0.22	47429	407	0.30	0.170	27429	375	0.26	0.22	47429
300	444	0.24	0.170	32000	410	0.21	0.22	53548	469	0.24	0.170	32000	433	0.21	0.22	53548
380	511	0.195	0.165	36923	472	0.170	0.22	59286	543	0.195	0.165	36923	502	0.170	0.22	59286
480	591	0.155	0.165	41739	546	0.140	0.22	63846	629	0.155	0.165	41739	582	0.140	0.22	63846
600	679	0.130	0.160	45714	626	0.110	0.22	69167	722	0.130	0.160	45714	669	0.110	0.22	69167
740	771	0.105	0.160	50526	709	0.094	0.21	72174	820	0.105	0.160	50526	761	0.094	0.21	72174
960	900	0.086	0.155	53333	823	0.077	0.21	74107	953	0.086	0.155	53333	886	0.077	0.21	74107
1200	1022	0.074	0.155	56471	926	0.066	0.21	75455	1073	0.074	0.155	56471	999	0.066	0.21	75455

Ambient temperature correction factors for p.v.c insulated cable

	For rewirable fuses only							For all other protective devices					
Ambient temperature °C	25	30	35	40	45	50	55	60	65	25	30	35	40	45	55	60	65
Correction factor	1.03	1.0	0.97	0.94	0.91	0.87	0.84	0.69	0.48	1.03	1.0	0.94	0.87	0.79	0.71	0.61	0.50	0.35

Note: Factor gives cable size for 4% voltage drop. Multiply load current by cable length, then select a factor equal to, or larger than the calculated value of load × length. For loads with a power factor where length is critical, use the values of r and x in the formulas given in the introduction, or in Part 2 Voltage drop. Cables need derating when BS 3036 fuses are used for overload protection.

(4K1)

Itab MULTICORE CABLES — CR 7 — COPPER CONDUCTORS

Non-armoured multicore p.v.c. insulated and sheathed copper cables to BS6004 or BS 6346

Full thermal ratings — Ambient temperature 30 °C — Conductor operating temperature 70 °C

Conductor cross-sectional area in sq.mm.	Enclosed in conduit, trunking, underground conduit (with sheath). Not in contact with thermal insulation						Open, clipped direct, or lying on a non-metallic surface or enclosed in non-thermal insulating plaster						
	One twin cable or one twin and cpc cable single-phase a.c.			One 3 core, or 3 core & cpc cable or 4 core cable three-phase a.c.			One twin cable or one twin and cpc cable single-phase a.c.			One 3 core, or 3 core & cpc cable or 4 core cable three-phase a.c.			
	Current rating amperes	Voltage drop mV/A/m		Factor length × amperes	Voltage drop mV/A/m		Factor length × amperes	Current rating amperes	Voltage drop mV/A/m		Factor length × amperes	Voltage drop mV/A/m		Factor length × amperes
		r	x		r	x			r	x		r	x	
1	2	3	4	5	6	7	8	9	10	11	12	13	14	15	16	17
1.0	13	44	-	218	11.5	38	-	437	15	44	-	218	13.5	38	-	437
1.5	16.5	29	-	331	15	25	-	664	19.5	29	-	331	17.5	25	-	664
2.5	23	18	-	533	20	15	-	1107	27	18	-	533	24	15	-	1107
4.0	30	11	-	873	27	9.5	-	1747	36	11	-	873	32	9.5	-	1747
6.0	38	7.3	-	1315	34	6.4	-	2594	46	7.3	-	1315	41	6.4	-	2594
10	52	4.4	-	2182	46	3.8	-	4368	63	4.4	-	2182	57	3.8	-	4368
16	69	2.8	-	3429	62	2.4	-	6917	85	2.8	-	3429	76	2.4	-	6917
25	90	1.75	0.170	5486	80	1.5	0.145	11067	112	1.75	0.170	5486	96	1.5	0.145	11067
35	111	1.25	0.165	7680	99	1.1	0.145	15091	138	1.25	0.165	7680	119	1.1	0.145	15091

Ambient temperature correction factors for p.v.c insulated cable

	For rewirable fuses only								For all other protective devices							
Ambient temperature °C	25	30	35	40	45	50	55	60	65	25	30	35	40	45	50	55	60	65
Correction factor	1.03	1.0	0.97	0.94	0.91	0.87	0.84	0.69	0.48	1.03	1.0	0.94	0.87	0.79	0.71	0.61	0.50	0.35

Note: Factor gives cable size for 4% voltage drop. Multiply load current by cable length, then select a factor equal to, or larger than the calculated value of load × length. For loads with a power factor where length is critical, use the values of r and x in the formulas given in the introduction, or in Part 2 Voltage drop. Cables need derating when BS 3036 fuses are used for overload protection.

(4D2)

Itab MULTICORE CABLES — CR 7 (a) — COPPER CONDUCTORS

Non-armoured multicore p.v.c insulated and sheathed copper cables to BS6004 or BS 6346

Full thermal ratings — Ambient temperature 30 °C — Conductor operating temperature 70 °C

Conductor cross-sectional area in sq.mm.	Enclosed in conduit, trunking, underground conduit (with sheath). Not in contact with thermal insulation					Open, clipped direct, or lying on a non-metallic surface or enclosed in non-thermal insulating plaster					
	One twin cable or one twin and cpc cable single-phase a.c.					One twin cable or one twin and cpc cable single-phase a.c.					
	Current rating amperes	Voltage drop mV/A/m		Factor length × amperes		Current rating amperes	Voltage drop mV/A/m		Factor length × amperes		
		r	x				r	x			
		One 3 core, or 3 core & cpc cable or 4 core cable three-phase a/c.					One 3 core, or 3 core & cpc cable or 4 core cable three-phase a/c.				
	Current rating amperes	Voltage drop mV/A/m		Factor length × amperes		Current rating amperes	Voltage drop mV/A/m		Factor length × amperes		
		r	x				r	x			
1	2	3	4	5	6	7	8	9	10	11	12	13	14	15	16	17

sq.mm	Col2	Col3	Col4	Col5	Col6	Col7	Col8	Col9	Col10	Col11	Col12	Col13	Col14	Col15	Col16	Col17
50	133	0.93	0.165	10213	118	0.80	0.140	20494	168	0.93	0.165	10213	144	0.80	0.140	20494
70	168	0.63	0.160	14769	149	0.55	0.140	29123	213	0.63	0.160	14769	184	0.55	0.140	29123
95	201	0.47	0.155	19200	179	0.41	0.135	38605	258	0.47	0.155	19200	223	0.41	0.135	38605
120	232	0.38	0.155	23415	206	0.33	0.135	47429	299	0.38	0.155	23415	259	0.33	0.135	47429
150	258	0.30	0.155	28235	225	0.26	0.130	57241	344	0.30	0.155	28235	299	0.26	0.130	57241
185	294	0.25	0.150	33103	255	0.21	0.130	66400	392	0.25	0.150	33103	341	0.21	0.130	66400
240	344	0.190	0.150	40000	297	0.165	0.130	79048	461	0.190	0.150	40000	403	0.165	0.130	79048
300	394	0.155	0.145	45714	339	0.135	0.130	89730	530	0.155	0.145	45714	464	0.135	0.130	89730
400	470	0.115	0.145	51892	402	0.100	0.125	103750	634	0.115	0.145	51892	557	0.100	0.125	103750

Ambient temperature correction factors for p.v.c insulated cable

	For rewirable fuses only								For all other protective devices						
Ambient temperature °C	25	30	35	40	45	50	55	60	65	25	30	35	40	45	50	55	60	65
Correction factor	1.03	1.0	0.97	0.94	0.91	0.87	.84	0.69	0.48	1.03	1.0	0.94	0.87	0.79	0.71	0.61	0.50	0.35

Note: Factor gives cable size for 4% voltage drop. Multiply load current by cable length, then select a factor equal to, or larger than the calculated value of load × length. For loads with a power factor where length is critical, use the values of r and x in the formulas given in the introduction, or in Part 2 Voltage drop. Cables need derating when BS 3036 fuses are used for overload protection.

(4D2)

I_tab MULTICORE CABLES   CR 8   COPPER CONDUCTORS

Non-armoured multicore p.v.c. insulated and sheathed copper cables to BS6004 or BS 6346

Full thermal ratings   Ambient temperature 30 °C   Conductor operating temperature 70 °C

Conductor cross-sectional area in sq.mm.	Installed direct, or in conduit, in a thermally insulated wall or ceiling, in contact with a 10 W/m²K thermally conductive surface on one side only					Installed in free air on perforated cable tray, where the perforations occupy 30% of tray area, or on brackets with a space of 0.3 x cable diameter between wall and cable, or spaced 2 x cable diameter when installed on ladder racking.					
	One twin cable or one twin and cpc cable single-phase a.c.					One 3 core, or 3 core & cpc cable or 4 core cable three-phase a.c.					
	Current rating amperes	Voltage drop mV/A/m		Factor length x amperes		Current rating amperes	Voltage drop mV/A/m		Factor length x amperes		
		r	x				r	x			
1	2	3	4	5		6	7	8	9		
1.0	11	44	-	218		10	38	-	437		
1.5	14	29	-	331		13	25	-	664		
2.5	18.5	18	-	533		17.5	15	-	1107		
4.0	25	11	-	873		23	9.5	-	1747		
6.0	32	7.3	-	1315		29	6.4	-	2594		
10	43	4.4	-	2182		39	3.8	-	4368		
16	57	2.8	-	3429		52	2.4	-	6917		
25	75	1.75	0.170	5486		68	1.5	0.145	11067		
35	92	1.25	0.165	7680		83	1.1	0.145	15091		

Conductor cross-sectional area in sq.mm.	Installed in free air on perforated cable tray, where the perforations occupy 30% of tray area, or on brackets with a space of 0.3 x cable diameter between wall and cable, or spaced 2 x cable diameter when installed on ladder racking.				
	One twin cable or one twin and cpc cable single-phase a.c.				
	Current rating amperes	Voltage drop mV/A/m		Factor length x amperes	
		r	x		
	10	11	12	13	
1.0	17	44	-	218	
1.5	22	29	-	331	
2.5	30	18	-	533	
4.0	40	11	-	873	
6.0	51	7.3	-	1315	
10	70	4.4	-	2182	
16	94	2.8	-	3429	
25	119	1.75	0.170	5486	
35	148	1.25	0.165	7680	

	One 3 core, or 3 core & cpc cable or 4 core cable three-phase a/c.				
	Current rating amperes	Voltage drop mV/A/m		Factor length x amperes	
		r	x		
	14	15	16	17	
1.0	14.5	38	-	437	
1.5	18.5	25	-	664	
2.5	25	15	-	1107	
4.0	34	9.5	-	1747	
6.0	43	6.4	-	2594	
10	60	3.8	-	4368	
16	80	2.4	-	6917	
25	101	1.5	0.145	11067	
35	126	1.1	0.145	15091	

Ambient temperature correction factors for p.v.c insulated cable

	For rewirable fuses only								For all other protective devices							
Ambient temperature °C	25	30	35	40	45	50	55	60	65	25	30	35	40	45	50	55	60	65
Correction factor	1.03	1.0	0.97	0.94	0.91	0.87	0.84	0.69	0.48	1.03	1.0	0.94	0.87	0.79	0.71	0.61	0.50	0.35

Note: Factor gives cable size for 4% voltage drop. Multiply load current by cable length, then select a factor equal to, or larger than the calculated value of load × length. For loads with a power factor where length is critical, use the values of r and x in the formulas given in the introduction, or in Part 2 Voltage drop. Cables need derating when BS 3036 fuses are used for overload protection.

(4D2)

Itab MULTICORE CABLES — CR 8 (a) — COPPER CONDUCTORS

Non-armoured multicore p.v.c. insulated and sheathed copper cables to BS6004 or BS 6346

Full thermal ratings — Ambient temperature 30 °C — Conductor operating temperature 70 °C

Conductor cross-sectional area in sq.mm.	Installed direct, or in conduit, in a thermally insulated wall or ceiling, in contact with a 10 W/m²K thermally conductive surface on one side only					Installed in free air on perforated cable tray, where the perforations occupy 30% of tray area, or on brackets with a space of 0.3 x cable diameter between wall and cable, or spaced 2 x cable diameter when installed on ladder racking.									
	One twin cable or one twin and cpc cable single-phase a.c.					One 3 core, or 3 core & cpc cable or 4 core cable three-phase a.c.					One twin cable or one twin and cpc cable single-phase a.c.			One 3 core, or 3 core & cpc cable or 4 core cable three-phase a/c.	
	Current rating amperes	Voltage drop mV/A/m		Factor length × amperes		Current rating amperes	Voltage drop mV/A/m		Factor length × amperes		Current rating amperes	Voltage drop mV/A/m	Factor length × amperes	Voltage drop mV/A/m	Factor length × amperes
		r	x				r	x							
1	2	3	4	5		6	7	8	9		10	11 12	13	14 15 16	17
50	110	0.93	0.165	10213		99	0.80	0.140	20494		180	0.93 0.165	10213	153 0.80 0.140	20494
70	139	0.63	0.160	14769		125	0.55	0.140	29123		232	0.63 0.160	14769	196 0.55 0.140	29123
95	167	0.47	0.155	19200		150	0.41	0.135	38605		282	0.47 0.155	19200	238 0.41 0.135	38605
120	192	0.38	0.155	23415		172	0.33	0.135	47429		328	0.38 0.155	23415	276 0.33 0.135	47429
150	219	0.30	0.155	28235		196	0.26	0.130	57241		379	0.30 0.155	28235	319 0.26 0.130	57241
185	248	0.25	0.150	33103		223	0.21	0.130	66400		434	0.25 0.150	33103	364 0.21 0.130	66400
240	291	0.190	0.150	40000		261	0.165	0.130	79048		514	0.190 0.150	40000	430 0.165 0.130	79048
300	334	0.155	0.145	45714		298	0.135	0.130	89730		593	0.155 0.145	45714	497 0.135 0.130	89730
400	-	-	-	-		-	-	-	-		715	0.115 0.145	51892	597 0.100 0.125	103750

Ambient temperature correction factors for p.v.c insulated cable

	For rewirable fuses only								For all other protective devices						
Ambient temperature °C	25	30	35	40	45	50	55	60	65	25	30	35	40	45	50	55	60	65
Correction factor	1.03	1.0	0.97	0.94	0.91	0.87	0.84	0.69	0.48	1.03	1.0	0.94	0.87	0.79	0.71	0.61	0.50	0.35

Note: Factor gives cable size for 4% voltage drop. Multiply load current by cable length, then select a factor equal to, or larger than the calculated value of load × length. For loads with a power factor where length is critical, use the values of r and x in the formulas given in the introduction, or in Part 2 Voltage drop. Cables need derating when BS 3036 fuses are used for overload protection.

(4D2)

I_tab ARMOURED CABLES

CR 9 COPPER CONDUCTORS

Armoured multicore p.v.c. insulated copper cables to BS 6346

Full thermal ratings

Ambient temperature 30 °C Conductor operating temperature 70 °C

Open, clipped direct, or lying on a non-metallic surface, or enclosed in non-thermal insulating plaster.

Installed in free air on perforated cable tray, where the perforations occupy 30% of tray area, or on brackets with a space of 0.3 x cable diameter between wall and cable, or spaced 2 x cable diameter when installed on ladder racking.

Conductor cross-sectional area in sq.mm.	One two core cables single-phase a.c.				One three or four core cable three-phase a.c.				One two core cables single-phase a.c.				One three or four core cable three-phase a.c.			
	Current rating amperes	Voltage drop mV/A/m		Factor length x amperes	Current rating amperes	Voltage drop mV/A/m		Factor length x amperes	Current rating amperes	Voltage drop mV/A/m		Factor length x amperes	Current rating amperes	Voltage drop mV/A/m		Factor length x amperes
		r	x			r	x			r	x			r	x	
1	2	3	4	5	6	7	8	9	10	11	12	13	14	15	16	17
1.5	21	29	-	331	18	25	-	664	22	29	-	331	19	25	-	664
2.5	28	18	-	533	25	15	-	1107	31	18	-	533	26	15	-	1107
4.0	38	11	-	873	33	9.5	-	1747	41	11	-	873	35	9.5	-	1747
6.0	49	7.3	-	1315	42	6.4	-	2594	53	7.3	-	1315	45	6.4	-	2594
10	67	4.4	-	2182	58	3.8	-	4368	72	4.4	-	2182	62	3.8	-	4368
16	89	2.8	-	3429	77	2.4	-	6917	97	2.8	-	3429	83	2.4	-	6917
25	118	1.75	0.170	5486	102	1.5	0.145	11067	128	1.75	0.170	5486	110	1.5	0.145	11067
35	145	1.25	0.165	7680	125	1.1	0.145	15091	157	1.25	0.165	7680	135	1.1	0.145	15091
50	175	0.93	0.165	10213	151	0.8	0.140	20494	190	0.93	0.165	10213	163	0.8	0.140	20494

Ambient temperature correction factors for p.v.c insulated cable

	For rewirable fuses only							For all other protective devices						
Ambient temperature °C	25	30	35	40	45	50	55	60	65	25	30	35	40	45	50	55	60	65
Correction factor	1.03	1.0	0.97	0.94	0.91	0.87	0.84	0.69	0.48	1.03	1.0	0.94	0.87	0.79	0.71	0.61	0.50	0.35

Note: Factor gives cable size for 4% voltage drop. Multiply load current by cable length, then select a factor equal to, or larger than the calculated value of load x length. For loads with a power factor where length is critical, use the values of r and x in the formulas given in the introduction, or in Part 2 Voltage drop. Cables need derating when BS 3036 fuses are used for overload protection.

(4D4)

314

Itab ARMOURED CABLES

CR 9 (a)

COPPER CONDUCTORS

Armoured multicore p.v.c. insulated copper cables to BS 6346

Full thermal ratings

Ambient temperature 30 °C — Conductor operating temperature 70 °C

Open, clipped direct, or lying on a non-metallic surface, or enclosed in non-thermal insulating plaster.

Installed in free air on perforated cable tray, where the perforations occupy 30% of tray area, or on brackets with a space of 0.3 x cable diameter between wall and cable, or spaced 2 x cable diameter when installed on ladder racking.

Conductor cross-sectional area in sq.mm.	One two core cables single-phase a.c.			One three or four core cable three-phase a.c.				One two core cables single-phase a.c.				One three or four core cables three-phase a.c.			
	Current rating amperes	Voltage drop mV/A/m		Factor length x amperes	Current rating amperes	Voltage drop mV/A/m		Factor length x amperes	Current rating amperes	Voltage drop mV/A/m		Factor length x amperes	Current rating amperes	Voltage drop mV/A/m		Factor length x amperes
		r	x			r	x			r	x			r	x	
1	2	3	4	5	6	7	8	9	10	11	12	13	14	15	16	17
70	222	0.63	0.160	14769	192	0.550	0.140	29123	241	0.63	0.160	14769	207	0.550	0.140	29123
95	269	0.47	0.155	19200	231	0.410	0.135	38605	291	0.47	0.155	19200	251	0.410	0.135	38605
120	310	0.38	0.155	23415	267	0.330	0.135	47429	336	0.38	0.155	23415	290	0.330	0.135	47429
150	356	0.30	0.155	28235	306	0.260	0.130	57241	386	0.30	0.155	28235	332	0.260	0.130	57241
185	405	0.25	0.150	33103	348	0.210	0.130	66400	439	0.25	0.150	33103	378	0.210	0.130	66400
240	476	0.19	0.150	40000	409	0.165	0.130	79048	516	0.19	0.150	40000	445	0.165	0.130	79048
300	547	0.155	0.145	45714	469	0.135	0.130	89730	592	0.155	0.145	45714	510	0.135	0.130	89730
400	621	0.115	0.145	51892	540	0.100	0.125	103750	683	0.115	0.145	51892	590	0.100	0.125	103750

Ambient temperature correction factors for p.v.c insulated cable

	For rewirable fuses only								For all other protective devices							
Ambient temperature °C	25	30	35	40	45	50	55	60	65	25	30	35	40	45	50	55	60	65
Correction factor	1.03	1.0	0.97	0.94	0.91	0.87	0.84	0.69	0.48	1.03	1.0	0.94	0.87	0.79	0.71	0.61	0.50	0.35

Note: Factor gives cable size for 4% voltage drop. Multiply load current by cable length, then select a factor equal to, or larger than the calculated value of load x length. For loads with a power factor where length is critical, use the values of r and x in the formulas given in the introduction, or in Part 2 Voltage drop. Cables need derating when BS 3036 fuses are used for overload protection.

(4D4)

Itab ARMOURED CABLES — CR 10 — COPPER CONDUCTORS

Armoured multicore thermosetting (XLPE) insulated copper cables to BS 5467 or BS 6724

Ambient temperature 30 °C   Conductor operating temperature 90 °C

Full thermal ratings

Open, clipped direct, or lying on a non-metallic surface, or enclosed in non-thermal insulating plaster.

Installed in free air on perforated cable tray, where the perforations occupy 30% of tray area, or on brackets with a space of 0.3 x cable diameter between wall and cable, or spaced 2 x cable diameter when installed on ladder racking.

Conductor cross-sectional area in sq.mm.	One two core cables single-phase a.c.			Factor length × amperes	One three or four core cable three-phase a.c.			Factor length × amperes	One two core cables single-phase a.c.			Factor length × amperes	One three or four core cable three-phase a.c.			Factor length × amperes
	Current rating amperes	Voltage drop mV/A/m			Current rating amperes	Voltage drop mV/A/m			Current rating amperes	Voltage drop mV/A/m			Current rating amperes	Voltage drop mV/A/m		
		r	x			r	x			r	x			r	x	
1	2	3	4	5	6	7	8	9	10	11	12	13	14	15	16	17
1.5	27	31	-	310	23	27	-	615	29	31	-	310	25	27	-	615
2.5	36	19	-	505	31	16	-	1038	39	19	-	505	33	16	-	1038
4	49	12	-	800	42	10	-	1660	52	12	-	800	44	10	-	1660
6	62	7.9	-	1215	53	6.8	-	2441	66	7.9	-	1215	56	6.8	-	2441
10	85	4.7	-	2043	73	4.0	-	4150	90	4.7	-	2043	78	4.0	-	4150
16	110	2.9	-	3310	94	2.5	-	6640	115	2.9	-	3310	99	2.5	-	6640
25	146	1.85	0.160	5053	124	1.60	0.140	10061	152	1.85	0.160	5053	131	1.60	0.140	10061
35	180	1.35	0.155	7111	154	1.15	0.135	14435	188	1.35	0.155	7111	162	1.15	0.135	14435
50	219	0.99	0.155	9600	187	0.86	0.135	19080	228	0.99	0.155	9600	197	0.86	0.135	19080

Ambient temperature correction factors for thermosetting insulation

For rewirable fuses only

Ambient temperature °C	25	30	35	40	45	50	55	60	65
Correction factor	1.02	1.0	0.98	0.95	0.93	0.91	0.89	0.87	0.85

For all other protective devices

Ambient temperature °C	25	30	35	40	45	50	55	60	65
Correction factor	1.02	1.0	0.96	0.91	0.87	0.82	0.76	0.71	0.65

Note: Factor gives cable size for 4% voltage drop. Multiply load current by cable length, then select a factor equal to, or larger than the calculated value of load × length. Factor gives cable size for 4% voltage drop. For loads with a power factor where length is critical, use the values of r and x in the formulas given in the introduction, or in Part 2 Voltage drop. Cables need derating when BS 3036 fuses are used for overload protection.

(4E4)

Itab ARMOURED CABLES   **CR 10 (a)**   **COPPER CONDUCTORS**

Armoured multicore thermosetting (XLPE) insulated copper cables to BS 5467 or BS 6724

Full thermal ratings   Ambient temperature 30 °C   Conductor operating temperature 90 °C

	Open, clipped direct, or lying on a non-metallic surface, or enclosed in non-thermal insulating plaster.							Installed in free air on perforated cable tray, where the perforations occupy 30% of tray area, or on brackets with a space of 0.3 x cable diameter between wall and cable, or spaced 2 x cable diameter when installed on ladder racking.							
	One two core cables single-phase a.c.				One three or four core cable three-phase a.c.				One two core cables single-phase a.c.				One three or four core cable three-phase a.c.			
Conductor cross-sectional area in sq.mm.	Current rating amperes	Voltage drop mV/A/m		Factor length × amperes	Current rating amperes	Voltage drop mV/A/m		Factor length × amperes	Current rating amperes	Voltage drop mV/A/m		Factor length × amperes	Current rating amperes	Voltage drop mV/A/m		Factor length × amperes
		r	x			r	x			r	x			r	x	
1	2	3	4	5	6	7	8	9	10	11	12	13	14	15	16	17
70	279	0.67	0.150	13913	238	0.59	0.130	27667	291	0.67	0.150	13913	251	0.59	0.130	27667
95	338	0.50	0.150	18462	289	0.43	0.130	36889	354	0.50	0.150	18462	304	0.43	0.130	36889
120	392	0.40	0.145	22857	335	0.34	0.130	44865	410	0.40	0.145	22857	353	0.34	0.130	44865
150	451	0.32	0.145	27429	386	0.28	0.125	55333	472	0.32	0.145	27429	406	0.28	0.125	55333
185	515	0.26	0.145	33103	441	0.22	0.125	63846	539	0.26	0.145	33103	463	0.22	0.125	63846
240	607	0.20	0.140	40000	520	0.175	0.125	79048	636	0.20	0.140	40000	546	0.175	0.125	79048
300	698	0.16	0.140	45714	599	0.140	0.120	89730	732	0.16	0.140	45714	628	0.140	0.120	89730
400	787	0.13	0.145	49231	673	0.115	0.125	97647	847	0.13	0.145	49231	728	0.115	0.125	97647

Ambient temperature correction factors for thermosetting insulation

	For rewirable fuses only								For all other protective devices							
Ambient temperature °C	25	30	35	40	45	50	55	60	65	25	30	35	40	45	50	55	60	65
Correction factor	1.02	1.0	0.98	0.95	0.93	0.91	0.89	0.87	0.85	1.02	1.0	0.96	0.91	0.87	0.82	0.76	0.71	0.65

Note: Factor gives cable size for 4% voltage drop. Multiply load current by cable length, then select a factor equal to, or larger than the calculated value of load × length. For loads with a power factor where length is critical, use the values of r and x in the formulas given in the introduction, or in Part 2 Voltage drop. Cables need derating when BS 3036 fuses are used for overload protection.

(4E4)

I$_{tab}$ ARMOURED CABLES CR 11 ALUMINIUM CONDUCTORS

Armoured multicore p.v.c. insulated aluminium cables to BS 6346

Full thermal ratings Ambient temperature 30 °C Conductor operating temperature 70 °C

	Open, clipped direct, or lying on a non-metallic surface, or enclosed in non-thermal insulating plaster.									Installed in free air on perforated cable tray, where the perforations occupy 30% of tray area, or on brackets with a space of 0.3 x cable diameter between wall and cable, or spaced 2 x cable diameter when installed on ladder racking.								
Conductor cross-sectional area in sq.mm.	One two core cables single-phase a.c.				One three or four core cable three-phase a.c.					One two core cables single-phase a.c.				One three or four core cable three-phase a.c.			
	Current rating amperes	Voltage drop mV/A/m		Factor length × amperes	Current rating amperes	Voltage drop mV/A/m		Factor length × amperes		Current rating amperes	Voltage drop mV/A/m		Factor length × amperes	Current rating amperes	Voltage drop mV/A/m		Factor length × amperes
		r	x			r	x				r	x			r	x	
1	2	3	4	5	6	7	8	9		10	11	12	13	14	15	16	17
16	68	4.50	-	2133	58	3.90	-	4256		71	4.50	-	2133	61	3.90	-	4256
25	89	2.90	0.175	3310	76	2.50	0.150	6640		94	2.90	0.175	3310	80	2.50	0.150	6640
35	109	2.10	0.170	4571	94	1.80	0.150	9222		115	2.10	0.170	4571	99	1.80	0.150	9222
50	131	1.55	0.170	6194	113	1.35	0.145	12296		139	1.55	0.170	6194	119	1.35	0.145	12296
70	165	1.05	0.165	9143	143	0.90	0.140	18043		175	1.05	0.165	9143	151	0.90	0.140	18043
95	199	0.77	0.160	12152	174	0.67	0.140	24412		211	0.77	0.160	12152	186	0.67	0.140	24412
120	-	-	-	-	202	0.53	0.135	30182		-	-	-	-	216	0.53	0.135	30182
150	-	-	-	-	232	0.42	0.135	37727		-	-	-	-	250	0.42	0.135	37727
185	-	-	-	-	265	0.34	0.135	44865		-	-	-	-	287	0.34	0.135	44865
240	-	-	-	-	312	0.26	0.130	55333		-	-	-	-	342	0.26	0.130	55333
300	-	-	-	-	360	0.21	0.130	66400		-	-	-	-	399	0.21	0.130	66400

Ambient temperature correction factors for p.v.c insulated cable

	For rewirable fuses only								For all other protective devices							
Ambient temperature °C	25	30	35	40	45	50	55	60	65	25	30	35	40	45	50	55	60	65
Correction factor	1.03	1.0	0.97	0.94	0.91	0.87	0.84	0.69	0.48	1.03	1.0	0.94	0.87	0.79	0.71	0.61	0.50	0.35

Note: Factor gives cable size for 4% voltage drop. Multiply load current by cable length, then select a factor equal to, or larger than the calculated value of load × length. For loads with a power factor where length is critical, use the values of r and x in the formulas given in the introduction, or in Part 2 Voltage drop. Cables need derating when BS 3036 fuses are used for overload protection.

(4K4)

I_tab ARMOURED CABLES — CR 12 — ALUMINIUM CONDUCTORS

Armoured multicore thermosetting (XLPE), insulated aluminium cables to BS 5467

Full thermal ratings

Ambient temperature 30 °C
Conductor operating temperature 90 °C

Conductor cross-sectional area in sq.mm.	Open, clipped direct, or lying on a non-metallic surface, or enclosed in non-thermal insulating plaster.								Installed in free air on perforated cable tray, where the perforations occupy 30% of tray area, or on brackets with a space of 0.3 × cable diameter between wall and cable, or spaced 2 × cable diameter when installed on ladder racking.							
	One two core cables single-phase a.c.				One three or four core cable three-phase a.c.				One two core cables single-phase a.c.				One three or four core cable three-phase a.c.			
	Current rating amperes	Voltage drop mV/A/m		Factor length × amperes	Current rating amperes	Voltage drop mV/A/m		Factor length × amperes	Current rating amperes	Voltage drop mV/A/m		Factor length × amperes	Current rating amperes	Voltage drop mV/A/m		Factor length × amperes
		r	x			r	x			r	x			r	x	
1	2	3	4	5	6	7	8	9	10	11	12	13	14	15	16	17
16	82	4.80	-	2000	71	4.20	-	3952	85	4.80	-	2000	74	4.20	-	3952
25	108	3.10	0.165	3097	92	2.70	0.140	6148	112	3.10	0.165	3097	98	2.70	0.140	6148
35	132	2.20	0.160	4364	113	1.90	0.140	8513	138	2.20	0.160	4364	120	1.90	0.140	8513
50	159	1.65	0.160	5818	137	1.40	0.135	11448	166	1.65	0.160	5818	145	1.40	0.135	11448
70	201	1.10	0.155	8348	174	0.96	0.135	17113	211	1.10	0.155	8348	185	0.96	0.135	17113
95	242	0.82	0.150	11429	214	0.71	0.130	23056	254	0.82	0.150	11429	224	0.71	0.130	23056
120	-	-	-	-	249	0.56	0.130	28621	-	-	-	-	264	0.56	0.130	28621
150	-	-	-	-	284	0.45	0.130	35319	-	-	-	-	305	0.45	0.130	35319
185	-	-	-	-	328	0.37	0.130	42564	-	-	-	-	350	0.37	0.130	42564
240	-	-	-	-	386	0.28	0.125	53548	-	-	-	-	418	0.28	0.125	53548
300	-	-	-	-	441	0.23	0.125	63846	-	-	-	-	488	0.23	0.125	63846

Ambient temperature correction factors for thermosetting insulation

Ambient temperature °C	For rewirable fuses only						For all other protective devices						
	25	30	35	40	45	50	55	60	65	25	30	35	40	45	50	55	60	65
Correction factor	1.02	1.0	0.98	0.95	0.93	0.91	0.89	0.87	0.85	1.02	1.0	0.96	0.91	0.87	0.82	0.76	0.71	0.65

Note: Factor gives cable size for 4% voltage drop. Multiply load current by cable length, then select a factor equal to, or larger than the calculated value of load × length. For loads with a power factor where length is critical, use the values of r and x in the formulas given in the introduction, or in Part 2 Voltage drop. Cables need derating when BS 3036 fuses are used for overload protection.

(4L.4)

Itab MICC PVC CABLES — CR 13 — COPPER SHEATH

Mineral insulated copper cables with p.v.c. sheath to BS 6207
Open and clipped direct or lying on a non-metallic surface

Full thermal ratings

Ambient temperature 30 °C
Sheath operating temperature 70 °C

Conductor cross sectional area in sq. mm.	Two single core, or one 2 core cable single-phase a.c.		Three single core in trefoil or 1-three core three-phase a.c.		Three single core in flat formation three-phase a.c.		One four core all cores loaded three-phase a.c.		One seven core all cores loaded single-phase a.c.		One twelve core all cores loaded single-phase a.c.	
	Current rating amps	Factor length x amps	Current rating amps	Factor length x amps	Current rating amps	Factor length x amps	Current rating amps	Factor length x amps	Current rating amps	Factor length x amps	Current rating amps	Factor length x amps
1	2	3	4	5	6	7	8	9	10	11	12	13	14	15
Light duty 500V												
1.0	18.5	229	15	461	17	461	15	461	10	229	9.5	229
1.5	23	343	19	692	21	692	19.5	692	13	343	12	343
2.5	31	565	26	1186	29	1186	26	1186	17.5	565	16	565
4.0	40	960	35	1824	38	1824	-	-	-	-	-	-
Heavy duty 750V												
1.0	19.5	229	16	461	18	461	14.5	461	11.5	229	-	-
1.5	25	343	21	692	23	692	18	692	14.5	343	-	-
2.5	34	565	28	1186	31	1186	25	1186	19.5	565	-	-
4.0	45	960	37	1824	41	1824	32	1824	26	960	-	-
6.0	57	1371	48	2767	52	2767	41	2767	-	-	-	-
10	77	2286	65	4611	70	4611	55	4611	-	-	-	-
16	102	3692	86	7217	92	7217	72	7217	-	-	-	-

Ambient temperature correction factors for m.i.c.c with 70 °C sheath

For rewirable fuses only

Ambient temperature °C	25	30	35	40	45	50	55	60	65
Correction factor	1.03	1.0	0.96	0.93	0.89	0.86	0.79	0.62	0.42

For all other protective devices

Ambient temperature °C	25	30	35	40	45	50	55	60
Correction factor	1.03	1.0	0.93	0.85	0.77	0.67	0.57	0.45

Note: Factor gives cable size for 4% voltage drop. Multiply load current by cable length, then select a factor equal to, or larger than the calculated value of load x length. Single core cable ratings only apply where cables are bonded at both ends. For unsheathed cables exposed to touch, see table CR 14. Cables need derating when BS 3036 fuses are used for overload protection

(4J1).

Itab MICC CABLES — CR 14 — COPPER SHEATH

Mineral insulated copper unsheathed cables exposed to touch to BS 6207
Open and clipped direct or lying on a non-metallic surface

Full thermal ratings

Ambient temperature 30 °C / Sheath operating temperature 70 °C

Conductor cross sectional area in sq. mm.	Two single core, or one 2 core cable single-phase a.c.		Three single core in trefoil or 1-three core three-phase a.c.		Three single core in flat formation three-phase a.c.		One four core with three cores loaded three-phase a.c.		One four core all cores loaded three-phase a.c.		One seven core all cores loaded single-phase a.c.		One twelve core all cores loaded single-phase a.c.	
	Current rating amps	Factor length x amps	Current rating amps	Factor length x amps	Current rating amps	Factor length x amps	Current rating amps	Factor length x amps	Current rating amps	Factor length x amps	Current rating amps	Factor length x amps	Current rating amps	Factor length x amps
1	2	3	4	5	6	7	8	9	10	11	12	13	14	15
Light duty 500V														
1.0	16.65	229	13.5	461	15.3	461	13.5	461	11.7	461	9	229	-	-
1.5	20.70	343	17.1	692	18.9	692	17.6	692	14.9	692	11.7	343	-	-
2.5	27.90	565	23.4	1186	26.1	1186	23.4	1186	19.8	1186	15.8	565	-	-
4.0	36.00	960	31.5	1824	34.2	1824	-	-	-	-	-	-	-	-
Heavy duty 750V														
1.0	17.55	229	14.4	461	16.2	461	14.9	461	13.1	461	10.4	229	8.6	229
1.5	22.50	343	18.9	692	20.7	692	18.9	692	16.2	692	13.1	343	10.8	343
2.5	30.60	565	25.2	1186	27.9	1186	25.2	1186	22.5	1186	17.6	565	14.4	565
4.0	40.50	960	33.3	1824	36.9	1824	33.3	1824	28.8	1824	23.4	960	-	-
6.0	51.30	1371	43.2	2767	46.8	2767	42.3	2767	36.9	2767	-	-	-	-
10	69.30	2286	58.5	4611	63	4611	57.6	4611	49.5	4611	-	-	-	-
16	91.80	3692	77.4	7217	82.8	7217	76.5	7217	64.8	7217	-	-	-	-

Ambient temperature correction factors for m.i.c.c with 70 °C sheath

	For rewirable fuses only						For all other protective devices						
Ambient temperature °C	25	30	35	40	45	50	55	25	30	35	40	45	50	55	60
Correction factor	1.03	1.0	0.96	0.93	0.89	0.86	0.79	1.03	1.0	0.93	0.85	0.77	0.67	0.57	0.45

(The temperature row also includes 60 and 65 for rewirable fuses: 0.62, 0.42)

Note: Factor gives cable size for 4% voltage drop. Multiply load current by cable length, then select a factor equal to, or larger than the calculated value of load × length. Single core cable ratings only apply where cables are bonded at both ends. Cables need derating when BS 3036 fuses are used for overload protection.

(4JI)

Itab MICC CABLES

CR 15 COPPER SHEATH

Mineral insulated copper cables pvc sheathed to BS 6207. Installed on perforated cable tray, horizontal or vertical. (See notes for unsheathed cables).

Full thermal ratings

Ambient temperature 30 °C / Sheath operating temperature 70 °C

Conductor cross sectional area in sq. mm.	One two core cable single-phase a.c. or dc.		One three core cable three-phase a.c.		Three single core in flat formation Three-phase a.c.		One four core with three cores loaded three-phase a.c.		One four core all cores loaded three-phase a.c.		One seven core all cores loaded single-phase a.c.		One twelve core all cores loaded single-phase a.c.	
	Current rating amps	Factor length x amps	Current rating amps	Factor length x amps	Current rating amps	Factor length x amps	Current rating amps	Factor length x amps	Current rating amps	Factor length x amps	Current rating amps	Factor length x amps	Current rating amps	Factor length x amps
1	2	3	4	5	6	7	8	9	10	11	12	13	14	15
Light duty 500V														
1.0	19.5	229	16.5	461	17	461	16	461	14	461	11	229	10	229
1.5	25	343	21	692	22	692	21	692	18	692	14	343	13	343
2.5	33	565	28	1186	29	1186	28	1186	24	1186	19	565	17	565
4.0	44	960	37	1824	39	1824	-	-	-	-	-	-	-	-
Heavy duty 750V														
1.0	21	229	17.5	461	19	461	18	461	16	461	12	229	10	229
1.5	26	343	22	692	25	692	23	692	20	692	15.5	343	13	343
2.5	36	565	30	1186	32	1186	30	1186	27	1186	21	565	17	565
4.0	47	960	40	1824	43	1824	40	1824	35	1824	28	960	-	-
6.0	60	1371	51	2767	54	2767	51	2767	44	2767	-	-	-	-
10	82	2286	69	4611	73	4611	68	4611	59	4611	-	-	-	-
16	109	3692	92	7217	97	7217	89	7217	78	7217	-	-	-	-

Ambient temperature correction factors for m.i.c.c with 70 °C sheath

For rewirable fuses only

Ambient temperature °C	25	30	35	40	45	50	55	60	65
Correction factor	1.03	1.0	0.96	0.93	0.89	0.86	0.79	0.62	0.42

For all other protective devices

Ambient temperature °C	25	30	35	40	45	50	55	60
Correction factor	1.03	1.0	0.93	0.85	0.77	0.67	0.57	0.45

Note: Factor gives cable size for 4% voltage drop. Multiply load current by cable length, then select a factor equal to, or larger than the calculated value of load × length. Single core cable ratings only apply where cables are bonded at both ends. Cable ratings to be derated by 0.9 for unsheathed cables exposed to touch. Cables need derating when BS 3036 fuses are used for overload protection.

(4J1).

Itab MICC CABLES   **CR 16**   **COPPER SHEATH**

Mineral insulated bare copper cables to BS 6207, not exposed to touch, or in contact with combustible materials. Open and clipped direct or lying on a non-metallic surface

Full thermal ratings   Ambient temperature 30 °C   Sheath operating temperature 105 °C

Conductor cross sectional area in sq. mm.	Two single core, or one 2 core cable single-phase a.c.		Three single core in trefoil or 1-three core three-phase a.c.		Three single core in flat formation three-phase a.c.		One four core with three cores loaded three-phase a.c.		One four core all cores loaded three-phase a.c.		One seven core all cores loaded single-phase a.c.		One twelve core all cores loaded single-phase a.c.	
	Current rating amps	Factor length × amps	Current rating amps	Factor length × amps	Current rating amps	Factor length × amps	Current rating amps	Factor length × amps	Current rating amps	Factor length × amps	Current rating amps	Factor length × amps	Current rating amps	Factor length × amps
1	2	3	4	5	6	7	8	9	10	11	12	13	14	15
Light duty 500V														
1.0	22	204	19	415	21	415	18.5	415	16.5	415	13	204		
1.5	28	310	24	615	27	615	24	615	21	615	16.5	310		
2.5	38	505	33	1038	36	1038	33	1038	28	1038	22	505		
4.0	51	800	44	1660	47	1660	-	-	-	-	-	-		
Heavy duty 750V														
1.0	24	204	20	415	24	415	20	415	17.5	415	14	204	12	204
1.5	31	310	26	615	30	615	26	615	22	615	17.5	310	15.5	310
2.5	42	505	35	1038	41	1038	35	1038	30	1038	24	505	20	505
4.0	55	800	47	1660	53	1660	46	1660	40	1660	32	800		
6.0	70	1231	59	2441	67	2441	58	2441	50	2441				
10	96	2043	81	4049	91	4049	78	4049	68	4049				
16	127	3200	107	6385	119	6385	103	6385	90	6385				

Ambient temperature correction factors for 105 °C sheath

For rewirable fuses only

Ambient temperature °C	25	30	35	40	45	50	55	60	65
Correction factor	1.02	1.0	0.98	0.96	0.93	0.91	0.89	0.86	0.84

NO CORRECTION FOR GROUPING REQUIRED

For all other protective devices

Ambient temperature °C	25	30	35	40	45	50	55	60	65
Correction factor	1.02	1.0	0.96	0.92	0.88	0.84	0.80	0.75	0.70

Note: Factor gives cable size for 4% voltage drop. Multiply load current by cable length, then select a factor equal to, or larger than the calculated value of load × length. Single core cable ratings only apply where cables are bonded at both ends. Cables need derating when BS 3036 fuses are used for overload protection. (4J2).

I_tab FLEXIBLE CORDS CR 17 COPPER CONDUCTORS

Flexible cords to BS 6500

Conductor cross-sectional area in sq. mm.	Current-carrying capacity		Maximum length in metres for 4% volt drop at rated current		Factor length × amperes		Maximum weight allowed (in kg) to be supported by a twin flexible cord
	Single-phase a.c.	Three-phase a.c.	Single-phase	Three-phase	Single-phase	Three-phase	
1	2	3	4	5	6	7	8
0.50	3	3	34.41	69.17	103	208	2
0.75	6	6	25.81	51.23	155	307	3
1.00	10	10	20.87	41.50	209	415	5
1.25	13	-	19.96	-	259	-	5
1.50	16	16	18.75	38.43	300	615	5
2.50	25	20	20.21	51.88	505	1038	5
4.00	32	25	25.00	66.40	800	1660	5

Ambient temperature correction factors

Ambient temperature °C	35	40	45	50	55	60	65	70	120	125	130	135	140	145	150	155	160	165	170
60 °C rubber and pvc cords	0.91	0.82	0.71	0.58	0.41														
85 °C rubber cords having HOFR or heat resisting pvc sheath	1.0	1.0	1.0	1.0	0.96	0.83	0.67	0.47											
150 °C rubber cords	1.0	1.0	1.0	1.0	1.0	1.0	1.0	1.0	1.0	0.96	0.85	0.74	0.60	0.42					
Glass fibre cords	1.0	1.0	1.0	1.0	1.0	1.0	1.0	1.0	1.0	1.0	1.0	1.0	1.0	1.0	1.0	0.92	0.82	0.71	0.57

Notes: To determine the voltage drop in mV/A/m for single phase, divide 9,600 by the factor for the cable size. For the mV/A/m for three phase, divide 16,600 by the factor for the cable size.

(4H3)

SHOCK PROTECTION

BS 88 HRC FUSES

ZS 1

Maximum design & testing values of earth loop impedance Z_S for BS 88 Parts 2 & 6 fuses when U_0 is 240V

DISCONNECTION TIME 0.4 SECONDS

Maximum values of Z_S in ohms for different HRC fuse sizes

Temperature degrees C	Fuse size in amperes for 0.4 second disconnection time															
	2	4	6	10	16	20	25	32	40	50	63	80	100	125	160	200	250	315	400	500

Maximum value of Z_S in ohms

| Design | 36 | 17 | 8.89 | 5.33 | 2.82 | 1.85 | 1.5 | 1.09 | 0.86 | 0.63 | 0.47 | 0.32 | 0.24 | 0.18 | 0.14 | 0.11 | 0.086 | 0.063 | 0.048 | 0.035 |

Testing — For testing the following maximum values of Z_S are based on a design temperature of 115 °C

Temp °C	2	4	6	10	16	20	25	32	40	50	63	80	100	125	160	200	250	315	400	500
30	26.87	12.69	6.63	3.98	2.10	1.38	1.12	0.81	0.64	0.47	0.358	0.239	0.179	0.134	0.104	0.082	0.064	0.047	0.036	0.026
25	26.47	12.50	6.54	3.92	2.07	1.36	1.10	0.80	0.63	0.46	0.353	0.235	0.176	0.132	0.103	0.081	0.063	0.046	0.035	0.026
20	26.09	12.32	6.44	3.86	2.04	1.34	1.09	0.79	0.62	0.46	0.348	0.232	0.174	0.130	0.101	0.080	0.062	0.046	0.035	0.025
15	25.71	12.14	6.35	3.81	2.01	1.32	1.07	0.78	0.61	0.45	0.343	0.229	0.171	0.129	0.100	0.079	0.061	0.045	0.034	0.025
10	25.35	11.97	6.26	3.75	1.99	1.30	1.06	0.77	0.61	0.44	0.338	0.225	0.169	0.127	0.099	0.077	0.061	0.044	0.034	0.025
5	25.00	11.81	6.17	3.70	1.96	1.28	1.04	0.76	0.60	0.44	0.333	0.222	0.167	0.125	0.097	0.076	0.060	0.044	0.033	0.024
0	24.66	11.64	6.09	3.65	1.93	1.27	1.03	0.75	0.59	0.43	0.329	0.219	0.164	0.123	0.096	0.075	0.059	0.043	0.033	0.024
-5	24.32	11.49	6.01	3.60	1.91	1.25	1.01	0.74	0.58	0.43	0.324	0.216	0.162	0.122	0.095	0.074	0.058	0.043	0.032	0.024

Notes: The above values for testing are based on the phase and cpc conductors, both being copper or aluminium. Calculations are based on the resistance-temperature coefficient of 0.004 per °C at 20 °C.

Where the voltage to earth is not 240V, multiply the Z_S from the table by the actual voltage to earth, and then divide by 240V, to give the revised Z_S value for the actual voltage to earth.

SHOCK PROTECTION

Z_S 1A

BS 88 HRC FUSES

Maximum design & testing values of earth loop impedance Z_S for BS 88 Part 2.2 fuses when U_0 is 240V

DISCONNECTION TIME 5 SECONDS

Maximum values of Z_S in ohms for different HRC fuse sizes

Temperature degrees C	Fuse size in amperes for 5 second disconnection time																
	2	4	6	10	16	20	25	32	40	50	63	80	100	125	160	200	250	315	400	500

Maximum value of Z_S in ohms

| Design | 47 | 23 | 14.1 | 7.74 | 4.36 | 3.04 | 2.4 | 1.92 | 1.41 | 1.09 | 0.86 | 0.6 | 0.44 | 0.35 | 0.27 | 0.2 | 0.16 | 0.12 | 0.09 | 0.065 |

Testing — For testing the following maximum values of Z_S are based on a design temperature of 115 °C

Temp																				
30	35.1	17.2	10.52	5.78	3.25	2.27	1.79	1.43	1.05	0.81	0.64	0.448	0.328	0.261	0.201	0.149	0.122	0.087	0.067	0.049
25	34.6	16.9	10.37	5.69	3.21	2.24	1.76	1.41	1.04	0.80	0.63	0.441	0.324	0.257	0.199	0.147	0.120	0.086	0.066	0.048
20	34.1	16.7	10.22	5.61	3.16	2.20	1.74	1.39	1.02	0.79	0.62	0.435	0.319	0.254	0.196	0.145	0.118	0.085	0.065	0.047
15	33.6	16.4	10.07	5.53	3.11	2.17	1.71	1.37	1.01	0.78	0.61	0.429	0.314	0.250	0.193	0.143	0.116	0.084	0.064	0.046
10	33.1	16.2	9.93	5.45	3.07	2.14	1.69	1.35	0.99	0.77	0.61	0.423	0.310	0.246	0.190	0.141	0.115	0.082	0.063	0.046
5	32.6	16.0	9.79	5.38	3.03	2.11	1.67	1.33	0.98	0.76	0.60	0.417	0.306	0.243	0.188	0.139	0.113	0.081	0.063	0.045
0	32.2	15.8	9.66	5.30	2.99	2.08	1.64	1.32	0.97	0.75	0.59	0.411	0.301	0.240	0.185	0.137	0.112	0.080	0.062	0.045
-5	31.8	15.5	9.53	5.23	2.95	2.05	1.62	1.30	0.95	0.74	0.58	0.405	0.297	0.236	0.182	0.135	0.110	0.079	0.061	0.044

Notes: The above values for testing are based on the phase and cpc conductors, both being copper or aluminium. Calculations are based on the resistance-temperature coefficient of 0.004 per °C at 20 °C.
Where the voltage to earth is not 240V, multiply the Z_S from the table by the actual voltage to earth, and then divide by 240V, to give the revised Z_S value for the actual voltage to earth.

SHOCK PROTECTION — ZS 2 — BS 1361 FUSES

Maximum design & testing values of earth loop impedance Z_S for BS 1361 fuses when U_o is 240V

Maximum values of Z_S in ohms for different fuse sizes

Temperature degrees C	Fuse size in amperes for 0.4 second disconnection time							Fuse size in amperes for 5 second disconnection time						
	5	15	20	30	45	60	80	100	5	15	20	30	45	60	80	100
	Maximum value of Z_S in ohms								Maximum value of Z_S in ohms							
Design	10.91	3.43	1.78	1.2	0.6	0.4	0.3	0.2	17.14	5.22	2.93	1.92	1.0	0.73	0.52	0.38
Testing	For testing the following maximum values of Z_S are based on a design temperature of 115 °C															
30	8.14	2.56	1.33	0.90	0.45	0.30	0.22	0.15	12.79	3.90	2.19	1.43	0.75	0.54	0.39	0.28
25	8.02	2.52	1.31	0.88	0.44	0.29	0.22	0.15	12.60	3.84	2.15	1.41	0.74	0.54	0.38	0.28
20	7.91	2.49	1.29	0.87	0.43	0.29	0.22	0.14	12.42	3.78	2.12	1.39	0.72	0.53	0.38	0.28
15	7.79	2.45	1.27	0.86	0.43	0.29	0.21	0.14	12.24	3.73	2.09	1.37	0.71	0.52	0.37	0.27
10	7.68	2.42	1.25	0.85	0.42	0.28	0.21	0.14	12.07	3.68	2.06	1.35	0.70	0.51	0.37	0.27
5	7.58	2.38	1.24	0.83	0.42	0.28	0.21	0.14	11.90	3.63	2.03	1.33	0.69	0.51	0.36	0.26
0	7.47	2.35	1.22	0.82	0.41	0.27	0.21	0.14	11.74	3.58	2.01	1.32	0.68	0.50	0.36	0.26
-5	7.37	2.32	1.20	0.81	0.41	0.27	0.20	0.14	11.58	3.53	1.98	1.30	0.68	0.49	0.35	0.26

Notes: The above values for testing are based on the phase and cpc conductors, both being copper or aluminium. Calculations are based on the resistance-temperature coefficient of 0.004 per °C at 20 °C.

Where the voltage to earth is not 240V, multiply the Z_S from the table by the actual voltage to earth, and then divide by 240V, to give the revised Z_S value for the actual voltage to earth.

SHOCK PROTECTION — ZS 3 — BS 3036 FUSES

Maximum design & testing values of earth loop impedance Z_S for BS 3036 fuses when U_0 is 240V

Maximum values of Z_S in ohms for different fuse sizes

Temperature degrees C	Fuse size in amperes for 0.4 second disconnection time							Fuse size in amperes for 5 second disconnection time					
	5	15	20	30	45	60	100	5	15	20	30	45	60	100
	Maximum value of Z_S in ohms							Maximum value of Z_S in ohms						
Design	2.67	1.85	1.14	0.85	0.62	0.44	0.2	18.46	5.58	4.0	2.76	1.66	1.17	0.558
Testing	For testing the following maximum values of Z_S are based on a design temperature of 115 °C													
30	7.46	1.99	1.38	0.85	0.46	0.328	0.149	13.81	4.16	2.99	2.06	1.239	0.873	0.416
25	7.35	1.96	1.36	0.84	0.46	0.324	0.147	13.60	4.10	2.94	2.03	1.221	0.860	0.410
20	7.25	1.93	1.34	0.83	0.45	0.319	0.145	13.41	4.04	2.90	2.00	1.203	0.848	0.404
15	7.14	1.91	1.32	0.81	0.44	0.314	0.143	13.21	3.99	2.86	1.97	1.186	0.836	0.399
10	7.04	1.88	1.30	0.80	0.44	0.310	0.141	13.03	3.93	2.82	1.94	1.169	0.824	0.393
5	6.94	1.85	1.28	0.79	0.43	0.306	0.139	12.85	3.88	2.78	1.92	1.153	0.813	0.388
0	6.85	1.83	1.27	0.78	0.42	0.301	0.137	12.67	3.82	2.74	1.89	1.137	0.801	0.382
-5	6.76	1.80	1.25	0.77	0.42	0.297	0.135	12.50	3.77	2.70	1.86	1.122	0.791	0.377

Notes: The above values for testing are based on the phase and cpc conductors, both being copper or aluminium. Calculations are based on the resistance-temperature coefficient of 0.004 per °C at 20 °C.

Where the voltage to earth is not 240V, multiply the Z_S from the table by the actual voltage to earth, and then divide by 240V, to give the revised Z_S value for the actual voltage to earth.

SHOCK PROTECTION

TYPE 1 MCB'S

ZS 4

Maximum design & testing values of earth loop impedance Z_S for BS 3871 Type 1 mcb's when U_0 is 240V

Maximum values of Z_S in ohms for different Type 1 mcb sizes

Mcb size for 0.1 second to 5 second disconnection time

Temperature degrees C	5	6	10	15	16	20	25	30	32	40	45	50	63	80	100	Other mcb sizes
Design	12	10	6	4	3.75	3	2.4	2	1.88	1.5	1.33	1.2	0.95	0.75	0.6	$60/I_n$ or $240V/4I_n$

Maximum value of Z_S in ohms

Testing — For testing the following maximum values of Z_S are based on a design temperature of 115 °C

30	8.96	7.46	4.48	2.99	2.80	2.24	1.79	1.49	1.40	1.12	0.99	0.90	0.71	0.56	0.45	$60/1.34I_n$
25	8.82	7.35	4.41	2.94	2.76	2.21	1.76	1.47	1.38	1.10	0.98	0.88	0.70	0.55	0.44	$60/1.36I_n$
20	8.70	7.25	4.35	2.90	2.72	2.17	1.74	1.45	1.36	1.09	0.96	0.87	0.69	0.54	0.43	$60/1.38I_n$
15	8.57	7.14	4.29	2.86	2.68	2.14	1.71	1.43	1.34	1.07	0.95	0.86	0.68	0.54	0.43	$60/1.40I_n$
10	8.45	7.04	4.23	2.82	2.64	2.11	1.69	1.41	1.32	1.06	0.94	0.85	0.67	0.53	0.42	$60/1.42I_n$
5	8.33	6.94	4.17	2.78	2.60	2.08	1.67	1.39	1.31	1.04	0.92	0.83	0.66	0.52	0.42	$60/1.44I_n$
0	8.22	6.85	4.11	2.74	2.57	2.05	1.64	1.37	1.29	1.03	0.91	0.82	0.65	0.51	0.41	$60/1.46I_n$
-5	8.11	6.76	4.05	2.70	2.53	2.03	1.62	1.35	1.27	1.01	0.90	0.81	0.64	0.51	0.41	$60/1.48I_n$

Notes: The above values for testing are based on the phase and cpc conductors, both being copper or aluminium. Calculations are based on the resistance-temperature coefficient of 0.004 per °C at 20 °C.
Where the voltage to earth is not 240V, multiply the Z_S from the table by the actual voltage to earth, and then divide by 240V, to give the revised Z_S value for the actual voltage to earth.

SHOCK PROTECTION

ZS 5

TYPE 2 MCB'S

Maximum design & testing values of earth loop impedance Z_S for BS 3871 Type 2 mcb's when U_0 is 240V

Maximum values of Z_S in ohms for different Type 2 mcb sizes

Temperature degrees C	\multicolumn{11}{c}{Mcb size for 0.1 second to 5 second disconnection time}	Other mcb sizes
	5	6	10	15	16	20	25	30	32	40	45	50	63	80	100	
	\multicolumn{15}{c}{Maximum value of Z_S in ohms}	
Design	6.86	5.71	3.43	2.29	2.14	1.71	1.37	1.14	1.07	0.86	0.76	0.69	0.54	0.43	0.34	$34.3/I_n$ or $240V/7I_n$
Testing	\multicolumn{15}{c}{For testing the following maximum values of Z_S are based on a design temperature of 115 °C}	
30	5.12	4.26	2.56	1.71	1.60	1.28	1.02	0.85	0.80	0.64	0.57	0.51	0.40	0.32	0.25	34.3/1.34I_n
25	5.04	4.20	2.52	1.68	1.57	1.26	1.01	0.84	0.79	0.63	0.56	0.51	0.40	0.32	0.25	34.3/1.36I_n
20	4.97	4.14	2.49	1.66	1.55	1.24	0.99	0.83	0.78	0.62	0.55	0.50	0.39	0.31	0.25	34.3/1.38I_n
15	4.90	4.08	2.45	1.64	1.53	1.22	0.98	0.81	0.76	0.61	0.54	0.49	0.39	0.31	0.24	34.3/1.40I_n
10	4.83	4.02	2.42	1.61	1.51	1.20	0.96	0.80	0.75	0.61	0.54	0.49	0.38	0.30	0.24	34.3/1.42I_n
5	4.76	3.97	2.38	1.59	1.49	1.19	0.95	0.79	0.74	0.60	0.53	0.48	0.38	0.30	0.24	34.3/1.44I_n
0	4.70	3.91	2.35	1.57	1.47	1.17	0.94	0.78	0.73	0.59	0.52	0.47	0.37	0.29	0.23	34.3/1.46I_n
-5	4.64	3.86	2.32	1.55	1.45	1.16	0.93	0.77	0.72	0.58	0.51	0.47	0.36	0.29	0.23	34.3/1.48I_n

Notes: The above values for testing are based on the phase and cpc conductors, both being copper or aluminium. Calculations are based on the resistance-temperature coefficient of 0.004 per °C at 20 °C.

Where the voltage to earth is not 240V, multiply the Z_S from the table by the actual voltage to earth, and then divide by 240V, to give the revised Z_S value for the actual voltage to earth.

329

ZS 6 — TYPE B MCB'S

Maximum design & testing values of earth loop impedance Z_S for BS 3871 Type B mcb's when U_0 is 240V

Maximum values of Z_S in ohms for different Type B mcb sizes

Temperature degrees C	5	6	10	15	16	20	25	30	32	40	45	50	63	80	100	Other mcb sizes
	Mcb size for 0.1 second to 5 second disconnection time															$48/I_n$ or $240V/5I_n$
	Maximum value of Z_S in ohms															
Design	9.6	8	4.8	3.2	3	2.4	1.92	1.6	1.5	1.2	1.07	0.96	0.76	0.6	0.48	
Testing	For testing the following maximum values of Z_S are based on a design temperature of 115 °C															
30	7.16	5.97	3.58	2.39	2.24	1.79	1.43	1.19	1.12	0.90	0.80	0.72	0.57	0.45	0.36	$48/1.34I_n$
25	7.06	5.88	3.53	2.35	2.21	1.76	1.41	1.18	1.10	0.88	0.79	0.71	0.56	0.44	0.35	$48/1.36I_n$
20	6.96	5.80	3.48	2.32	2.17	1.74	1.39	1.16	1.09	0.87	0.78	0.70	0.55	0.43	0.35	$48/1.38I_n$
15	6.86	5.71	3.43	2.29	2.14	1.71	1.37	1.14	1.07	0.86	0.76	0.69	0.54	0.43	0.34	$48/1.40I_n$
10	6.76	5.63	3.38	2.25	2.11	1.69	1.35	1.13	1.06	0.85	0.75	0.68	0.54	0.42	0.34	$48/1.42I_n$
5	6.67	5.56	3.33	2.22	2.08	1.67	1.33	1.11	1.04	0.83	0.74	0.67	0.53	0.42	0.33	$48/1.44I_n$
0	6.58	5.48	3.29	2.19	2.05	1.64	1.32	1.10	1.03	0.82	0.73	0.66	0.52	0.41	0.33	$48/1.46I_n$
-5	6.49	5.41	3.24	2.16	2.03	1.62	1.30	1.08	1.01	0.81	0.72	0.65	0.51	0.41	0.32	$48/1.48I_n$

Notes: The above values for testing are based on the phase and cpc conductors, both being copper or aluminium. Calculations are based on the resistance-temperature coefficient of 0.004 per °C at 20 °C.

Where the voltage to earth is not 240V, multiply the Z_S from the table by the actual voltage to earth, and then divide by 240V, to give the revised Z_S value for the actual voltage to earth.

SHOCK PROTECTION

ZS 7

TYPE 3 & C MCB'S

Maximum design & testing values of earth loop impedance Z_S for BS 3871 Type 3 & C mcb's when U_0 is 240V

Temperature degrees C	Maximum values of Z_S in ohms for different Type 3 & C mcb sizes												Other mcb sizes
	Mcb size for 0.1 second to 5 second disconnection time											
	5	6	10	15	16	20	25	30	32	40	45	50	63	80	100	
	Maximum value of Z_S in ohms															$24/I_n$ or $240V/10I_n$
Design	4.8	4	2.4	1.6	1.5	1.2	0.96	0.8	0.75	0.6	0.53	0.48	0.38	0.3	0.24	
Testing	For testing the following maximum values of Z_S are based on a design temperature of 115 °C															
30	3.58	2.99	1.79	1.19	1.12	0.90	0.72	0.60	0.56	0.45	0.40	0.36	0.28	0.22	0.18	$24/1.34I_n$
25	3.53	2.94	1.76	1.18	1.10	0.88	0.71	0.59	0.55	0.44	0.39	0.35	0.28	0.22	0.18	$24/1.36I_n$
20	3.48	2.90	1.74	1.16	1.09	0.87	0.70	0.58	0.54	0.43	0.38	0.35	0.28	0.22	0.17	$24/1.38I_n$
15	3.43	2.86	1.71	1.14	1.07	0.86	0.69	0.57	0.54	0.43	0.38	0.34	0.27	0.21	0.17	$24/1.40I_n$
10	3.38	2.82	1.69	1.13	1.06	0.85	0.68	0.56	0.53	0.42	0.37	0.34	0.27	0.21	0.17	$24/1.42I_n$
5	3.33	2.78	1.67	1.11	1.04	0.83	0.67	0.56	0.52	0.42	0.37	0.33	0.26	0.21	0.17	$24/1.44I_n$
0	3.29	2.74	1.64	1.10	1.03	0.82	0.66	0.55	0.51	0.41	0.36	0.33	0.26	0.21	0.16	$24/1.46I_n$
-5	3.24	2.70	1.62	1.08	1.01	0.81	0.65	0.54	0.51	0.41	0.36	0.32	0.26	0.20	0.16	$24/1.48I_n$

Notes: The above values for testing are based on the phase and cpc conductors, both being copper or aluminium. Calculations are based on the resistance-temperature coefficient of 0.004 per °C at 20 °C.

Where the voltage to earth is not 240V, multiply the Z_S from the table by the actual voltage to earth, and then divide by 240V, to give the revised Z_S value for the actual voltage to earth.

SHOCK PROTECTION

ZS 8

REDUCED VOLTAGE

Maximum design values of earth loop impedance for different types of protective device, when disconnection time required is 5 seconds and U_o is 55 volts

MAXIMUM DISCONNECTION TIME 5 SECONDS

Protective device type	\multicolumn{17}{c}{Protective device size in amperes}																
	2A	4A	5A	6A	10A	15A	16A	20A	25A	30A	32A	40A	45A	50A	60A	63A	80A	100A	125A	160A	200A
	\multicolumn{21}{c}{Maximum design value of earth loop impedance Z_S in ohms}
BS 88 Fuse	10.77	5.27	-	3.23	1.77	-	1.00	0.697	0.55	-	0.44	0.323	-	0.25	-	0.197	0.138	0.101	0.08	0.062	0.046
BS1361 fuse	-	-	3.928	-	-	1.196	-	0.671	-	0.440	-	-	0.229	-	0.167	-	0.119	0.087	-	-	-
BS 3036 fuse	-	-	4.230	-	-	1.279	-	0.917	-	0.633	-	-	0.380	-	0.268	-	-	0.128	-	-	-
Type 1 mcb	6.9	3.45	2.76	2.3	1.38	0.92	0.863	0.69	0.552	0.46	0.431	0.345	0.307	0.276	0.23	0.219	0.173	0.138	0.11	0.086	0.069
Type 2 mcb	3.95	1.975	1.58	1.317	0.79	0.527	0.494	0.395	0.316	0.263	0.247	0.198	0.176	0.158	0.132	0.125	0.099	0.079	0.063	0.049	0.04
Type B mcb	5.50	2.75	2.2	1.833	1.1	0.733	0.688	0.55	0.44	0.367	0.344	0.275	0.244	0.22	0.183	0.175	0.138	0.11	0.088	0.069	0.055
Type 3&C mcb	2.75	1.375	1.1	0.917	0.55	0.367	0.344	0.275	0.22	0.183	0.172	0.138	0.122	0.110	0.092	0.087	0.069	0.055	0.044	0.034	0.028

Notes: To obtain the value of earth loop impedance allowed for a U_o voltage of 63.5 V, divide the values in the above table by 55 and multiply by 63.5

471A

SHOCK PROTECTION   ZS 9   SPECIAL SITES

Maximum design values of earth loop impedance for different types of protective device, when disconnection time required is 0.2 seconds and U_0 is 240 volts

MAXIMUM DISCONNECTION TIME 0.2 SECONDS

Protective device size in amperes
Maximum design value of earth loop impedance Z_S in ohms

Protective device type	2A	4A	5A	6A	10A	15A	16A	20A	25A	30A	32A	40A	45A	50A	60A	63A	80A	100A	125A	160A	200A
BS 88 Fuse	33.8	15.19	-	7.74	4.71	-	2.53	1.60	1.33	-	0.92	0.71	-	0.53	-	0.39	0.27	0.20	0.17	0.12	0.09
BS1361 fuse	-	-	9.6	-	-	3.0	-	1.55	-	1.0	-	-	0.51	-	0.33	-	0.28	0.17	-	-	-
BS 3036 fuse	-	-	7.5	-	-	1.92	-	1.33	-	0.80	-	-	0.40	-	0.30	-	-	0.13	-	-	-
Type 1 mcb	30	15	12	10	6.0	4.0	3.75	3.0	2.4	2.0	1.88	1.50	1.33	1.20	1.0	0.95	0.75	0.60	0.48	0.38	0.30
Type 2 mcb	17.14	8.57	6.86	5.71	3.43	2.29	2.14	1.71	1.37	1.14	1.07	0.86	0.76	0.69	0.57	0.54	0.43	0.34	0.27	0.21	0.17
Type B mcb	24	12	9.6	8.0	4.8	3.2	3.0	2.4	1.92	1.6	1.5	1.2	1.07	0.96	0.80	0.76	0.60	0.48	0.384	0.30	0.24
Type 3&C mcb	12	6.0	4.8	4.0	2.4	1.6	1.5	1.2	0.96	0.80	0.75	0.60	0.53	0.48	0.40	0.38	0.30	0.24	0.19	0.15	0.12

Notes: To obtain the value of earth loop impedance allowed for any other voltage between 220 V and 277 V, divide the values in the above table by 240, and multiply by the voltage being used. For motor fuses, check that the characteristic is the same as that for the standard fuse; if not, consult the manufacturer; the characteristics at the rear of this book are the same, so the Z_S figures given can be used for motor fuses as well as for general purpose fuses, when using this type and make of fuse

604/605B1/B2

C.P.C. PROTECTION

PCZ 1

BS 88 HRC FUSES

Maximum values of Z_S for thermal protection of the protective conductor, when $U_0 = 240V$

Size of copper c.p.c. in sq. mm.	Size of H.R.C. fuse in amperes															
	2	4	6	10	16	20	25	32	40	50	63	80	100	125	160	200	250	315	400	500
	Maximum value of Z_S at design temperature																
1	47	23	14	7.7	4.1	2.3	1.5	0.9	-	-	-	-	-	-	-	-	-	-	-	-
1.5	47	23	14	7.7	4.4	2.9	2.0	1.3	0.81	0.43	-	-	-	-	-	-	-	-	-	-
2.5	47	23	14	7.7	4.4	3.0	2.4	1.9	1.16	0.73	0.40	-	-	-	-	-	-	-	-	-
4	47	23	14	7.7	4.4	3.0	2.4	1.9	1.41	1.09	0.65	0.33	0.175	-	-	-	-	-	-	-
6	47	23	14	7.7	4.4	3.0	2.4	1.9	1.41	1.09	0.86	0.49	0.269	0.180	-	-	-	-	-	-
10	47	23	14	7.7	4.4	3.0	2.4	1.9	1.41	1.09	0.86	0.60	0.434	0.275	0.205	0.090	-	-	-	-
16	47	23	14	7.7	4.4	3.0	2.4	1.9	1.41	1.09	0.86	0.60	0.436	0.348	0.251	0.151	0.096	0.051	-	-
25	47	23	14	7.7	4.4	3.0	2.4	1.9	1.41	1.09	0.86	0.60	0.436	0.348	0.260	0.200	0.146	0.078	0.045	-
35	47	23	14	7.7	4.4	3.0	2.4	1.9	1.41	1.09	0.86	0.60	0.436	0.348	0.260	0.200	0.163	0.111	0.090	0.033
50	47	23	14	7.7	4.4	3.0	2.4	1.9	1.41	1.09	0.86	0.60	0.436	0.348	0.260	0.200	0.163	0.117	0.090	0.051
70	47	23	14	7.7	4.4	3.0	2.4	1.9	1.41	1.09	0.86	0.60	0.436	0.348	0.260	0.200	0.163	0.117	0.090	0.065
95 or larger	47	23	14	7.7	4.4	3.0	2.4	1.9	1.41	1.09	0.86	0.60	0.436	0.348	0.260	0.200	0.163	0.117	0.090	0.065

Notes: The impedances in the table are based on an initial conductor temperature of 70 °C and a k factor of 115, for a maximum disconnection time of 5 seconds, for thermal protection of the cpc. Where the voltage to earth is not 240V, multiply the Z_S from the table by the actual voltage to earth, and then divide by 240V, to give the revised Z_S value for the actual voltage to earth.

How to use: Work out the Z_S for the circuit at the design temperature for the conductors, then compare this with the Z_S given in the table under the size of the protective device for the size of circuit protective conductor being used. If the Z_S in the table is larger than the Z_S worked out for the circuit, the protective conductor is thermally protected. If the circuit Z_S is greater than the table Z_S, recalculate the circuit Z_S using a larger cpc., until the circuit Z_S is less than that given in the table. Now check the calculated Z_S with the appropriate ZS table, to ensure that the circuit gives protection against indirect contact.

C.P.C. PROTECTION   PCZ 2   BS 1361 FUSES

Maximum values of Z_S for thermal protection of the protective conductor, when $U_o = 240V$

Size of copper c.p.c. in sq. mm.	Size of BS 1361 fuse in amperes						
	5	15	20	30	45	60	80	100
	Maximum value of Z_S at design temperature							
1.0	17.14	5.22	2.38	1.00	0.286	-	-	-
1.5	17.14	5.22	2.79	1.48	0.462	-	-	-
2.5	17.14	5.22	2.93	1.92	0.686	0.286	-	-
4.0	17.14	5.22	2.93	1.92	1.000	0.453	0.267	-
6.0	17.14	5.22	2.93	1.92	1.000	0.686	0.393	0.218
10	17.14	5.22	2.93	1.92	1.000	0.727	0.522	0.324
16	17.14	5.22	2.93	1.92	1.000	0.727	0.522	0.381
25 or larger	17.14	5.22	2.93	1.92	1.000	0.727	0.522	0.381

Notes: The impedances in the table are based on an initial conductor temperature of 70 °C and a k factor of 115, for a maximum disconnection time of 5 seconds, for thermal protection of the cpc. Where the voltage to earth is not 240V, multiply the Z_S from the table by the actual voltage to earth, and then divide by 240V, to give the revised Z_S value for the actual voltage to earth.

How to use: Work out the Z_S for the circuit at the design temperature for the conductors, then compare this with the Z_S given in the table under the size of the protective device for the size of circuit protective conductor being used. If the Z_S in the table is larger than the Z_S worked out for the circuit, the protective conductor is thermally protected. If the circuit Z_S is greater than the table Z_S, recalculate the circuit Z_S using a larger cpc, until the circuit Z_S is less than that given in the table. Now check the calculated Z_S with the appropriate ZS table, to ensure that the circuit gives protection against indirect contact.

C.P.C. PROTECTION PCZ 3 BS 3036 FUSES

Maximum values of Z_S for thermal protection of the protective conductor, when $U_o = 240V$

Size of copper c.p.c. in sq. mm.	Size of BS 3036 fuse in amperes					
	5	15	20	30	45	60	100
	Maximum value of Z_S at design temperature						
1.0	18.46	5.58	3.44	-	-	-	-
1.5	18.46	5.58	4.00	2.39	-	-	-
2.5	18.46	5.58	4.00	2.76	1.38	1.17	-
4.0	18.46	5.58	4.00	2.76	1.66	1.17	-
6.0	18.46	5.58	4.00	2.76	1.66	1.17	-
10	18.46	5.58	4.00	2.76	1.66	1.17	0.558
16	18.46	5.58	4.00	2.76	1.66	1.17	0.558
25	18.46	5.58	4.00	2.76	1.66	1.17	0.558
35 or larger	18.46	5.58	4.00	2.76	1.66	1.17	0.558

Notes: The impedances in the table are based on an initial conductor temperature of 70 °C and a k factor of 115, for a maximum disconnection time of 5 seconds, for thermal protection of the cpc. Where the voltage to earth is not 240V, multiply the Z_S from the table by the actual voltage to earth, and then divide by 240V, to give the revised Z_S value for the actual voltage to earth.

How to use: Work out the Z_S for the circuit at the design temperature for the conductors, then compare this with the Z_S given in the table under the size of the protective device for the size of circuit protective conductor being used. If the Z_S in the table is larger than the Z_S worked out for the circuit, the protective conductor is thermally protected. If the circuit Z_S is greater than the table Z_S, recalculate the circuit Z_S using a larger cpc., until the circuit Z_S is less than that given in the table. Now check the calculated Z_S with the appropriate ZS table, to ensure that the circuit gives protection against indirect contact.

THERMAL PROTECTION

PCZ4

BS 3871 TYPE 1 MCB

Maximum values of Z_S for thermal protection of the protective conductor, when $U_0 = 240V$

Size of copper c.p.c. in sq. mm.	Size of BS 3871 TYPE 1 mcb in amperes											Minimum value of impedance for all mcb sizes
	5	6	10	15	16	20	25	30	32	40	45	50	63	80	100	
	Maximum value of Z_S or Z_{pn} at design temperature											
1.0	12.0	10.0	6.0	4.0	3.75	3.0	2.4	2.0	1.88	1.5	1.3	1.2	0.95	0.75	0.60	0.2200
1.5	12.0	10.0	6.0	4.0	3.75	3.0	2.4	2.0	1.88	1.5	1.3	1.2	0.95	0.75	0.60	0.1500
2.5	12.0	10.0	6.0	4.0	3.75	3.0	2.4	2.0	1.88	1.5	1.3	1.2	0.95	0.75	0.60	0.0880
4.0	12.0	10.0	6.0	4.0	3.75	3.0	2.4	2.0	1.88	1.5	1.3	1.2	0.95	0.75	0.60	0.0550
6.0	12.0	10.0	6.0	4.0	3.75	3.0	2.4	2.0	1.88	1.5	1.3	1.2	0.95	0.75	0.60	0.0365
10	12.0	10.0	6.0	4.0	3.75	3.0	2.4	2.0	1.88	1.5	1.3	1.2	0.95	0.75	0.60	0.0219
16	12.0	10.0	6.0	4.0	3.75	3.0	2.4	2.0	1.88	1.5	1.3	1.2	0.95	0.75	0.60	0.0142
25	12.0	10.0	6.0	4.0	3.75	3.0	2.4	2.0	1.88	1.5	1.3	1.2	0.95	0.75	0.60	0.0088
35	12.0	10.0	6.0	4.0	3.75	3.0	2.4	2.0	1.88	1.5	1.3	1.2	0.95	0.75	0.60	0.0063
50 or larger	12.0	10.0	6.0	4.0	3.75	3.0	2.4	2.0	1.88	1.5	1.3	1.2	0.95	0.75	0.60	0.0044

Notes: The impedances in the table are based on an initial conductor temperature of 70 °C and a k factor of 115, for a maximum disconnection time of 5 seconds, for thermal protection of the cpc. Where the voltage to earth is not 240V, multiply the Z_S from the table by the actual voltage to earth, and then divide by 240V, to give the revised Z_S value for the actual voltage to earth.

How to use: Work out the Z_S for the circuit at the design temperature for the conductors, then compare this with the Z_S given in the table under the size of the protective device for the size of circuit protective conductor being used. If the Z_S in the table is larger than the Z_S worked out for the circuit, the protective conductor is thermally protected. If the circuit Z_S is greater than the table Z_S, recalculate the circuit Z_S using a larger cpc, until the circuit Z_S is less than that given in the table. Now check the calculated Z_S with the appropriate ZS table, to ensure that the circuit gives protection against indirect contact.

THERMAL PROTECTION PCZ 5 BS 3871 TYPE 2 MCB

Maximum values of Z_S for thermal protection of the protective conductor, when U_o = 240V

Size of copper c.p.c. in sq. mm.	Size of BS 3871 TYPE 2 mcb in amperes												Minimum value of impedance for all mcb sizes
	5	6	10	15	16	20	25	30	32	40	45	50	63	80	100	
	Maximum value of Z_S or Z_{pn} at design temperature															
1.0	6.86	5.71	3.43	2.29	2.14	1.71	1.37	1.14	1.07	0.86	0.76	0.69	0.54	0.43	0.34	0.2191
1.5	6.86	5.71	3.43	2.29	2.14	1.71	1.37	1.14	1.07	0.86	0.76	0.69	0.54	0.43	0.34	0.1462
2.5	6.86	5.71	3.43	2.29	2.14	1.71	1.37	1.14	1.07	0.86	0.76	0.69	0.54	0.43	0.34	0.0877
4.0	6.86	5.71	3.43	2.29	2.14	1.71	1.37	1.14	1.07	0.86	0.76	0.69	0.54	0.43	0.34	0.0548
6.0	6.86	5.71	3.43	2.29	2.14	1.71	1.37	1.14	1.07	0.86	0.76	0.69	0.54	0.43	0.34	0.0365
10	6.86	5.71	3.43	2.29	2.14	1.71	1.37	1.14	1.07	0.86	0.76	0.69	0.54	0.43	0.34	0.0219
16	6.86	5.71	3.43	2.29	2.14	1.71	1.37	1.14	1.07	0.86	0.76	0.69	0.54	0.43	0.34	0.0142
25	6.86	5.71	3.43	2.29	2.14	1.71	1.37	1.14	1.07	0.86	0.76	0.69	0.54	0.43	0.34	0.0088
35	6.86	5.71	3.43	2.29	2.14	1.71	1.37	1.14	1.07	0.86	0.76	0.69	0.54	0.43	0.34	0.0063
50 or larger	6.86	5.71	3.43	2.29	2.14	1.71	1.37	1.14	1.07	0.86	0.76	0.69	0.54	0.43	0.34	0.0044

Notes: The impedances in the table are based on an initial conductor temperature of 70 °C and a k factor of 115, for a maximum disconnection time of 5 seconds, for thermal protection of the cpc. Where the voltage to earth is not 240V, multiply the Z_S from the table by the actual voltage to earth, and then divide by 240V, to give the revised Z_S value for the actual voltage to earth.

How to use: Work out the Z_S for the circuit at the design temperature for the conductors, then compare this with the Z_S given in the table under the size of the protective device for the size of circuit protective conductor being used. If the Z_S in the table is larger than the Z_S worked out for the circuit, the protective conductor is thermally protected. If the circuit Z_S is greater than the table Z_S, recalculate the circuit Z_S using a larger cpc., until the circuit Z_S is less than that given in the table. Now check the calculated Z_S with the appropriate ZS table, to ensure that the circuit gives protection against indirect contact.

THERMAL PROTECTION

PCZ 6 — BS 3871 TYPE B MCB

Maximum values of Z_S for thermal protection of the protective conductor, when $U_o = 240V$

Size of copper c.p.c. in sq. mm.	Size of BS 3871 TYPE B mcb in amperes												Minimum value of impedance for all mcb sizes
	5	6	10	15	16	20	25	30	32	40	45	50	63	80	100	
	Maximum value of Z_S or Z_{pn} at design temperature												
1.0	9.6	8.0	4.8	3.2	3.0	2.4	1.92	1.60	1.50	1.20	1.07	0.96	0.76	0.60	0.48	0.2191
1.5	9.6	8.0	4.8	3.2	3.0	2.4	1.92	1.60	1.50	1.20	1.07	0.96	0.76	0.60	0.48	0.1462
2.5	9.6	8.0	4.8	3.2	3.0	2.4	1.92	1.60	1.50	1.20	1.07	0.96	0.76	0.60	0.48	0.0877
4.0	9.6	8.0	4.8	3.2	3.0	2.4	1.92	1.60	1.50	1.20	1.07	0.96	0.76	0.60	0.48	0.0548
6.0	9.6	8.0	4.8	3.2	3.0	2.4	1.92	1.60	1.50	1.20	1.07	0.96	0.76	0.60	0.48	0.0365
10	9.6	8.0	4.8	3.2	3.0	2.4	1.92	1.60	1.50	1.20	1.07	0.96	0.76	0.60	0.48	0.0219
16	9.6	8.0	4.8	3.2	3.0	2.4	1.92	1.60	1.50	1.20	1.07	0.96	0.76	0.60	0.48	0.0142
25	9.6	8.0	4.8	3.2	3.0	2.4	1.92	1.60	1.50	1.20	1.07	0.96	0.76	0.60	0.48	0.0088
35	9.6	8.0	4.8	3.2	3.0	2.4	1.92	1.60	1.50	1.20	1.07	0.96	0.76	0.60	0.48	0.0063
50 or larger	9.6	8.0	4.8	3.2	3.0	2.4	1.92	1.60	1.50	1.20	1.07	0.96	0.76	0.60	0.48	0.0044

Notes: The impedances in the table are based on an initial conductor temperature of 70 °C and a k factor of 115, for a maximum disconnection time of 5 seconds, for thermal protection of the cpc. Where the voltage to earth is not 240V, multiply the Z_S from the table by the actual voltage to earth, and then divide by 240V, to give the revised Z_S value for the actual voltage to earth.

How to use: Work out the Z_S for the circuit at the design temperature for the conductors, then compare this with the Z_S given in the table under the size of the protective device for the size of circuit protective conductor being used. If the Z_S in the table is larger than the Z_S worked out for the circuit, the protective conductor is thermally protected. If the circuit Z_S is greater than the table Z_S, recalculate the circuit Z_S using a larger cpc, until the circuit Z_S is less than that given in the table. Now check the calculated Z_S with the appropriate ZS table, to ensure that the circuit gives protection against indirect contact.

C.P.C. PROTECTION PCZ 7 BS 3871 TYPE 3 & C MCB

Maximum values of Z_S for thermal protection of the protective conductor, when $U_o = 240V$

Size of copper c.p.c. in sq. mm.	Size of BS 3871 TYPE 3 & C mcb in amperes												Minimum value of impedance for all mcb sizes
	5	6	10	15	16	20	25	30	32	40	45	50	63	80	100	
	Maximum value of Z_S or Z_{pn} at design temperature												
1.0	4.8	4.0	2.4	1.6	1.5	1.2	0.96	0.80	0.75	0.60	0.53	0.48	0.38	0.30	0.24	0.2191
1.5	4.8	4.0	2.4	1.6	1.5	1.2	0.96	0.80	0.75	0.60	0.53	0.48	0.38	0.30	0.24	0.1462
2.5	4.8	4.0	2.4	1.6	1.5	1.2	0.96	0.80	0.75	0.60	0.53	0.48	0.38	0.30	0.24	0.0877
4.0	4.8	4.0	2.4	1.6	1.5	1.2	0.96	0.80	0.75	0.60	0.53	0.48	0.38	0.30	0.24	0.0548
6.0	4.8	4.0	2.4	1.6	1.5	1.2	0.96	0.80	0.75	0.60	0.53	0.48	0.38	0.30	0.24	0.0365
10	4.8	4.0	2.4	1.6	1.5	1.2	0.96	0.80	0.75	0.60	0.53	0.48	0.38	0.30	0.24	0.0219
16	4.8	4.0	2.4	1.6	1.5	1.2	0.96	0.80	0.75	0.60	0.53	0.48	0.38	0.30	0.24	0.0142
25	4.8	4.0	2.4	1.6	1.5	1.2	0.96	0.80	0.75	0.60	0.53	0.48	0.38	0.30	0.24	0.0088
35	4.8	4.0	2.4	1.6	1.5	1.2	0.96	0.80	0.75	0.60	0.53	0.48	0.38	0.30	0.24	0.0063
50 or larger	4.8	4.0	2.4	1.6	1.5	1.2	0.96	0.80	0.75	0.60	0.53	0.48	0.38	0.30	0.24	0.0044

Notes: The impedances in the table are based on an initial conductor temperature of 70 °C and a k factor of 115, for a maximum disconnection time of 5 seconds, for thermal protection of the cpc. Where the voltage to earth is not 240V, multiply the Z_S from the table by the actual voltage to earth, and then divide by 240V, to give the revised Z_S value for the actual voltage to earth.

How to use: Work out the Z_S for the circuit at the design temperature for the conductors, then compare this with the Z_S given in the table under the size of the protective device for the size of circuit protective conductor being used. If the Z_S in the table is larger than the Z_S worked out for the circuit, the protective conductor is thermally protected. If the circuit Z_S is greater than the table Z_S, recalculate the circuit Z_S using a larger cpc., until the circuit Z_S is less than that given in the table. Now check the calculated Z_S with the appropriate ZS table, to ensure that the circuit gives protection against indirect contact.

SHOCK PROTECTION — PCZ 8 — SOCKET CIRCUITS

Maximum design values of circuit protective conductor impedance for different types of protective device, for use with socket outlet circuits, Class I circuits, or circuits for portable equipment intended to be moved.

MAXIMUM DISCONNECTION TIME 5 SECONDS

Protective device type	Protective device size in amperes — Maximum design value of c.p.c. impedance in ohms																				
	2A	4A	5A	6A	10A	15A	16A	20A	25A	30A	32A	40A	45A	50A	60A	63A	80A	100A	125A	160A	200A
BS 88 Fuse	9.79	4.79	-	2.94	1.61	-	0.91	0.63	0.50	-	0.40	0.29	-	0.23	-	0.179	0.125	0.092	0.073	0.056	0.042
BS1361 fuse	-	-	3.57	-	-	1.09	-	0.61	-	0.40	-	-	0.21	-	0.15	-	0.108	0.079	-	-	-
BS 3036 fuse	-	-	3.85	-	-	1.16	-	0.83	-	0.58	-	-	0.35	-	0.24	-	-	0.116	-	-	-
Type 1 mcb	6.25	3.13	2.5	2.08	1.25	0.83	0.78	0.63	0.5	0.42	0.39	0.31	0.28	0.25	0.21	0.198	0.156	0.125	0.10	0.078	0.063
Type 2 mcb	3.57	1.79	1.43	1.19	0.71	0.48	0.45	0.36	0.29	0.24	0.22	0.18	0.16	0.14	0.12	0.113	0.089	0.071	0.057	0.045	0.036
Type B mcb	5.0	2.5	2.0	1.67	1.0	0.67	0.63	0.50	0.40	0.33	0.31	0.25	0.22	0.20	0.17	0.16	0.13	0.10	0.08	0.06	0.05
Type 3&C mcb	2.50	1.25	1.00	0.83	0.50	0.33	0.31	0.25	0.20	0.17	0.16	0.13	0.11	0.10	0.08	0.08	0.06	0.05	0.04	0.03	0.03

Notes: The disconnection time for socket outlet circuits, or for Class 1 hand held equipment circuits, can be increased to 5 seconds, provided that the impedance of the protective conductor does not exceed the maximum value given in the above table. The impedance of the protective conductor is to be taken all the way back to the point where the protective conductor is connected to the main equipotential bonding conductors. The point of equipotential bonding can be at a local distribution board, providing it is bonded to the same types of extraneous conductive parts to which the main equipotential bonding conductors are connected.

41C

SHORT-CIRCUIT PROTECTION   SCZ 1   BS 3871 TYPE 1 MCB

Maximum values of Z_{pn} for copper conductors, for thermal protection of the live conductors, when U_o = 240V

Size of copper live conductors in sq.mm.	Size of BS 3871 TYPE 1 mcb in amperes												Minimum value of impedance for all mcb sizes
	5	6	10	15	16	20	25	30	32	40	45	50	63	80	100	
	Maximum value of Z_S or Z_{pn} at design temperature															
1.0	12.0	10.0	6.0	4.0	3.75	3.0	2.4	2.0	1.88	1.5	1.3	1.2	0.95	0.75	0.60	0.2200
1.5	12.0	10.0	6.0	4.0	3.75	3.0	2.4	2.0	1.88	1.5	1.3	1.2	0.95	0.75	0.60	0.1500
2.5	12.0	10.0	6.0	4.0	3.75	3.0	2.4	2.0	1.88	1.5	1.3	1.2	0.95	0.75	0.60	0.0880
4.0	12.0	10.0	6.0	4.0	3.75	3.0	2.4	2.0	1.88	1.5	1.3	1.2	0.95	0.75	0.60	0.0550
6.0	12.0	10.0	6.0	4.0	3.75	3.0	2.4	2.0	1.88	1.5	1.3	1.2	0.95	0.75	0.60	0.0365
10	12.0	10.0	6.0	4.0	3.75	3.0	2.4	2.0	1.88	1.5	1.3	1.2	0.95	0.75	0.60	0.0219
16	12.0	10.0	6.0	4.0	3.75	3.0	2.4	2.0	1.88	1.5	1.3	1.2	0.95	0.75	0.60	0.0142
25	12.0	10.0	6.0	4.0	3.75	3.0	2.4	2.0	1.88	1.5	1.3	1.2	0.95	0.75	0.60	0.0088
35	12.0	10.0	6.0	4.0	3.75	3.0	2.4	2.0	1.88	1.5	1.3	1.2	0.95	0.75	0.60	0.0063
50 or larger	12.0	10.0	6.0	4.0	3.75	3.0	2.4	2.0	1.88	1.5	1.3	1.2	0.95	0.75	0.60	0.0044

Notes: The impedances in the table are based on an initial conductor temperature of 70 °C and a k factor of 115. For protection of conductors, the maximum impedance of the conductors is required. With single phase circuits, or three phase and neutral circuits, take the total impedance of the phase and neutral (Z_{pn}) at the design temperature, and compare it with the impedance given in the table. For safety reasons, the circuit impedance should be less than the maximum allowed by the table. For three-phase three-wire circuits, take the total impedance of two phases (Z_{pp}), and divide by 1.732; compare the resultant impedance with the value from the table. As before, the circuits impedance should be less than the figure in the table.

SHORT-CIRCUIT PROTECTION SCZ 2 BS 3871 TYPE 2 MCB

Maximum values of Z_{pn} for copper conductors, for thermal protection of the live conductors, when $U_0 = 240V$

Size of copper live conductors in sq.mm.	Size of BS 3871 TYPE 2 mcb in amperes											Minimum value of impedance for all mcb sizes
	5	6	10	15	16	20	25	30	32	40	45	50	63	80	100	

Maximum value of Z_S or Z_{pn} at design temperature

Size	5	6	10	15	16	20	25	30	32	40	45	50	63	80	100	Min
1.0	6.86	5.71	3.43	2.29	2.14	1.71	1.37	1.14	1.07	0.86	0.76	0.69	0.54	0.43	0.34	0.2191
1.5	6.86	5.71	3.43	2.29	2.14	1.71	1.37	1.14	1.07	0.86	0.76	0.69	0.54	0.43	0.34	0.1462
2.5	6.86	5.71	3.43	2.29	2.14	1.71	1.37	1.14	1.07	0.86	0.76	0.69	0.54	0.43	0.34	0.0877
4.0	6.86	5.71	3.43	2.29	2.14	1.71	1.37	1.14	1.07	0.86	0.76	0.69	0.54	0.43	0.34	0.0548
6.0	6.86	5.71	3.43	2.29	2.14	1.71	1.37	1.14	1.07	0.86	0.76	0.69	0.54	0.43	0.34	0.0365
10	6.86	5.71	3.43	2.29	2.14	1.71	1.37	1.14	1.07	0.86	0.76	0.69	0.54	0.43	0.34	0.0219
16	6.86	5.71	3.43	2.29	2.14	1.71	1.37	1.14	1.07	0.86	0.76	0.69	0.54	0.43	0.34	0.0142
25	6.86	5.71	3.43	2.29	2.14	1.71	1.37	1.14	1.07	0.86	0.76	0.69	0.54	0.43	0.34	0.0088
35	6.86	5.71	3.43	2.29	2.14	1.71	1.37	1.14	1.07	0.86	0.76	0.69	0.54	0.43	0.34	0.0063
50 or larger	6.86	5.71	3.43	2.29	2.14	1.71	1.37	1.14	1.07	0.86	0.76	0.69	0.54	0.43	0.34	0.0044

Notes: The impedances in the table are based on an initial conductor temperature of 70 °C and a k factor of 115. For protection of conductors, the maximum impedance of the conductors is required. With single phase circuits, or three phase and neutral circuits, take the total impedance of the phase and neutral (Z_{pn}) at the design temperature, and compare it with the impedance given in the table. For safety reasons, the circuit impedance should be less than the maximum allowed by the table. For three-phase three-wire circuits, take the total impedance of two phases (Z_{pp}), and divide by 1.732; compare the resultant impedance with the value from the table. As before, the circuits impedance should be less than the figure in the table.

SHORT-CIRCUIT PROTECTION   SCZ 3   BS 3871 TYPE B MCB

Maximum values of Z_{pn} for copper conductors, for thermal protection, of the live conductors, when U_o = 240V

Size of copper live conductors in sq.mm.	Size of BS 3871 TYPE B mcb in amperes													Minimum value of impedance for all mcb sizes
	5	6	10	15	16	20	25	30	32	40	45	50	63	80	100	
	Maximum value of Z_S or Z_{pn} at design temperature														
1.0	9.6	8.0	4.8	3.2	3.0	2.4	1.92	1.60	1.50	1.20	1.07	0.96	0.76	0.60	0.48	0.2191
1.5	9.6	8.0	4.8	3.2	3.0	2.4	1.92	1.60	1.50	1.20	1.07	0.96	0.76	0.60	0.48	0.1462
2.5	9.6	8.0	4.8	3.2	3.0	2.4	1.92	1.60	1.50	1.20	1.07	0.96	0.76	0.60	0.48	0.0877
4.0	9.6	8.0	4.8	3.2	3.0	2.4	1.92	1.60	1.50	1.20	1.07	0.96	0.76	0.60	0.48	0.0548
6.0	9.6	8.0	4.8	3.2	3.0	2.4	1.92	1.60	1.50	1.20	1.07	0.96	0.76	0.60	0.48	0.0365
10	9.6	8.0	4.8	3.2	3.0	2.4	1.92	1.60	1.50	1.20	1.07	0.96	0.76	0.60	0.48	0.0219
16	9.6	8.0	4.8	3.2	3.0	2.4	1.92	1.60	1.50	1.20	1.07	0.96	0.76	0.60	0.48	0.0142
25	9.6	8.0	4.8	3.2	3.0	2.4	1.92	1.60	1.50	1.20	1.07	0.96	0.76	0.60	0.48	0.0088
35	9.6	8.0	4.8	3.2	3.0	2.4	1.92	1.60	1.50	1.20	1.07	0.96	0.76	0.60	0.48	0.0063
50 or larger	9.6	8.0	4.8	3.2	3.0	2.4	1.92	1.60	1.50	1.20	1.07	0.96	0.76	0.60	0.48	0.0044

Notes: The impedances in the table are based on an initial conductor temperature of 70°C and a k factor of 115. For protection of conductors, the maximum impedance of the conductors is required. With single phase circuits, or three phase and neutral circuits, take the total impedance of the phase and neutral (Z_{pn}) at the design temperature, and compare it with the impedance given in the table. For safety reasons, the circuit impedance should be less than the maximum allowed by the table. For three-phase three-wire circuits, take the total impedance of two phases (Z_{pp}), and divide by 1.732; compare the resultant impedance with the value from the table. As before, the circuits impedance should be less than the figure in the table.

SHORT-CIRCUIT PROTECTION SCZ 4 BS 3871 TYPE 3 & C MCB

Maximum values of Z_{pn} for copper conductors, for thermal protection, of the live conductors, when $U_o = 240V$

Size of copper live conductors in sq.mm.	Size of BS 3871 TYPE 3 & C mcb in amperes												Minimum value of impedance for all mcb sizes
	5	6	10	15	16	20	25	30	32	40	45	50	63	80	100	
	Maximum value of Z_S or Z_{pn} at design temperature															
1.0	4.8	4.0	2.4	1.6	1.5	1.2	0.96	0.80	0.75	0.60	0.53	0.48	0.38	0.30	0.24	0.2191
1.5	4.8	4.0	2.4	1.6	1.5	1.2	0.96	0.80	0.75	0.60	0.53	0.48	0.38	0.30	0.24	0.1462
2.5	4.8	4.0	2.4	1.6	1.5	1.2	0.96	0.80	0.75	0.60	0.53	0.48	0.38	0.30	0.24	0.0877
4.0	4.8	4.0	2.4	1.6	1.5	1.2	0.96	0.80	0.75	0.60	0.53	0.48	0.38	0.30	0.24	0.0548
6.0	4.8	4.0	2.4	1.6	1.5	1.2	0.96	0.80	0.75	0.60	0.53	0.48	0.38	0.30	0.24	0.0365
10	4.8	4.0	2.4	1.6	1.5	1.2	0.96	0.80	0.75	0.60	0.53	0.48	0.38	0.30	0.24	0.0219
16	4.8	4.0	2.4	1.6	1.5	1.2	0.96	0.80	0.75	0.60	0.53	0.48	0.38	0.30	0.24	0.0142
25	4.8	4.0	2.4	1.6	1.5	1.2	0.96	0.80	0.75	0.60	0.53	0.48	0.38	0.30	0.24	0.0088
35	4.8	4.0	2.4	1.6	1.5	1.2	0.96	0.80	0.75	0.60	0.53	0.48	0.38	0.30	0.24	0.0063
50 or larger	4.8	4.0	2.4	1.6	1.5	1.2	0.96	0.80	0.75	0.60	0.53	0.48	0.38	0.30	0.24	0.0044

Notes: The impedances in the table are based on an initial conductor temperature of 70 °C and a k factor of 115. For protection of conductors, the maximum impedance of the conductors is required. With single phase circuits, or three phase and neutral circuits, take the total impedance of the phase and neutral (Z_{pn}) at the design temperature, and compare it with the impedance given in the table. For safety reasons, the circuit impedance should be less than the maximum allowed by the table. For three-phase three-wire circuits, take the total impedance of two phases (Z_{pp}), and divide by 1.732; compare the resultant impedance with the value from the table. As before, the circuits impedance should be less than the figure in the table.

SHORT-CIRCUIT PROTECTION

SCZ 5

BS 88 HRC FUSES

Maximum values of Z_{pn} for copper conductors, for thermal protection, of the live conductors, when $U_o = 240V$

Size of copper live conductors in sq.mm.	Size of BS 88 H.R.C. fuse in amperes																	
	2	4	6	10	16	20	25	32	40	50	63	80	100	125	160	200	250	315	400	500
	Maximum value of Z_{pn} in ohms with a 'k' factor of 115																			
1	60.5	27.8	16.6	8.0	3.6	2.1	1.3	0.8	-	-	-	-	-	-	-	-	-	-	-	-
1.5	64.2	29.0	18.6	9.2	4.3	2.6	1.7	1.2	0.72	0.38	-	-	-	-	-	-	-	-	-	-
2.5	64.2	30.9	20.9	9.9	5.2	3.2	2.3	1.7	1.03	0.65	0.36	-	-	-	-	-	-	-	-	-
4	64.2	30.9	21.4	11.3	6.0	4.0	3.0	1.7	1.45	0.95	0.58	0.30	0.156	-	-	-	-	-	-	-
6	64.2	30.9	21.4	12.3	6.5	4.6	3.5	1.7	1.74	1.21	0.80	0.43	0.240	0.161	-	-	-	-	-	-
10	64.2	30.9	21.4	12.3	7.5	5.5	4.3	1.7	2.20	1.58	1.10	0.63	0.386	0.246	0.145	0.080	-	-	-	-
16	64.2	30.9	21.4	12.3	7.5	6.5	4.9	1.7	2.61	1.90	1.35	0.85	0.557	0.360	0.240	0.135	0.085	-	-	-
25	64.2	30.9	21.4	12.3	7.5	6.5	4.9	1.7	2.98	2.26	1.74	1.04	0.730	0.497	0.334	0.193	0.130	0.045	-	-
35	64.2	30.9	21.4	12.3	7.5	6.5	4.9	1.7	2.98	2.46	1.81	1.30	0.835	0.580	0.417	0.255	0.180	0.070	0.041	-
50	64.2	30.9	21.4	12.3	7.5	6.5	4.9	1.7	2.98	2.46	2.05	1.36	0.949	0.696	0.522	0.326	0.232	0.099	0.060	0.030
70	64.2	30.9	21.4	12.3	7.5	6.5	4.9	1.7	2.98	2.46	2.05	1.49	1.098	0.803	0.614	0.398	0.286	0.137	0.087	0.046
95	64.2	30.9	21.4	12.3	7.5	6.5	4.9	1.7	2.98	2.46	2.05	1.49	1.193	0.907	0.673	0.474	0.334	0.174	0.116	0.061
120	64.2	30.9	21.4	12.3	7.5	6.5	4.9	1.7	2.98	2.46	2.05	1.49	1.193	0.994	0.727	0.522	0.379	0.209	0.137	0.085
150	64.2	30.9	21.4	12.3	7.5	6.5	4.9	1.7	2.98	2.46	2.05	1.49	1.193	0.994	0.773	0.549	0.401	0.246	0.159	0.104
185	64.2	30.9	21.4	12.3	7.5	6.5	4.9	1.7	2.98	2.46	2.05	1.49	1.193	0.994	0.773	0.580	0.435	0.268	0.174	0.123
240	64.2	30.9	21.4	12.3	7.5	6.5	4.9	1.7	2.98	2.46	2.05	1.49	1.193	0.994	0.773	0.614	0.474	0.298	0.199	0.139
300	64.2	30.9	21.4	12.3	7.5	6.5	4.9	1.7	2.98	2.46	2.05	1.49	1.193	0.994	0.773	0.614	0.497	0.334	0.232	0.168
																		0.363	0.246	0.186

Notes: The impedances in the table are based on an initial conductor temperature of 70 °C and a k factor of 115. For protection of conductors, the maximum impedance of the conductors is required. With single phase circuits, or three phase and neutral circuits, take the total impedance of the phase and neutral (Z_{pn}) at the design temperature, and compare it with the impedance given in the table. For safety reasons, the circuit impedance should be less than the maximum allowed by the table. For three-phase three-wire circuits, take the total impedance of two phases (Z_{pp}), and divide by 1.732; compare the resultant impedance with the value from the table. As before, the circuits impedance should be less than the figure in the table.

SHORT-CIRCUIT PROTECTION

SCZ 6

BS 1361 FUSES

Maximum values of Z_{pn} for copper conductors, for thermal protection, of the live conductors, when $U_o = 240V$

Size of copper live conductors in sq.mm.	Size of BS 1361 fuse in amperes							
	5	15	20	30	45	60	80	100
	Maximum value of Z_{pn} in ohms with a 'k' factor of 115							
1	20.87	4.54	2.07	0.87	0.25	-	-	-
1.5	20.87	5.22	2.43	1.28	0.40	-	-	-
2.5	23.19	5.64	3.02	1.67	0.60	0.248	-	-
4	23.19	6.14	3.48	2.24	0.77	0.394	0.232	-
6	24.27	6.73	4.17	2.43	0.97	0.596	0.342	0.190
10	24.27	7.32	4.85	2.90	1.35	0.835	0.469	0.282
16	24.27	7.59	5.42	3.26	1.61	1.128	0.652	0.409
25	24.27	8.03	5.80	3.60	2.04	1.391	0.835	0.522
35	24.27	8.03	6.14	3.94	2.27	1.670	0.994	0.673
50 or larger	24.27	8.03	6.14	4.17	2.55	1.988	1.128	0.773

Notes: The impedances in the table are based on an initial conductor temperature of 70 °C and a k factor of 115. For protection of conductors, the maximum impedance of the conductors is required. With single phase circuits, or three phase and neutral circuits, take the total impedance of the phase and neutral (Z_{pn}) at the design temperature, and compare it with the impedance given in the table. For safety reasons, the circuit impedance should be less than the maximum allowed by the table. For three-phase three-wire circuits, take the total impedance of two phases (Z_{pp}), and divide by 1.732; compare the resultant impedance with the value from the table. As before, the circuits impedance should be less than the figure in the table.

SHORT-CIRCUIT PROTECTION SCZ 7 BS 3036 FUSES

Maximum values of Z_{pn} for copper conductors, for thermal protection, of the live conductors, when $U_o = 240V$

Size of copper live conductors in sq.mm.	Size of BS 3036 fuse in amperes						
	5	15	20	30	45	60	100
	Maximum value of Z_{pn} in ohms with a 'k' factor of 115						
1	20.87	5.22	3.07	-	-	-	-
1.5	21.74	5.96	3.86	2.13	-	-	-
2.5	21.74	6.42	4.35	2.90	1.23	-	-
4	21.74	6.73	4.64	3.31	1.74	1.043	-
6	21.74	6.73	4.85	3.54	1.90	1.391	-
10	21.74	7.07	5.09	3.73	2.13	1.739	0.835
16	21.74	7.20	5.22	3.73	2.13	1.815	0.773
25	21.74	7.26	5.22	3.73	2.13	1.897	0.870
35	21.74	7.26	5.22	3.73	2.13	1.988	0.928
50 or larger	21.74	7.26	5.22	3.73	2.13	2.036	0.971

Notes: The impedances in the table are based on an initial conductor temperature of 70 °C and a k factor of 115. For protection of conductors, the maximum impedance of the conductors is required. With single phase circuits, or three phase and neutral circuits, take the total impedance of the phase and neutral (Z_{pn}) at the design temperature, and compare it with the impedance given in the table. For safety reasons, the circuit impedance should be less than the maximum allowed by the table. For three-phase three-wire circuits, take the total impedance of two phases (Z_{pp}), and divide by 1.732; compare the resultant impedance with the value from the table. As before, the circuits impedance should be less than the figure in the table.

GROUPED SINGLE CORE CABLES   CSG 1   ENCLOSED

Cable sizes required for three phase & single phase circuits using p.v.c insulated copper single core cable with or without sheath, grouped and enclosed in conduit or trunking, not in contact with thermal insulation..

Size of I_n or I_b in amperes, excluding BS 3036 fuses when they are used for overload protection

Number of circuits grouped together	Single-phase circuits											Three-phase circuits											
	2 to 5	6	10	15	16	20	25	30	32	35	40	2 to 4	5	6	10	15	16	20	25	30	32	35	40
	Size of p.v.c. single core cable required in sq.mm.											Size of p.v.c. single core cable required in sq.mm.											
2	1.0	1.0	1.0	2.5	2.5	4	4	6	6	10	10	1.0	1.0	1.0	1.5	2.5	2.5	4	6	10	10	10	10
3	1.0	1.0	1.5	2.5	2.5	4	6	10	10	10	16	1.0	1.0	1.0	1.5	4	4	6	6	10	10	16	16
4	1.0	1.0	1.5	4	4	4	6	10	10	10	16	1.0	1.0	1.0	1.5	4	4	6	10	10	16	16	16
5	1.0	1.0	1.5	4	4	6	10	10	10	16	16	1.0	1.0	1.0	2.5	4	4	6	10	10	16	16	16
6	1.0	1.0	1.5	4	4	6	10	10	10	16	16	1.0	1.0	1.0	2.5	4	4	6	10	16	16	16	25
7	1.0	1.0	2.5	4	4	6	10	10	16	16	16	1.0	1.0	1.0	2.5	4	6	10	10	16	16	25	25
8	1.0	1.0	2.5	4	4	6	10	16	16	16	25	1.0	1.0	1.0	2.5	6	6	10	10	16	16	25	25
9	1.0	1.0	2.5	4	4	6	10	16	16	25	25	1.0	1.0	1.5	2.5	6	6	10	16	16	25	25	25
10	1.0	1.0	2.5	6	6	10	10	16	16	25	25	1.0	1.0	1.5	4	6	6	10	16	16	25	25	25
11	1.0	1.0	2.5	6	6	10	10	16	16	25	25	1.0	1.0	1.5	4	6	6	10	16	25	25	25	25
12	1.0	1.0	2.5	6	6	10	10	16	16	25	25	1.0	1.0	1.5	4	6	10	10	16	25	25	25	35
13	1.0	1.5	2.5	6	6	10	16	16	25	25	25	1.0	1.0	1.5	4	6	10	10	16	25	25	25	35
14	1.0	1.5	2.5	6	6	10	16	16	25	25	25	1.0	1.0	1.5	4	6	10	10	16	25	25	25	35
15	1.0	1.5	2.5	6	6	10	16	16	25	25	25	1.0	1.0	1.5	4	6	10	10	16	25	25	25	35
16	1.0	1.5	4	6	6	10	16	16	25	25	25	1.0	1.5	1.5	4	10	10	10	16	25	25	25	35
17	1.0	1.5	4	6	6	10	16	16	25	25	25	1.0	1.5	1.5	4	10	10	10	16	25	25	35	35
18	1.0	1.5	4	6	6	10	16	25	25	25	35	1.0	1.5	1.5	4	10	10	16	16	25	25	35	35
19	1.0	1.5	4	6	10	10	16	25	25	25	35	1.0	1.5	2.5	4	10	10	16	16	25	25	35	35
20	1.0	1.5	4	6	10	10	16	25	25	25	35	1.0	1.5	2.5	4	10	10	16	16	25	25	35	35

Note: Where a cable or circuit will not carry more than 30% of the tabulated current rating of its grouped cable size, it can be ignored when counting the number grouped together for the remaining circuits. Where the distance between adjacent conductors or cables is more than twice the diameter of the larger cable, no derating for grouping need be applied. Circuit groupings can be a mixture of three phase and single phase circuits. Where the circuit is not likely to carry overload current, use the full load current I_b of the circuit, and not the protective device size I_n, when sizing the cable from the above table.

GROUPED SINGLE CORE CABLES — CSG 2 — BUNCHED

Cable sizes required for three phase & single phase circuits using p.v.c insulated copper single core cable with or without sheath, bunched and clipped direct to a non-metallic surface, not in contact with thermal insulation.

Size of I_n or I_b in amperes, excluding BS 3036 fuses when they are used for overload protection

Number of circuits grouped together	Single-phase circuits — Size of p.v.c. single core cable required in sq.mm.											Three-phase circuits — Size of p.v.c. single core cable required in sq.mm.											
	2 to 5	6	10	15	16	20	25	30	32	35	40	2 to 5	6	10	15	16	20	25	30	32	35	40	45
2	1.0	1.0	1.0	1.5	1.5	2.5	4	6	6	6	10	1.0	1.0	1.0	2.5	2.5	2.5	4	6	6	10	10	10
3	1.0	1.0	1.0	2.5	2.5	4	4	6	6	10	10	1.0	1.0	1.5	2.5	2.5	4	6	6	10	10	16	16
4	1.0	1.0	1.5	2.5	2.5	4	6	6	10	10	10	1.0	1.0	1.5	2.5	4	4	6	10	10	10	16	16
5	1.0	1.0	1.5	2.5	2.5	4	6	10	10	10	16	1.0	1.0	1.5	2.5	4	6	6	10	10	16	16	16
6	1.0	1.5	1.5	2.5	4	4	6	10	10	10	16	1.0	1.0	1.5	4	4	6	10	10	10	16	16	16
7	1.0	1.5	1.5	4	4	4	6	10	10	16	16	1.0	1.0	2.5	4	4	6	10	10	16	16	16	25
8	1.0	1.5	1.5	4	4	6	10	10	10	16	16	1.0	1.0	2.5	4	4	6	10	16	16	16	16	25
9	1.0	1.5	1.5	4	4	6	10	10	16	16	16	1.0	1.0	2.5	4	4	6	10	16	16	16	25	25
10	1.0	1.0	2.5	4	4	6	10	10	16	16	16	1.0	1.5	2.5	4	6	6	10	16	16	16	25	25
11	1.0	1.0	2.5	4	4	6	10	16	16	16	25	1.0	1.0	2.5	4	6	6	10	16	16	16	25	25
12	1.0	1.0	2.5	4	4	6	10	16	16	16	25	1.0	1.0	2.5	6	6	10	10	16	16	25	25	25
13	1.0	1.0	2.5	4	4	6	10	16	16	16	25	1.0	1.0	2.5	6	6	10	10	16	16	25	25	25
14	1.0	1.0	2.5	4	6	6	10	16	16	16	25	1.0	1.5	2.5	6	6	10	10	16	16	25	25	35
15	1.0	1.0	2.5	4	6	10	10	16	16	16	25	1.0	1.5	2.5	6	6	10	16	16	16	25	25	35
16	1.0	1.0	2.5	4	6	10	10	16	16	16	25	1.0	1.5	2.5	6	6	10	16	16	16	25	25	35
17	1.0	1.0	2.5	6	6	10	10	16	16	25	25	1.0	1.5	4	6	6	10	16	16	25	25	25	35
18	1.0	1.0	2.5	6	6	10	10	16	16	25	25	1.0	1.5	4	6	6	10	16	16	25	25	25	35
19	1.0	1.5	2.5	6	6	10	16	16	16	25	25	1.0	1.5	4	6	6	10	16	16	25	25	35	35
20	1.0	1.5	2.5	6	6	10	16	16	16	25	25	1.0	1.5	4	6	6	10	16	16	25	25	35	35

Note: Where a cable or circuit will not carry more than 30% of the tabulated current rating of its grouped cable size, it can be ignored when counting the number grouped together for the remaining circuits. Where the distance between adjacent conductors or cables is more than twice the diameter of the larger cable, no derating for grouping need be applied. Circuit groupings can be a mixture of three phase and single phase circuits. Where the circuit is not likely to carry overload current, use the full load current I_b of circuit, and not the protective device size I_n, when sizing the cable from the above table.

GROUPED SINGLE CORE CABLES   CSG 3   ENCLOSED/THERMAL

Cable sizes required for three phase & single phase circuits using p.v.c insulated copper single core cable with or without sheath, grouped and enclosed, in conduit or trunking in contact with thermal insulation, one side being in contact with a thermally conductive surface.

Size of I_n or I_b in amperes, excluding BS 3036 fuses when they are used for overload protection

Number of circuits grouped together	Single-phase circuits											Three-phase circuits											
	2 TO 4	5	6	10	15	16	20	25	30	32	35	2 to 4	5	6	10	15	16	20	25	30	32	35	40
	Size of p.v.c. single core cable required in sq.mm.											Size of p.v.c. single core cable required in sq.mm.											
2	1.0	1.0	1.0	1.5	2.5	4	4	6	10	10	10	1.0	1.0	1.0	1.5	4	4	6	10	10	10	16	16
3	1.0	1.0	1.0	1.5	4	4	6	10	10	10	16	1.0	1.0	1.0	2.5	4	4	6	10	16	16	16	25
4	1.0	1.0	1.0	2.5	4	4	6	10	16	16	16	1.0	1.0	1.0	2.5	4	6	6	10	16	16	16	25
5	1.0	1.0	1.0	2.5	4	6	6	10	16	16	16	1.0	1.0	1.0	2.5	6	6	10	10	16	16	25	25
6	1.0	1.0	1.0	2.5	6	6	10	10	16	16	25	1.0	1.0	1.5	2.5	6	6	10	16	16	16	25	25
7	1.0	1.0	1.0	2.5	6	6	10	16	16	16	25	1.0	1.0	1.5	4	6	6	10	16	16	25	25	35
8	1.0	1.0	1.5	4	6	6	10	16	25	25	25	1.0	1.0	1.5	4	6	10	10	16	25	25	25	35
9	1.0	1.0	1.5	4	6	6	10	16	25	25	25	1.0	1.0	1.5	4	6	10	10	16	25	25	25	35
10	1.0	1.0	1.5	4	10	10	16	16	25	25	25	1.0	1.0	1.5	4	10	10	10	16	25	25	25	35
11	1.0	1.0	1.5	4	6	10	10	16	25	25	25	1.0	1.5	1.5	4	10	10	16	16	25	25	35	35
12	1.0	1.0	1.5	4	6	10	10	16	25	25	25	1.0	1.5	2.5	4	10	10	16	16	25	25	35	35
13	1.0	1.5	1.5	4	6	10	16	25	25	25	35	1.0	1.5	2.5	4	10	10	16	25	25	35	35	50
14	1.0	1.5	1.5	4	10	10	16	25	25	25	35	1.0	1.5	2.5	4	10	10	16	25	25	35	35	50
15	1.0	1.5	1.5	4	10	10	16	25	25	25	35	1.0	1.5	2.5	4	10	10	16	25	25	35	35	50
16	1.0	1.5	2.5	4	10	10	16	16	25	25	35	1.0	1.5	2.5	6	10	10	16	25	35	35	35	50
17	1.0	1.5	2.5	4	10	10	16	25	25	35	35	1.0	1.5	2.5	6	10	10	16	25	35	35	35	50
18	1.0	1.5	2.5	4	10	10	16	25	35	35	35	1.0	1.5	2.5	6	10	10	16	25	35	35	50	50
19	1.0	1.5	2.5	4	10	10	16	25	35	35	35	1.0	1.5	2.5	6	10	10	16	25	35	35	50	50
20	1.0	1.5	2.5	6	10	10	16	25	35	35	35	1.0	1.5	2.5	6	10	10	16	25	35	35	50	50

Note: Where a cable or circuit will not carry more than 30% of the tabulated current rating of its grouped cable size, it can be ignored when counting the number grouped together for the remaining circuits. Where the distance between adjacent conductors or cables is more than twice the diameter of the larger cable, no derating for grouping need be applied.   Circuit groupings can be a mixture of three phase and single phase circuits. Where the circuit is not likely to carry overload current, use the full load current I_b of the circuit, and not the protective device size I_n, when sizing the cable from the above table.

GROUPED MULTICORE CABLES   CSG 4   ENCLOSED

Cable sizes required for three phase & single phase circuits using p.v.c insulated multicore copper cable or twin & cpc copper cable, grouped and enclosed in conduit or trunking, not in contact with thermal insulation.

Size of I_n or I_b in amperes, excluding BS 3036 fuses when they are used for overload protection

Number of grouped multicore cables	Single-phase circuits — Size of p.v.c. single core cable required in sq.mm.											Three-phase circuits — Size of p.v.c. single core cable required in sq.mm.											
	2 TO 4	5	6	10	15	16	20	25	30	32	35	2 to 4	5	6	10	15	16	20	25	30	32	35	40
2	1.0	1.0	1.0	1.5	2.5	2.5	4	6	6	10	10	1.0	1.0	1.0	1.5	2.5	2.5	4	6	10	10	10	16
3	1.0	1.0	1.0	1.5	2.5	2.5	4	6	10	10	10	1.0	1.0	1.0	1.5	4	4	6	10	10	10	16	16
4	1.0	1.0	1.0	2.5	2.5	4	6	6	10	10	16	1.0	1.0	1.0	2.5	4	4	6	10	16	16	16	16
5	1.0	1.0	1.0	2.5	4	4	6	10	10	16	16	1.0	1.0	1.0	2.5	4	6	6	10	16	16	16	25
6	1.0	1.0	1.0	2.5	4	4	6	10	16	16	16	1.0	1.0	1.0	2.5	4	6	10	10	16	16	16	25
7	1.0	1.0	1.0	2.5	4	4	6	10	16	16	16	1.0	1.0	1.0	2.5	6	6	10	16	16	16	25	25
8	1.0	1.0	1.5	2.5	4	6	10	10	16	16	16	1.0	1.0	1.5	2.5	6	6	10	16	16	25	25	25
9	1.0	1.0	1.0	2.5	6	6	10	10	16	16	25	1.0	1.0	1.0	4	6	6	10	16	16	25	25	25
10	1.0	1.0	1.0	2.5	6	6	10	10	16	16	25	1.0	1.0	1.5	4	6	6	10	16	25	25	25	35
11	1.0	1.0	1.0	2.5	6	6	10	16	16	16	25	1.0	1.0	1.5	4	6	10	10	16	25	25	25	35
12	1.0	1.0	1.5	2.5	6	6	10	16	16	25	25	1.0	1.0	1.5	4	6	10	10	16	25	25	25	35
13	1.0	1.0	1.5	4	6	6	10	16	16	25	25	1.0	1.0	1.5	4	6	10	10	16	25	25	35	35
14	1.0	1.0	1.5	4	6	6	10	16	16	25	25	1.0	1.5	1.5	4	10	10	16	16	25	25	35	35
15	1.0	1.0	1.5	4	6	6	10	16	25	25	25	1.0	1.5	1.5	4	10	10	16	16	25	35	35	35
16	1.0	1.0	1.5	4	6	10	10	16	25	25	25	1.0	1.5	1.5	4	10	10	16	16	25	25	35	35
17	1.0	1.0	1.5	4	6	10	16	16	25	25	35	1.0	1.5	1.5	4	10	10	16	25	25	35	35	50
18	1.0	1.0	1.5	4	10	10	16	16	25	25	35	1.0	1.5	2.5	4	10	10	16	25	25	35	35	50
19	1.0	1.0	1.5	4	10	10	16	16	25	25	35	1.0	1.5	1.5	4	10	10	16	25	25	35	35	50
20	1.0	1.5	1.5	4	10	10	16	16	25	35	35	1.0	1.5	2.5	4	10	10	16	25	25	35	35	50

Note: Where a cable or circuit will not carry more than 30% of the tabulated current rating of its grouped cable size, it can be ignored when counting the number grouped together for the remaining circuits. Where the distance between adjacent conductors or cables is more than twice the diameter of the larger cable, no derating for grouping need be applied. Circuit groupings can be a mixture of three phase and single phase circuits. Where the circuit is not likely to carry overload current, use the full load current I_b of the circuit, and not the protective device size I_n, when sizing the cable from the above table.

GROUPED MULTICORE CABLES   CSG 5   BUNCHED

Cable sizes required for three phase & single phase circuits using p.v.c insulated multicore copper cable or twin & cpc copper cable, bunched and clipped direct to a non-metallic surface, not in contact with thermal insulation.

Size of I_n or I_b in amperes, excluding BS 3036 fuses when they are used for overload protection

Number of grouped multicore cables	Single-phase circuits — Size of p.v.c. multicore cable required in sq.mm.											Three-phase circuits — Size of p.v.c. multicore cable required in sq.mm.												
	2 TO 5	6	10	15	16	20	25	30	32	35	40	2 to 5	6	10	15	16	20	25	30	32	35	40	45
2	1.0	1.0	1.0	1.5	2.5	2.5	4	6	6	6	10	1.0	1.0	1.0	2.5	2.5	4	4	6	6	10	10	10
3	1.0	1.0	1.0	2.5	2.5	4	4	6	6	10	10	1.0	1.0	1.5	2.5	2.5	4	6	10	10	10	16	16
4	1.0	1.0	1.5	2.5	2.5	4	6	10	10	10	10	1.0	1.0	1.5	2.5	4	4	6	10	10	16	16	16
5	1.0	1.0	1.5	2.5	2.5	4	6	10	10	10	16	1.0	1.0	1.5	4	4	6	10	10	10	16	16	16
6	1.0	1.0	1.5	2.5	4	4	6	10	10	16	16	1.0	1.0	1.5	4	4	6	10	10	10	16	16	25
7	1.0	1.0	1.5	4	4	6	10	10	10	16	16	1.0	1.0	2.5	4	4	6	10	16	16	16	25	25
8	1.0	1.0	1.5	4	4	6	10	10	16	16	16	1.0	1.0	2.5	4	4	10	10	16	16	16	25	25
9	1.0	1.0	2.5	4	4	6	10	10	16	16	16	1.0	1.0	2.5	4	6	10	10	16	16	25	25	25
10	1.0	1.0	2.5	4	4	6	10	10	16	16	16	1.0	1.0	2.5	4	6	10	10	16	16	25	25	25
11	1.0	1.0	2.5	4	4	6	10	16	16	16	25	1.0	1.0	2.5	6	6	10	10	16	16	16	25	35
12	1.0	1.0	2.5	4	4	6	10	16	16	16	25	1.0	1.0	2.5	6	6	10	10	16	16	25	25	35
13	1.0	1.0	2.5	4	6	6	10	16	16	16	25	1.0	1.5	2.5	6	6	10	16	16	16	25	25	35
14	1.0	1.0	2.5	4	6	10	10	16	16	16	25	1.0	1.5	2.5	6	6	10	16	16	25	25	25	35
15	1.0	1.0	2.5	4	6	10	10	16	16	16	25	1.0	1.5	2.5	6	6	10	16	16	25	25	25	35
16	1.0	1.0	2.5	6	6	10	10	16	16	25	25	1.5	1.5	4	6	6	10	16	16	25	25	35	35
17	1.0	1.0	2.5	6	6	10	16	16	16	25	25	1.0	1.5	4	6	6	10	16	16	25	25	35	35
18	1.0	1.5	2.5	6	6	10	16	16	16	25	25	1.0	1.5	4	6	6	10	16	25	25	25	35	35
19	1.0	1.5	2.5	6	6	10	16	16	16	25	25	1.0	1.5	4	6	10	10	16	25	25	25	35	35
20	1.0	1.5	2.5	6	6	10	16	16	16	25	25	1.0	1.5	4	6	10	10	16	25	25	25	35	35

Note: Where a cable or circuit will not carry more than 30% of the tabulated current rating of its grouped cable size, it can be ignored when counting the number grouped together for the remaining circuits. Where the distance between adjacent conductors or cables is more than twice the diameter of the larger cable, no derating for grouping need be applied. Circuit groupings can be a mixture of three phase and single phase circuits. Where the circuit is not likely to carry overload current, use the full load current I_b of the circuit, and not the protective device size I_n, when sizing the cable from the above table.

GROUPED MULTICORE CABLES — CSG 6 — BUNCHED/THERMAL

Cable sizes required for three phase & single phase circuits using p.v.c insulated multicore copper cable, or twin & cpc copper cable, bunched and clipped direct to a non-metallic surface, in contact with thermal insulation, one side being in contact with a thermally conductive surface.

Size of I_n or I_b in amperes, excluding BS 3036 fuses when they are used for overload protection

Size of p.v.c. multicore cable required in sq.mm.

Number of grouped multicore cables	Single-phase circuits											Three-phase circuits											
	2 TO 4	5	6	10	15	16	20	25	30	32	35	2 to 4	5	6	10	15	16	20	25	30	32	35	40
2	1.0	1.0	1.0	1.5	4	4	4	6	10	10	16	1.0	1.0	1.0	1.5	4	4	6	10	10	16	16	16
3	1.0	1.0	1.0	2.5	4	4	6	10	16	16	16	1.0	1.0	1.0	2.5	4	4	6	10	16	16	16	25
4	1.0	1.0	1.0	2.5	4	4	6	10	16	16	16	1.0	1.0	1.0	2.5	6	6	10	16	16	16	25	25
5	1.0	1.0	1.0	2.5	6	6	10	10	16	16	25	1.0	1.0	1.0	2.5	6	6	10	16	16	25	25	25
6	1.0	1.0	1.0	2.5	6	6	10	16	16	16	25	1.0	1.0	1.5	4	6	6	10	16	25	25	25	35
7	1.0	1.0	1.0	2.5	6	6	10	16	16	25	25	1.0	1.0	1.5	4	6	10	10	16	25	25	25	35
8	1.0	1.0	1.5	4	6	6	10	16	25	25	25	1.0	1.0	1.5	4	10	10	16	25	25	25	35	35
9	1.0	1.0	1.5	4	6	6	10	16	25	25	25	1.0	1.0	1.5	4	10	10	16	25	25	25	35	35
10	1.0	1.0	1.5	4	6	10	10	16	25	25	25	1.0	1.5	1.5	4	10	10	16	25	25	35	35	50
11	1.0	1.0	1.5	4	10	10	10	16	25	25	35	1.0	1.5	1.5	4	10	10	16	25	25	35	35	50
12	1.0	1.0	1.5	4	10	10	16	16	25	25	35	1.0	1.5	2.5	4	10	10	16	25	35	35	35	50
13	1.0	1.5	1.5	4	10	10	16	25	25	25	35	1.0	1.5	2.5	6	10	10	16	25	35	35	35	50
14	1.0	1.5	1.5	4	10	10	16	25	35	35	35	1.0	1.5	2.5	6	10	10	16	25	35	35	50	50
15	1.0	1.5	2.5	4	10	10	16	25	35	35	35	1.0	1.5	2.5	6	10	10	16	25	35	35	50	50
16	1.0	1.5	2.5	4	10	10	16	25	25	35	35	1.0	1.5	2.5	6	10	16	16	25	35	35	50	50
17	1.0	1.5	2.5	4	10	10	16	25	35	35	35	1.0	1.5	2.5	6	10	16	16	25	35	35	50	70
18	1.0	1.5	2.5	6	10	10	16	25	35	35	35	1.5	1.5	2.5	6	10	16	16	25	35	35	50	70
19	1.0	1.5	2.5	6	10	10	16	25	35	35	35	1.5	1.5	2.5	6	16	16	16	25	35	50	50	70
20	1.0	1.5	2.5	6	10	10	16	25	35	35	35	1.5	2.5	2.5	6	16	16	16	25	35	50	50	70

Note: Where a cable or circuit will not carry more than 30% of the tabulated current rating of its grouped cable size, it can be ignored when counting the number grouped together for the remaining circuits. Where the distance between adjacent conductors or cables is more than twice the diameter of the larger cable, no derating for grouping need be applied. Circuit groupings can be a mixture of three phase and single phase circuits. Where the circuit is not likely to carry overload current, use the full load current I_b of the circuit, and not the protective device size I_n, when sizing the cable from the above table.

GROUPED MULTICORE CABLES

CSG 7

CLIPPED DIRECT

Cable sizes required for three phase & single phase circuits using p.v.c insulated multicore copper cable or twin & cpc copper cable, clipped direct in a single layer to a non-metallic surface with cable sheaths touching, not in contact with thermal insulation.

Size of I_n or I_b in amperes, excluding BS 3036 fuses when they are used for overload protection

Number of grouped multicore cables	Single-phase circuits											Three-phase circuits									
	2 TO 10	15	16	20	25	30	32	35	40	45	50	2 to 6	10	15	16	20	25	30	32	35	40	45	50
	Size of p.v.c. multicore cable required in sq.mm.											Size of p.v.c. multicore cable required in sq.mm.											
2	1.0	1.5	2.5	2.5	4	4	6	6	10	10	10	1.0	1.0	2.5	2.5	2.5	4	6	6	10	10	10	16
3	1.0	1.5	2.5	2.5	4	6	6	6	10	10	16	1.0	1.0	2.5	2.5	4	4	6	6	10	10	10	16
4	1.0	2.5	2.5	2.5	4	6	6	10	10	10	16	1.0	1.0	2.5	2.5	4	6	6	6	10	10	16	16
5	1.0	2.5	2.5	4	4	6	6	10	10	10	16	1.0	1.5	2.5	2.5	4	6	6	6	10	10	16	16
6	1.0	2.5	2.5	4	4	6	6	10	10	10	16	1.0	1.5	2.5	2.5	4	6	6	10	10	10	16	16
7	1.0	2.5	2.5	4	4	6	6	10	10	10	16	1.0	1.5	2.5	2.5	4	6	6	10	10	10	16	16
8	1.0	2.5	2.5	4	4	6	6	10	10	16	16	1.0	1.5	2.5	2.5	4	6	6	10	10	10	16	16
9	1.0	2.5	2.5	4	4	6	6	10	10	16	16	1.0	1.5	2.5	2.5	4	6	6	10	10	10	16	16

Note: Where a cable or circuit will not carry more than 30% of the tabulated current rating of its grouped cable size, it can be ignored when counting the number grouped together for the remaining circuits. Where the distance between adjacent conductors or cables is more than twice the diameter of the larger cable, no derating for grouping need be applied. Circuit groupings can be a mixture of three phase and single phase circuits. Where the circuit is not likely to carry overload current, use the full load current I_b of the circuit, and not the protective device size I_n, when sizing the cable from the above table.

GROUPED MULTICORE CABLES — CSG 8 — CABLE TRAY

Cable sizes required for three phase & single phase circuits using p.v.c insulated multicore copper cable or twin & cpc copper cable, installed in a single layer on perforated metal cable tray.

Size of I_n or I_b in amperes, excluding BS 3036 fuses when they are used for overload protection

Number of grouped multicore cables	Single-phase circuits — Size of p.v.c. multicore cable required in sq.mm.											Three-phase circuits — Size of p.v.c. multicore cable required in sq.mm.										
	2 TO 10	15	16	20	25	32	35	40	45	50		2 to 10	15	16	20	25	30	32	35	40	45	50	60
2	1.0	1.5	1.5	2.5	2.5	4	6	6	10	10		1.0	1.5	2.5	2.5	4	6	6	6	10	10	10	16
3	1.0	1.5	1.5	2.5	4	4	6	6	10	10		1.0	1.5	2.5	2.5	4	6	6	6	10	10	16	16
4	1.0	1.5	1.5	2.5	4	4	6	10	10	10		1.0	2.5	2.5	4.0	4	6	6	10	10	10	16	16
5	1.0	1.5	1.5	2.5	4	6	6	10	10	10		1.0	2.5	2.5	4.0	4	6	6	10	10	10	16	16
6	1.0	1.5	1.5	2.5	4	6	6	10	10	10		1.0	2.5	2.5	4.0	4	6	10	10	10	16	16	25
7	1.0	1.5	1.5	2.5	4	6	6	10	10	10		1.0	2.5	2.5	4.0	6	6	10	10	10	16	16	25
8	1.0	1.5	1.5	2.5	4	6	6	10	10	10		1.0	2.5	2.5	4.0	6	6	10	10	10	16	16	25
9	1.0	1.5	2.5	2.5	4	6	6	10	10	10		1.0	2.5	2.5	4.0	6	6	10	10	10	16	16	25
10	1.0	1.5	2.5	2.5	4	6	6	10	10	16		1.0	2.5	2.5	4.0	6	6	10	10	10	16	16	25
11	1.0	1.5	2.5	2.5	4	6	6	10	10	16		1.0	2.5	2.5	4.0	6	6	10	10	10	16	16	25
12	1.0	1.5	2.5	2.5	4	6	6	10	10	16		1.0	2.5	2.5	4.0	6	6	10	10	10	16	16	25

Note: Where a cable or circuit will not carry more than 30% of the tabulated current rating of its grouped cable size, it can be ignored when counting the number grouped together for the remaining circuits. Where the distance between adjacent conductors or cables is more than twice the diameter of the larger cable, no derating for grouping need be applied. Circuit groupings can be a mixture of three phase and single phase circuits. Where the circuit is not likely to carry overload current, use the full load current I_b of the circuit, and not the protective device size I_n, when sizing the cable from the above table.

GROUPED ARMOURED CABLES   CSG 9   BUNCHED

Cable sizes required for three phase & single phase circuits using p.v.c insulated multicore armoured copper cable, bunched and clipped direct to a non-metallic surface, not in contact with thermal insulation.

Size of I_n or I_b in amperes, excluding BS 3036 fuses when they are used for overload protection

Number of grouped armoured cables	Single-phase circuits										Three-phase circuits											
	2 TO 6	10	15	16	20	30	32	35	40	45	2 to 6	10	15	16	20	25	30	32	35	40	45	50
	Size of armoured p.v.c. cable required in sq.mm.										Size of armoured p.v.c. cable required in sq.mm.											
2	1.5	1.5	1.5	1.5	2.5	4	6	6	10	10	1.5	1.5	2.5	2.5	2.5	4	6	6	10	10	10	16
3	1.5	1.5	2.5	2.5	4	4	6	10	10	10	1.5	1.5	2.5	2.5	4	6	10	10	10	16	16	16
4	1.5	1.5	2.5	2.5	4	6	10	10	10	16	1.5	1.5	2.5	2.5	4	6	10	10	10	16	16	16
5	1.5	1.5	2.5	2.5	4	6	10	10	10	16	1.5	1.5	2.5	4	6	6	10	10	16	16	16	25
6	1.5	1.5	2.5	2.5	4	6	10	10	16	16	1.5	2.5	4	4	6	10	10	16	16	16	25	25
7	1.5	1.5	2.5	4	4	6	10	10	16	16	1.5	2.5	4	4	6	10	10	16	16	16	25	25
8	1.5	1.5	4	4	6	10	10	10	16	16	1.5	2.5	4	4	6	10	10	16	16	16	25	25
9	1.5	1.5	4	4	6	10	10	16	16	25	1.5	2.5	4	4	6	10	16	16	16	25	25	25
10	1.5	1.5	4	4	6	10	10	16	16	25	1.5	2.5	4	6	6	10	16	16	16	25	25	35
11	1.5	2.5	4	4	6	10	16	16	16	25	1.5	2.5	4	6	10	10	16	16	16	25	25	35
12	1.5	2.5	4	4	6	10	16	16	16	25	1.5	2.5	4	6	10	10	16	16	25	25	25	35
13	1.5	2.5	4	4	6	10	16	16	25	25	1.5	2.5	4	6	10	10	16	16	25	25	35	35
14	1.5	2.5	4	4	6	10	16	16	25	25	1.5	2.5	6	6	10	10	16	16	25	25	35	35
15	1.5	2.5	4	4	6	10	16	16	25	25	1.5	2.5	6	6	10	16	16	16	25	25	35	35
16	1.5	2.5	4	6	6	10	16	16	25	25	1.5	2.5	6	6	10	16	16	25	25	25	35	35
17	1.5	2.5	4	6	10	10	16	16	25	25	1.5	2.5	6	6	10	16	16	25	25	25	35	35
18	1.5	2.5	6	6	10	16	16	25	25	25	1.5	4	6	6	10	16	25	25	25	35	35	50
19	1.5	2.5	6	6	10	16	16	25	25	25	1.5	4	6	6	10	16	25	25	25	35	35	50
20	1.5	2.5	6	6	10	16	16	25	25	35	1.5	4	6	6	10	16	25	25	25	35	35	50

Note: Where a cable or circuit will not carry more than 30% of the tabulated current rating of its grouped cable size, it can be ignored when counting the number grouped together for the remaining circuits. Where the distance between adjacent conductors or cables is more than twice the diameter of the larger cable, no derating for grouping need be applied. Circuit groupings can be a mixture of three phase and single phase circuits. Where the circuit is not likely to carry overload current, use the full load current I_b of the circuit, and not the protective device size I_n, when sizing the cable from the above table.

GROUPED ARMOURED CABLES CSG 10 CLIPPED DIRECT

Cable sizes required for three phase & single phase circuits using armoured multicore p.v.c insulated copper cable, clipped direct in a single layer to a non-metallic surface with cable sheaths touching, not in contact with thermal insulation.

Size of I_n or I_b in amperes, excluding BS 3036 fuses when they are used for overload protection

Number of grouped armoured cables	Single-phase circuits											Three-phase circuits										
	2 TO 10	15	16	20	25	30	32	35	40	45	50	2 to 10	15	16	20	25	30	32	35	40	45	50	60
	Size of armoured p.v.c. cable required in sq.mm.											Size of armoured p.v.c. cable required in sq.mm.										
2	1.5	1.5	1.5	2.5	4	4	4	6	6	10	10	1.5	1.5	2.5	2.5	4	6	6	6	10	10	16	16
3	1.5	1.5	1.5	2.5	4	4	6	6	10	10	10	1.5	2.5	2.5	4	4	6	6	10	10	10	16	16
4	1.5	1.5	2.5	2.5	4	4	6	6	10	10	10	1.5	2.5	2.5	4	6	6	6	10	10	16	16	25
5	1.5	1.5	2.5	2.5	4	4	6	6	10	10	16	1.5	2.5	2.5	4	6	6	10	10	10	16	16	25
6	1.5	1.5	2.5	2.5	4	4	6	6	10	10	16	1.5	2.5	2.5	4	6	6	6	10	10	16	16	25
7	1.5	1.5	2.5	2.5	4	6	6	6	10	10	16	1.5	2.5	2.5	4	6	6	10	10	10	16	16	25
8	1.5	1.5	2.5	4	4	6	6	10	10	10	16	1.5	2.5	2.5	4	6	10	10	10	10	16	16	25
9	1.5	2.5	2.5	4	4	6	6	10	10	10	16	1.5	2.5	2.5	4	6	10	10	10	10	16	16	25

Note: Where a cable or circuit will not carry more than 30% of the tabulated current rating of its grouped cable size, it can be ignored when counting the number grouped together for the remaining circuits. Where the distance between adjacent conductors or cables is more than twice the diameter of the larger cable, no derating for grouping need be applied. Circuit groupings can be a mixture of three phase and single phase circuits. Where the circuit is not likely to carry overload current, use the full load current I_b of the circuit, and not the protective device size I_n, when sizing the cable from the above table.

GROUPED ARMOURED CABLES CSG 11 CABLE TRAY

Cable sizes required for three phase & single phase circuits using armoured multicore p.v.c insulated copper cable, installed in a single layer on perforated metal cable tray with cable sheaths touching, not in contact with thermal insulation.

Size of I_n or I_b in amperes, excluding BS 3036 fuses when they are used for overload protection

Number of grouped armoured cables	Single-phase circuits											Three-phase circuits											
	2 TO 15	16	20	25	30	32	35	40	45	50	60	2 to 10	15	16	20	25	30	32	35	40	45	50	60
	Size of armoured p.v.c cable required in sq.mm.											Size of armoured p.v.c cable required in sq.mm.											
2	1.5	1.5	2.5	2.5	4	4	4	6	6	10	10	1.5	1.5	1.5	2.5	4	4	6	6	10	10	10	16
3	1.5	1.5	2.5	2.5	4	4	6	6	10	10	16	1.5	1.5	2.5	2.5	4	4	6	6	10	10	10	16
4	1.5	1.5	2.5	4	4	6	6	6	10	10	16	1.5	1.5	2.5	2.5	4	6	6	6	10	10	16	16
5	1.5	1.5	2.5	4	4	6	6	10	10	10	16	1.5	1.5	2.5	2.5	4	6	6	10	10	10	16	16
6	1.5	1.5	2.5	4	4	6	6	10	10	10	16	1.5	1.5	2.5	2.5	4	6	6	10	10	10	16	16
7	1.5	1.5	2.5	4	4	6	6	10	10	10	16	1.5	1.5	2.5	2.5	4	6	6	10	10	10	16	16
8	1.5	1.5	2.5	4	4	6	6	10	10	10	16	1.5	1.5	2.5	2.5	4	6	6	10	10	10	16	16
9	1.5	2.5	2.5	4	6	6	6	10	10	10	16	1.5	1.5	2.5	4	4	6	6	10	10	16	16	25
10	1.5	2.5	2.5	4	6	6	6	10	10	10	16	1.5	1.5	2.5	4	6	6	10	10	10	16	16	25
11	1.5	2.5	2.5	4	6	6	6	10	10	10	16	1.5	1.5	2.5	4	6	6	10	10	10	16	16	25
12	1.5	2.5	2.5	4	6	6	6	10	10	10	16	1.5	2.5	2.5	4	6	6	10	10	10	16	16	25

Note: Where a cable or circuit will not carry more than 30% of the tabulated current rating of its grouped cable size, it can be ignored when counting the number grouped together for the remaining circuits. Where the distance between adjacent conductors or cables is more than twice the diameter of the larger cable, no derating for grouping need be applied. Circuit groupings can be a mixture of three phase and single phase circuits. Where the circuit is not likely to carry overload current, use the full load current I_b of the circuit, and not the protective device size I_n, when sizing the cable from the above table.

GROUPED ARMOURED CABLES

CSG 12

BUNCHED XLPE CABLES

Cable sizes required for three phase & single phase circuits using multicore armoured copper cable with thermosetting (XLPE) insulation, bunched and clipped direct to a non-metallic surface, not in contact with thermal insulation.

Size of I_n or I_b in amperes, excluding BS 3036 fuses when they are used for overload protection

Number of grouped armoured cables	Single-phase circuits														Three-phase circuits												
	2 to 10	15	16	20	25	30	32	35	40	45	50	60	63	80	2 to 6	10	15	16	20	25	32	35	45	50	60	63	80
	Size of armoured XLPE cable required in sq.mm.														Size of armoured XLPE cable required in sq.mm.												
2	1.5	1.5	1.5	2.5	2.5	4	4	4	6	6	10	10	10	16	1.5	1.5	1.5	1.5	2.5	4	4	6	6	10	10	16	25
3	1.5	1.5	1.5	2.5	2.5	4	4	6	6	10	10	16	16	25	1.5	1.5	1.5	2.5	2.5	4	6	6	10	10	16	16	25
4	1.5	1.5	2.5	2.5	4	4	6	6	6	10	10	16	16	25	1.5	1.5	2.5	2.5	2.5	4	6	6	10	10	16	16	25
5	1.5	1.5	2.5	2.5	4	4	6	6	10	10	10	16	16	25	1.5	1.5	2.5	2.5	4	4	6	10	10	10	16	16	25
6	1.5	2.5	2.5	4	4	6	6	6	10	10	16	16	16	25	1.5	2.5	2.5	2.5	4	4	6	10	10	16	16	16	25
7	1.5	2.5	2.5	4	4	6	6	10	10	10	16	25	25	35	1.5	2.5	2.5	2.5	4	6	10	10	16	16	25	25	35
8	1.5	2.5	2.5	4	4	6	6	10	10	16	16	25	25	35	1.5	2.5	2.5	4	4	6	10	10	16	16	25	25	35
9	1.5	2.5	2.5	4	6	6	10	10	10	16	16	25	25	35	1.5	2.5	2.5	4	4	6	10	10	16	16	25	25	35
10	1.5	2.5	2.5	4	6	10	10	10	16	16	16	25	25	35	1.5	2.5	4	4	4	6	10	10	16	25	25	35	50
11	1.5	2.5	2.5	4	6	10	10	10	16	16	25	25	25	35	1.5	2.5	4	4	6	6	10	10	16	25	25	35	50
12	1.5	2.5	4	4	6	10	10	10	16	16	25	25	35	50	1.5	2.5	4	4	6	10	10	16	16	25	35	35	50
13	1.5	2.5	4	4	6	10	10	16	16	16	25	35	35	50	1.5	2.5	4	4	6	10	10	16	25	25	35	35	50
14	1.5	2.5	4	4	6	10	10	16	16	25	25	35	35	50	1.5	2.5	4	4	6	10	16	16	25	25	35	35	50
15	1.5	2.5	4	4	6	10	10	16	16	25	25	35	35	50	1.5	2.5	4	4	6	10	16	16	25	25	35	35	50
16	1.5	4	4	6	10	10	10	16	16	25	25	35	35	50	2.5	4	4	6	10	10	16	16	25	25	35	35	70
17	1.5	4	4	6	10	10	10	16	16	25	25	35	35	50	2.5	4	4	6	10	10	16	16	25	35	35	50	70
18	1.5	4	4	6	10	10	16	16	25	25	25	35	35	50	2.5	4	4	6	10	16	16	16	25	35	35	50	70
19	1.5	4	4	6	10	10	16	16	25	25	25	35	35	50	2.5	4	6	6	10	16	16	25	25	35	50	50	70
20	1.5	4	4	6	10	10	16	16	25	25	25	35	35	50	2.5	4	6	6	10	16	16	25	35	35	50	50	70

Note: Where a cable or circuit will not carry more than 30% of the tabulated current rating of its grouped cable size, it can be ignored when counting the number grouped together for the remaining circuits. Where the distance between adjacent conductors or cables is more than twice the diameter of the larger cable, no derating for grouping need be applied. Circuit groupings can be a mixture of three phase and single phase circuits. Where the circuit is not likely to carry overload current, use the full load current I_b of the circuit, and not the protective device size I_n, when sizing the cable from the above table.

GROUPED ARMOURED CABLES

CSG 13

XLPE CLIPPED DIRECT

Cable sizes required for three phase & single phase circuits using multicore armoured copper cable with thermosetting (XLPE) insulation, clipped direct in a single layer to a non-metallic surface with cable sheaths touching, not in contact with thermal insulation.

Number of grouped armoured cables	Size of I_n or I_b in amperes, excluding BS 3036 fuses when they are used for overload protection																							
	Single-phase circuits												Three-phase circuits											
	2 to 16	20	25	30	32	35	40	45	50	60	63	80	100	125	2 to 16	20	25	30	32	35	40	45	50	63	80	100	125	160
	Size of armoured XLPE cable required in sq. mm.														Size of armoured XLPE cable required in sq.mm.													
2	1.5	1.5	2.5	2.5	4	4	4	6	6	10	10	16	25	35	1.5	2.5	2.5	4	4	4	6	6	10	16	25	35	50	70
3	1.5	1.5	2.5	4	4	4	4	6	6	10	10	16	25	35	1.5	2.5	4	4	4	6	6	10	10	16	25	35	50	70
4	1.5	2.5	2.5	4	4	4	6	6	6	10	10	16	25	35	1.5	2.5	4	4	6	6	6	10	10	16	25	35	50	70
5	1.5	2.5	2.5	4	4	4	6	6	10	10	16	16	25	35	1.5	2.5	4	4	6	6	10	10	10	16	25	35	50	70
6	1.5	2.5	2.5	4	4	4	6	6	10	10	16	25	25	35	1.5	2.5	4	4	6	6	10	10	10	16	25	35	50	70
7	1.5	2.5	2.5	4	4	4	6	6	10	10	16	25	25	35	1.5	2.5	4	4	6	6	10	10	10	16	25	35	50	70
8	1.5	2.5	2.5	4	4	6	6	10	10	10	16	25	25	35	1.5	2.5	4	4	6	6	10	10	16	16	25	35	50	70
9	1.5	2.5	2.5	4	4	6	6	10	10	16	16	25	25	35	1.5	2.5	4	4	6	6	10	10	16	16	25	35	50	70

Note: Where a cable or circuit will not carry more than 30% of the tabulated current rating of its grouped cable size, it can be ignored when counting the number grouped together for the remaining circuits. Where the distance between adjacent conductors or cables is more than twice the diameter of the larger cable, no derating for grouping need be applied. Circuit groupings can be a mixture of three phase and single phase circuits. Where the circuit is not likely to carry overload current, use the full load current I_b of the circuit, and not the protective device size I_n, when sizing the cable from the above table.

GROUPED ARMOURED CABLES   CSG 14   XLPE ON TRAY

Cable sizes required for three phase & single phase circuits using multicore armoured copper cable with thermosetting (XLPE) insulation, installed in a single layer on perforated metal cable tray with cable sheaths touching, not in contact with thermal insulation.

Size of I_n or I_b in amperes, excluding BS 3036 fuses when they are used for overload protection

Number of grouped armoured cables	Single-phase circuits — Size of armoured XLPE cable required in sq.mm.																Three-phase circuits — Size of armoured XLPE cable required in sq. mm.															
	2 to 20	25	30	32	35	40	45	50	60	63	80	100	125	160			2 to 16	20	25	30	32	35	40	45	50	60	63	80	100	125	160	
2	1.5	1.5	2.5	2.5	4	4	6	6	10	10	16	25	25	35			1.5	1.5	2.5	4	4	4	6	6	10	10	10	16	25	35	50	
3	1.5	2.5	2.5	4	4	4	6	6	10	10	16	25	35	50			1.5	1.5	2.5	4	4	4	6	6	10	10	10	16	25	35	70	
4	1.5	2.5	2.5	4	4	4	6	6	10	10	16	25	35	50			1.5	2.5	2.5	4	4	4	6	6	10	10	16	25	35	50	70	
5	1.5	2.5	4	4	4	6	6	10	10	10	16	25	35	50			1.5	2.5	4	4	4	6	6	10	10	10	16	25	35	50	70	
6	1.5	2.5	4	4	4	6	6	10	10	10	16	25	35	50			1.5	2.5	4	4	4	6	6	10	10	16	16	25	35	50	70	
7	1.5	2.5	4	4	4	6	6	10	10	10	16	25	35	50			1.5	2.5	4	4	6	6	6	10	10	16	16	25	35	50	70	
8	1.5	2.5	4	4	4	6	6	10	10	10	16	25	35	50			1.5	2.5	4	4	6	6	6	10	10	16	16	25	35	50	70	
9	1.5	2.5	4	4	4	6	6	10	10	10	16	25	35	50			1.5	2.5	4	4	6	6	6	10	10	16	16	25	35	50	70	
10	1.5	2.5	4	4	6	6	6	10	10	10	16	25	35	50			1.5	2.5	4	4	6	6	10	10	10	16	16	25	35	50	70	
11	1.5	2.5	4	4	4	6	6	10	10	10	16	25	35	50			1.5	2.5	4	4	6	6	10	10	10	16	16	25	35	50	70	
12	1.5	2.5	4	4	4	6	6	10	10	10	16	25	35	70			1.5	2.5	4	4	6	6	10	10	10	16	16	25	35	50	70	

Note: Where a cable or circuit will not carry more than 30% of the tabulated current rating of its grouped cable size, it can be ignored when counting the number grouped together for the remaining circuits. Where the distance between adjacent conductors or cables is more than twice the diameter of the larger cable, no derating for grouping need be applied. Circuit groupings can be a mixture of three phase and single phase circuits. Where the circuit is not likely to carry overload current, use the full load current I_b of the circuit, and not the protective device size I_n, when sizing the cable from the above table.

GROUPED MICC CABLES — CSG 15 — CLIPPED DIRECT

Cable sizes required for three phase & single phase circuits using micc insulated copper cable p.v.c. sheathed, clipped direct in a single layer to a non-metallic surface with cable sheaths touching.

Size of I_n or I_b in amperes, excluding BS 3036 fuses when they are used for overload protection

Number of cables grouped together	One two core cable					One three core cable					One 4 core cable 3 cores loaded					One 4 core cable 4 cores loaded				
	2 to 10	15	16	20	25	2 to 10	15	16	20	25	2 to 10	15	16	20	25	2 to 6	10	15	16	20	25

Light duty 500V

2	1.0	1.0	1.5	2.5	2.5	1.0	1.5	1.5	2.5	4	1.0	1.5	1.5	2.5	–	1.0	1.0	2.5	2.5	–	–
3	1.0	1.5	1.5	2.5	4	1.0	1.5	2.5	2.5	4	1.0	1.5	2.5	2.5	–	1.0	1.0	2.5	2.5	–	–
4	1.0	1.5	1.5	2.5	4	1.0	2.5	2.5	4	4	1.0	2.5	2.5	–	–	1.0	1.5	2.5	2.5	–	–
5	1.0	1.5	1.5	2.5	4	1.0	2.5	2.5	4	4	1.0	2.5	2.5	–	–	1.0	1.5	2.5	2.5	–	–
6	1.0	1.5	1.5	2.5	4	1.0	2.5	2.5	4	4	1.0	2.5	2.5	–	–	1.0	1.5	2.5	–	–	–
7	1.0	1.5	1.5	2.5	4	1.0	2.5	2.5	4	4	1.0	2.5	2.5	–	–	1.0	1.5	2.5	–	–	–
8	1.0	1.5	1.5	2.5	4	1.0	2.5	2.5	4	–	1.0	2.5	2.5	–	–	1.0	1.5	2.5	–	–	–
9	1.0	1.5	1.5	2.5	4	1.0	2.5	2.5	4	–	1.0	2.5	2.5	–	–	1.0	1.5	2.5	–	–	–

Heavy duty 750V

2	1.0	1.0	1.0	1.5	2.5	1.0	1.5	1.5	2.5	4	1.0	1.5	1.5	2.5	4	1.0	1.0	1.5	2.5	2.5	4
3	1.0	1.0	1.5	2.5	2.5	1.0	1.5	1.5	2.5	4	1.0	1.5	1.5	2.5	4	1.0	1.0	2.5	2.5	4	4
4	1.0	1.5	1.5	2.5	2.5	1.0	1.5	2.5	2.5	4	1.0	1.5	2.5	2.5	4	1.0	1.0	2.5	2.5	4	6
5	1.0	1.5	1.5	2.5	4	1.0	1.5	2.5	2.5	4	1.0	1.5	2.5	2.5	4	1.0	1.0	2.5	2.5	4	6
6	1.0	1.5	1.5	2.5	4	1.0	1.5	2.5	2.5	4	1.0	1.5	2.5	2.5	4	1.0	1.0	2.5	2.5	4	6
7	1.0	1.5	1.5	2.5	4	1.0	1.5	2.5	2.5	4	1.0	1.5	2.5	2.5	4	1.0	1.0	2.5	2.5	4	6
8	1.0	1.5	1.5	2.5	4	1.0	1.5	2.5	4	4	1.0	1.5	2.5	4.0	4	1.0	1.0	2.5	2.5	4	6
9	1.0	1.5	1.5	2.5	4	1.0	2.5	2.5	4	4	1.0	2.5	2.5	4.0	4	1.0	1.0	2.5	2.5	4	6

Note: Where a cable or circuit will not carry more than 30% of the tabulated current rating of its grouped cable size, it can be ignored when counting the number grouped together for the remaining circuits. Where the distance between adjacent conductors or cables is more than twice the diameter of the larger cable, no derating for grouping need be applied. Circuit groupings can be a mixture of three phase and single phase circuits. Where the circuit is not likely to carry overload current, use the full load current I_b of the circuit, and not the protective device size I_n when sizing the cable from the above table.

GROUPED MICC CABLES — CSG 16 — ON TRAY

Cable sizes required for three phase & single phase circuits using micc pvc sheathed cable installed on horizontal or vertical (see notes) perforated metal cable tray with cable sheaths touching.

Size of I_n or I_b in amperes, excluding BS 3036 fuses when they are used for overload protection

Number of cables grouped together	One two core cable				One three core cable				One 4 core cable 3 cores loaded				One 4 core cable 4 cores loaded			
	2 to 10	15	16	20	25	2 to 10	15	16	20	25	2 to 10	15	16	20	25	2 to 10	15	16	20	25

Light duty 500V

2	1.0	1.0	1.0	1.5	2.5	1.0	1.5	1.5	2.5	2.5	1.0	1.5	1.5	2.5	2.5	1.0	1.5	1.5	2.5	-
3	1.0	1.0	1.5	1.5	2.5	1.0	1.5	1.5	2.5	4	1.0	1.5	1.5	2.5	-	1.0	2.5	2.5	-	-
4	1.0	1.0*	1.5	1.5*	2.5*	1.0	1.5	1.5*	2.5	4	1.0	1.5	1.5*	2.5	-	1.0	2.5	2.5	-	-
5	1.0	1.0*	1.5	2.5	2.5*	1.0	1.5	1.5*	2.5	4	1.0	1.5	1.5*	2.5	-	1.0	2.5	2.5	-	-
6	1.0	1.5	1.5	2.5	4	1.0	1.5	2.5	2.5	4	1.0	1.5	2.5	2.5	-	1.0	2.5	2.5	-	-
7	1.0	1.5	1.5	2.5	4	1.0	1.5	2.5	2.5	4	1.0	1.5	2.5	2.5	-	1.0	2.5	2.5	-	-
8	1.0	1.5	1.5	2.5	4	1.0	1.5	2.5	2.5	4	1.0	1.5	2.5	2.5	-	1.0	2.5	2.5	-	-
9	1.0	1.5	1.5	2.5	4	1.0	1.5*	2.5	2.5*	4	1.0	1.5*	2.5	2.5†	-	1.0*	2.5	2.5	-	-

Heavy duty 750V

2	1.0	1.0	1.0	1.5	2.5	1.0	1	1.5	2.5	2.5	1.0	1.0	1.0	1.5	2.5	1.0	1.5	1.5	2.5	4
3	1.0	1.0	1.5	1.5	2.5	1.0	1.5	1.5	2.5	4	1.0	1.5	1.5	2.5	4	1.0	1.5	1.5	2.5	4
4	1.0	1.0*	1.0*	1.5*	2.5	1.0	1.5	1.5	2.5	4	1.0	1.5	1.5*	2.5	4	1.0	1.5	1.5*	2.5	4
5	1.0	1.0*	1.0*	1.5*	2.5	1.0	1.5	1.5	2.5	4	1.0	1.5	1.5	2.5	4	1.0	1.5	2.5	2.5	4
6	1.0	1.0	1.5	2.5	2.5	1.0	1.5	1.5	2.5	4	1.0	1.5	2.5	2.5	4	1.0	1.5	2.5	2.5	4
7	1.0	1.5	1.5	2.5	2.5	1.0	1.5	1.5	2.5	4	1.0	1.5	2.5	2.5	4	1.0	1.5	2.5	2.5	4
8	1.0	1.5	1.5	2.5	2.5	1.0	1.5	1.5	2.5	4	1.0	1.5	2.5	2.5	4	1.0	1.5	2.5	2.5	4
9	1.0	1.0*	1.5	2.5	2.5	1.0	1.5	1.5*	2.5	4	1.0	1.5	2.5	2.5	4	1.0	1.5*	2.5	2.5*	4*

Note: Cable sizes marked with * should be increased to the next larger size cable when installed on a vertical cable tray. Cable sizes marked with † cannot be used on a vertical cable tray. Where a cable or circuit will not carry more than 30% of the tabulated current rating of its grouped cable size, it can be ignored when counting the number of cables grouped together for the remaining circuits. Where the distance between adjacent conductors or cables is more than twice the diameter of the larger cable, no derating for grouping need be applied. Circuit groupings can be a mixture of three phase and single phase circuits. Where the circuit is not likely to carry overload current, use the full load current I_b of the circuit, and not the protective device size I_n, when sizing the cable from the above table.

GROUPED ARMOURED CABLES   CSG 17   CLIPPED DIRECT
ALUMINIUM

Cable sizes required for three phase & single phase circuits using armoured multicore aluminium p.v.c insulated cable, clipped direct in a single layer to a non-metallic surface with cable sheaths touching, not in contact with thermal insulation.

Number of grouped armoured cables	Size of I_n or I_b in amperes, excluding BS 3036 fuses when they are used for overload protection																
	Single-phase circuits									Three-phase circuits								
	2 TO 45	50	60	63	80	100	125	160		2 to 40	45	50	60	63	80	100	125	160	200	250
	Size of armoured p.v.c. cable required in sq.mm.									Size of armoured p.v.c. cable required in sq.mm								
2	16	16	25	25	35	50	70	95		16	16	25	25	25	35	50	95	120	185	240
3	16	16	25	25	35	50	70	-		16	16	25	25	35	50	70	95	150	185	300
4	16	16	25	25	35	70	95	-		16	25	25	35	35	50	70	95	150	240	300
5	16	25	25	25	50	70	95	-		16	25	25	35	35	50	70	95	150	240	300
6	16	25	25	25	50	70	95	-		16	25	25	35	35	50	70	95	150	240	300
7	16	25	25	25	50	70	95	-		16	25	25	35	35	50	70	95	150	240	300
8	16	25	25	25	50	70	95	-		16	25	25	35	35	50	70	120	150	240	300
9	16	25	25	35	50	70	95	-		16	25	25	35	35	70	70	120	150	240	300

Note: Where a cable or circuit will not carry more than 30% of the tabulated current rating of its grouped cable size, it can be ignored when counting the number of cables grouped together for the remaining circuits. Where the distance between adjacent conductors or cables is more than twice the diameter of the larger cable, no derating for grouping need be applied. Circuit groupings can be a mixture of three phase and single phase circuits. Where the circuit is not likely to carry overload current, use the full load current I_b of the circuit, and not the protective device size I_n, when sizing the cable from the above table.

365

GROUPED ARMOURED CABLES

CSG 18 ALUMINIUM

CABLE TRAY

Cable sizes required for three phase & single phase circuits using armoured multicore aluminium p.v.c insulated cable, installed in a single layer on perforated metal cable tray with cable sheaths touching, not in contact with thermal insulation.

Size of I_n or I_b in amperes, excluding BS 3036 fuses when they are used for overload protection

Number of grouped armoured cables	Single-phase circuits Size of armoured p.v.c. cable required in sq.mm.								Three-phase circuits Size of armoured p.v.c. cable required in sq.mm										
	2 TO 45	50	60	63	80	100	125	160	2 to 40	45	50	60	63	80	100	125	160	200	250
2	16	16	16	25	25	50	70	95	16	16	16	25	25	35	50	70	95	150	240
3	16	16	25	25	35	50	70	95	16	16	25	25	25	35	70	95	120	150	240
4	16	16	25	25	35	50	70	95	16	16	25	25	35	50	70	95	120	185	240
5	16	16	25	25	35	50	70	-	16	16	25	25	35	50	70	95	120	185	240
6	16	16	25	25	35	50	70	-	16	16	25	35	35	50	70	95	150	185	240
7	16	16	25	25	35	50	70	-	16	25	25	35	35	50	70	95	150	185	300
8	16	16	25	25	35	50	70	-	16	25	25	35	35	50	70	95	150	185	300
9	16	16	25	25	35	50	70	-	16	25	25	35	35	50	70	95	150	185	300
10	16	16	25	25	35	70	95	-	16	25	25	35	35	50	70	95	150	185	300
11	16	16	25	25	35	70	95	-	16	25	25	35	35	50	70	95	150	185	300
12	16	25	25	25	35	70	95	-	16	25	25	35	35	50	70	95	150	185	300

Note: Where a cable or circuit will not carry more than 30% of the tabulated current rating of its grouped cable size, it can be ignored when counting the number of cables grouped together for the remaining circuits. Where the distance between adjacent conductors or cables is more than twice the diameter of the larger cable, no derating for grouping need be applied. Circuit groupings can be a mixture of three phase and single phase circuits. Where the circuit is not likely to carry overload current, use the full load current I_b of the circuit, and not the protective device size I_n, when sizing the cable from the above table.

GROUPED ARMOURED CABLES   CSG 19   XLPE CLIPPED DIRECT
ALUMINIUM

Cable sizes required for three phase & single phase circuits using armoured multicore aluminium cable with thermosetting (XLPE) insulation, clipped direct in a single layer to a non-metallic surface with cable sheaths touching, not in contact with thermal insulation.

Number of grouped armoured cables	Size of I_n or I_b in amperes, excluding BS 3036 fuses when they are used for overload protection																
	Single-phase circuits								Three-phase circuits								
	2 TO 50	60	63	80	100	125	160	200	2 to 45	50	60	63	80	100	125	160	200	250	315
	Size of armoured XLPE cable required in sq.mm.								Size of armoured XLPE cable required in sq.mm.								
2	16	16	16	25	35	50	70	95	16	16	16	25	35	50	70	95	120	185	240
3	16	16	16	25	35	50	95	-	16	16	25	25	35	50	70	95	150	185	300
4	16	16	25	25	50	70	95	-	16	16	25	25	35	50	70	95	150	240	300
5	16	25	25	35	50	70	95	-	16	16	25	25	35	50	70	120	150	240	300
6	16	25	25	35	50	70	95	-	16	16	25	25	35	70	70	120	150	240	300
7	16	25	25	35	50	70	95	-	16	16	25	25	35	70	70	120	150	240	300
8	16	25	25	35	50	70	95	-	16	16	25	25	35	70	95	120	150	240	-
9	16	25	25	35	50	70	95	-	16	25	25	25	50	70	95	120	185	240	-

Note: Where a cable or circuit will not carry more than 30% of the tabulated current rating of its grouped cable size, it can be ignored when counting the number of cables grouped together for the remaining circuits. Where the distance between adjacent conductors or cables is more than twice the diameter of the larger cable, no derating for grouping need be applied. Circuit groupings can be a mixture of three phase and single phase circuits. Where the circuit is not likely to carry overload current, use the full load current I_b of the circuit, and not the protective device size I_n, when sizing the cable from the above table.

GROUPED ARMOURED CABLES

CSG 20
ALUMINIUM

XLPE ON TRAY

Cable sizes required for three phase & single phase circuits using multicore armoured aluminium cable with thermosetting (XLPE) insulation, installed in a single layer on perforated metal cable tray with cable sheaths touching, not in contact with thermal insulation.

Size of I_n or I_b in amperes, excluding BS 3036 fuses when they are used for overload protection

Number of grouped armoured cables	Single-phase circuits											Three-phase circuits											
	2 TO 50	60	63	80	100	125	160	200				2 to 50	60	63	80	100	125	160	200	250	315	355
	Size of armoured XLPE cable required in sq. mm.											Size of armoured XLPE cable required in sq.mm.										
2	16	16	16	25	35	50	70	95				16	16	16	25	35	70	95	120	150	240	240
3	16	16	16	25	35	50	70	95				16	16	16	25	35	70	95	120	185	240	300
4	16	16	16	25	35	50	70	-				16	16	25	35	50	70	95	120	185	240	300
5	16	16	16	25	35	70	95	-				16	25	25	35	50	70	95	150	185	300	300
6	16	16	16	25	35	70	95	-				16	25	25	35	50	70	95	150	185	300	300
7	16	16	25	25	35	70	95	-				16	25	25	35	50	70	95	150	185	300	300
8	16	16	25	25	50	70	95	-				16	25	25	35	50	70	95	150	185	300	300
9	16	16	25	25	50	70	95	-				16	25	25	35	50	70	95	150	185	300	-
10	16	16	25	35	50	70	95	-				16	25	25	35	50	70	120	150	240	300	-
11	16	16	25	35	50	70	95	-				16	25	25	35	50	70	120	150	240	300	-
12	16	25	25	35	50	70	95	-				16	25	25	35	50	70	120	150	240	300	-

Note: Where a cable or circuit will not carry more than 30% of the tabulated current rating of its grouped cable size, it can be ignored when counting the number of cables grouped together for the remaining circuits. Where the distance between adjacent conductors or cables is more than twice the diameter of the larger cable, no derating for grouping need be applied. Circuit groupings can be a mixture of three phase and single phase circuits. Where the circuit is not likely to carry overload current, use the full load current I_b of the circuit, and not the protective device size I_n, when sizing the cable from the above table.

GROUPED MULTICORE CABLES   CSG 21   BUNCHED

OVERLOAD PROTECTION BY BS 3036 FUSES

Cable sizes required for three phase & single phase circuits using p.v.c insulated multicore copper cable or twin & cpc copper cable, bunched and clipped direct to a non-metallic surface, not in contact with thermal insulation.

Number of grouped multicore cables	Size of BS 3036 protective device in amperes when BS 3036 fuse is used for overload protection																			
	Single-phase circuits										Three-phase circuits									
	5	10	15	20	30	45	60	80	100		5	10	15	20	30	45	60	80	100
	Size of p.v.c. multicore cable required in sq.mm.										Size of p.v.c. multicore cable required in sq.mm.									
2	1.0	1.5	2.5	4	10	16	25	35	70		1.0	1.5	4	6	10	25	35	50	70
3	1.0	2.5	4	6	10	25	35	50	70		1.0	2.5	4	6	16	25	35	70	95
4	1.0	2.5	4	6	16	25	35	70	70		1.0	2.5	4	10	16	25	50	70	95
5	1.0	2.5	4	6	16	25	35	70	95		1.0	2.5	6	10	16	35	50	70	120
6	1.0	2.5	6	10	16	25	50	70	95		1.0	4	6	10	16	35	70	95	120
7	1.0	2.5	6	10	16	35	50	70	95		1.0	4	6	10	25	35	70	95	150
8	1.0	2.5	6	10	16	35	50	70	120		1.0	4	6	10	25	50	70	95	150
9	1.0	4	6	10	16	35	50	95	120		1.5	4	10	16	25	50	70	95	150
10	1.0	4	6	10	25	35	70	120	150		1.5	4	10	16	25	50	70	120	150
11	1.0	4	6	10	25	35	70	95	120		1.5	4	10	16	25	50	70	120	150
12	1.5	4	6	10	25	35	70	95	150		1.5	4	10	16	25	50	70	120	185
13	1.5	4	10	10	25	50	70	95	150		1.5	4	10	16	25	50	95	120	185
14	1.5	4	10	16	25	50	70	95	150		1.5	6	10	16	35	70	95	120	185
15	1.5	4	10	16	25	50	70	120	150		1.5	6	10	16	35	70	95	150	185
16	1.5	4	10	16	25	50	70	120	150		1.5	6	10	16	35	70	95	150	185
17	1.5	4	10	16	25	50	70	120	185		1.5	6	10	16	35	70	95	150	240
18	1.5	4	10	25	25	70	95	120	185		2.5	6	10	16	35	70	95	150	240
19	1.5	4	10	16	25	50	95	120	185		2.5	6	10	16	35	70	95	150	240
20	1.5	6	10	16	25	50	95	120	185		2.5	6	10	16	35	70	95	150	240

Note: Where a cable or circuit will not carry more than 30% of the tabulated current rating of its grouped cable size, it can be ignored when counting the number grouped together for the remaining circuits. Where the distance between adjacent conductors or cables is more than twice the diameter of the larger cable, no derating for grouping need be applied. Circuit groupings can be a mixture of three phase and single phase circuits.

GROUPED MULTICORE CABLES

CSG 22

CLIPPED DIRECT

OVERLOAD PROTECTION BY BS 3036 FUSES

Cable sizes required for three phase & single phase circuits using p.v.c insulated multicore copper cable or twin & cpc copper cable, clipped direct in a single layer to a non-metallic surface with cable sheaths touching, not in contact with thermal insulation.

Number of grouped multicore cables	Size of BS 3036 protective device in amperes, when BS 3036 fuse is used for overload protection															
	Single-phase circuits								Three-phase circuits							
	5	10	15	20	30	45	60	80	100	5	10	15	20	30	45	60	80	100
	Size of p.v.c. multicore cable required in sq.mm.								Size of p.v.c. multicore cable required in sq.mm.							
2	1.0	1.5	2.5	4	10	16	25	35	50	1.0	1.5	4.0	6	10	16	35	50	70
3	1.0	1.5	2.5	4	10	16	25	50	70	1.0	1.5	4.0	6	10	25	35	50	70
4	1.0	1.5	4.0	6	10	16	25	50	70	1.0	2.5	4.0	6	10	25	35	70	70
5	1.0	1.5	4.0	6	10	16	35	50	70	1.0	2.5	4.0	6	10	25	35	70	95
6	1.0	1.5	4.0	6	10	25	35	50	70	1.0	2.5	4.0	6	16	25	35	70	95
7	1.0	1.5	4.0	6	10	25	35	50	70	1.0	2.5	4.0	6	16	25	35	70	95
8	1.0	1.5	4.0	6	10	25	35	50	70	1.0	2.5	4.0	6	16	25	35	70	95
9	1.0	2.5	4.0	6	10	25	35	50	70	1.0	2.5	4.0	6	16	25	35	70	95

Note: Where a cable or circuit will not carry more than 30% of the tabulated current rating of its grouped cable size, it can be ignored when counting the number grouped together for the remaining circuits. Where the distance between adjacent conductors or cables is more than twice the diameter of the larger cable, no derating for grouping need be applied. Circuit groupings can be a mixture of three phase and single phase circuits.

GROUPED MULTICORE CABLES — CSG 23 — CABLE TRAY

OVERLOAD PROTECTION BY BS 3036 FUSES

Cable sizes required for three phase & single phase circuits using p.v.c insulated multicore copper cable or twin & cpc copper cable, installed in a single layer on perforated metal cable tray with cable sheaths touching, not in contact with thermal insulation.

Number of grouped multicore cables	Size of BS 3036 protective device in amperes, when BS 3036 fuse is used for overload protection																		
	Single-phase circuits									Three-phase circuits								
	5	10	15	20	30	45	60	80	100	5	10	15	20	30	45	60	80	100
	Size of p.v.c. multicore cable required in sq.mm.									Size of p.v.c. multicore cable required in sq.mm.								
2	1.0	1.0	2.5	4	6	16	25	35	50	1.0	1.5	2.5	4	10	16	25	50	70
3	1.0	1.0	2.5	4	6	16	25	35	50	1.0	1.5	4.0	4	10	16	35	50	70
4	1.0	1.5	2.5	4	10	16	25	35	50	1.0	1.5	4.0	6	10	25	35	50	70
5	1.0	1.5	2.5	4	10	16	25	35	70	1.0	1.5	4.0	6	10	25	35	50	70
6	1.0	1.5	2.5	4	10	16	25	50	70	1.0	2.5	4.0	6	10	25	35	50	70
7	1.0	1.5	2.5	4	10	16	25	50	70	1.0	2.5	4.0	6	10	25	35	50	70
8	1.0	1.5	2.5	4	10	16	25	50	70	1.0	2.5	4.0	6	10	25	35	50	70
9	1.0	1.5	2.5	4	10	16	25	50	70	1.0	2.5	4.0	6	10	25	35	50	70
10	1.0	1.5	2.5	4	10	16	25	50	70	1.0	2.5	4.0	6	10	25	35	70	70
11	1.0	1.5	2.5	4	10	16	25	50	70	1.0	2.5	4.0	6	10	25	35	70	70
12	1.0	1.5	2.5	4	10	16	25	50	70	1.0	2.5	4.0	6	10	25	35	70	95

Note: Where a cable or circuit will not carry more than 30% of the tabulated current rating of its grouped cable size, it can be ignored when counting the number grouped together for the remaining circuits. Where the distance between adjacent conductors or cables is more than twice the diameter of the larger cable, no derating for grouping need be applied. Circuit groupings can be a mixture of three phase and single phase circuits.

GROUPED MULTICORE CABLES   CSG 24   ENCLOSED

OVERLOAD PROTECTION BY BS 3036 FUSES

Cable sizes required for three phase & single phase circuits using p.v.c insulated multicore copper cable or twin & cpc copper cable, grouped and enclosed in conduit or trunking, not in contact with thermal insulation.

Number of grouped multicore cables	Size of BS 3036 protective device in amperes, when BS 3036 fuse is used for overload protection														
	Single-phase circuits									Three-phase circuits						
	5	10	15	20	30	45	60	80	100	5	10	15	20	30	45	60	80	100
	Size of p.v.c. multicore cable required in sq.mm.									Size of p.v.c. multicore cable required in sq.mm.								
2	1.0	2.5	4	6	10	25	35	70	95	1.0	2.5	4	10	16	25	50	70	95
3	1.0	2.5	4	10	16	25	50	70	95	1.0	2.5	6	10	16	35	70	95	120
4	1.0	2.5	6	10	16	35	50	95	120	1.0	4	6	10	25	35	70	95	150
5	1.0	2.5	6	10	16	35	70	95	120	1.0	4	10	10	25	50	70	120	185
6	1.0	4	6	10	25	35	70	95	150	1.5	4	10	16	25	50	70	120	185
7	1.0	4	10	10	25	50	70	120	150	1.5	4	10	16	25	50	95	120	240
8	1.5	4	10	16	25	50	70	120	185	1.5	4	10	16	25	70	95	150	240
9	1.5	4	10	16	25	50	70	120	185	1.5	6	10	16	35	70	95	150	240
10	1.5	4	10	16	25	50	95	120	185	1.5	6	10	16	35	70	95	185	240
11	1.5	4	10	16	25	70	95	150	240	1.5	6	10	16	35	70	95	185	240
12	1.5	6	10	16	35	70	95	150	240	2.5	6	10	16	35	70	120	185	300
13	1.5	6	10	16	35	70	95	150	240	2.5	6	16	25	35	70	120	185	300
14	1.5	6	10	16	35	70	95	150	240	2.5	6	16	25	35	70	120	240	300
15	1.5	6	10	16	35	70	95	185	240	2.5	6	16	25	35	70	120	240	300
16	2.5	6	10	16	35	70	120	185	240	2.5	6	16	25	50	95	120	240	300
17	2.5	6	10	16	35	70	120	185	300	2.5	10	16	25	50	95	150	240	400
18	2.5	6	16	25	35	70	120	185	300	2.5	10	16	25	50	95	150	240	400
19	2.5	6	16	25	35	70	120	185	300	2.5	10	16	25	50	95	150	240	400
20	2.5	6	16	25	35	70	120	185	300	2.5	10	16	25	50	95	150	240	400

Note: Where a cable or circuit will not carry more than 30% of the tabulated current rating of its grouped cable size, it can be ignored when counting the number grouped together for the remaining circuits. Where the distance between adjacent conductors or cables is more than twice the diameter of the larger cable, no derating for grouping need be applied.   Circuit groupings can be a mixture of three phase and single phase circuits.

CABLE CAPACITY

CCC 1

CONDUIT

Cable capacity of conduit

Multiply the quantity of each size of cable by the cable factor from Table 1; add the total factors obtained for each each cable size together, then select a factor from the appropriate column of Table 2 which factor is equal to, or greater than, the total of the cable factors. The conduit size required is given at the top of each column of factors.

Table 1

Cable size sq.mm	Cable factor
1.0	16
1.5	22
2.5	30
4.0	43
6.0	58
10	105
16	145
25	218
35	275
50	383

Table 2

Conduit length in metres between draw-in boxes	Select column for straight or number of bends between draw in boxes																	
	Straight runs						One bend					Two bends					Three bends			
	Conduit size						Conduit size					Conduit size					Conduit size			
	16	20	25	32	38		16	20	25	32	38	16	20	25	32	16	20	25	32
1	207	329	571	1000	1234		188	303	543	947	1128	177	286	514	900	158	256	463	818
2	207	329	571	1000	1234		177	286	514	900	1072	158	256	463	818	130	213	388	692
3	207	329	571	1000	1234		167	270	487	857	1020	143	233	422	750	111	182	333	600
4	177	286	514	900	1110		158	256	463	818	973	130	213	388	692	97	159	292	529
5	171	278	500	878	1082		150	244	442	783	931	120	196	356	643	86	141	260	474
6	167	270	487	857	1056		143	233	422	750	891	111	182	333	600				
7	162	263	475	837	1030		136	222	404	720	855	103	169	311	563				
8	158	256	463	818	1005		130	213	388	692	822	97	159	292	529				
9	154	250	452	800	980		125	204	373	667	792	91	149	275	500				
10	150	244	442	783	956		120	196	358	643	763	86	141	260	474				

Note: The 38mm column is for 1.5 inch conduit.

CCT 1

Cable capacity of trunking

Multiply each size of cable to be used by the factor for the cable size from table 1; add together the total obtained for each size of cable, then compare the total with the factors for trunking from table 2, selecting a trunking factor equal to, or greater than, the total factor obtained for the cables.

Table 1

Cable factors

Type	Cable size sq.mm	Factor
Solid	1.5	7.1
"	2.5	10.2
Stranded	1.5	8.1
"	2.5	11.4
"	4	15.2
"	6	22.9
"	10	36.3
"	16	50.3
"	25	75.4
"	35	95.0
"	50	132.7
"	70	176.7
"	95	227.0
"	120	284.0
"	150	346.0

Table 2

Trunking factors

Trunking size mm	Factor
75 x 25	738
50 x 37.5	767
100 x 25	993
50 x 50	1037
75 x 37.5	1146
100 x 37.5	1542
75 x 50	1555
100 x 50	2091
75 x 75	2371
150 x 50	3161
100 x 75	3189
100 x 100	4252
150 x 75	4787
150 x 100	6414
150 x 150	9575

PART FOUR

DESIGN TABLES

SPACE FOR NOTES OR AMENDMENTS:

Contents

Breaking capacity of equipment
MISC 3   Breaking capacity of fuses and mcbs 406

Cosine conversion
MISC 2   Table giving the sine of power factor angle for use with voltage drop calculations 329

Cross-sectional area of protective conductors
ELI 1   PVC and XLPE armoured cables up to 16 mm^2 399
ELI 2   PVC SWA & PVC copper and aluminium cables, 25mm^2 to 300 mm^2 400
ELI 3   XLPE SWA copper and aluminium cables, 25mm^2 to 300 mm^2 401
ELI 4   PVC & XLPE aluminium strip armoured solidal cables 16 mm^2 to 300 mm^2 402
ELI 5   Sheath area of micc cables 403
MISC 1   Table M54G giving cross-sectional areas required 404
MISC 1   Cross sectional area of steel conduit, trunking, and the cpc in twin & cpc cable 404

Current-carrying capacity required for grouped cables:
Current carrying capacity required for single phase and three phase circuits, or multicore cables when grouped together. (Excluding BS 3036 fuses if used for overload protection.)
CRG 1   Enclosed in conduit, trunking, or bunched and clipped direct 379
CRG 3   Single layer, clipped direct to a non-metallic surface, cable sheaths touching 381
CRG 5   Single layer, clipped direct to a non-metallic surface, 1 cable dia. between cables 383
CRG 6   Multicore cable in a single layer on metal cable tray , cable sheaths touching 384
CRG 8   Multicore cable in a single layer on metal cable tray, 1 cable dia. between cables 386
CRG 9   Multicore cable in a single layer on ladder support, cable sheaths touching 387
CRG 11   Micc cables installed on cable tray 389

Current carrying capacity required for single phase and three phase circuits or multicore cables when grouped together, & protected against overload by a BS 3036 fuse.
CRG 2   Enclosed in conduit, trunking, or bunched and clipped direct 380
CRG 4   Single layer, clipped direct to a non-metallic surface, cable sheaths touching 382
CRG 5   Single layer, clipped direct to a non-metallic surface, 1 cable dia. between cables 383
CRG 7   Multicore cable in a single layer on metal cable tray, cable sheaths touching 385
CRG 8   Multicore cable in a single layer on metal cable tray, 1 cable dia. between cables 386
CRG 10   Multicore cable in a single layer on ladder support, cable sheaths touching 388

Derating factors
GF 1   Derating factors for single core and multicore cables grouped together 390

Earth loop impedance of armoured cables and micc cables

ELI 1	Earth fault loop impedance for PVC and XLPE armoured cables up to 16 mm^2	399
ELI 2	Earth loop impedance of PVC SWA & PVC copper and aluminium cables	400
ELI 3	Earth loop impedance of XLPE SWA & PVC copper and aluminium cables	401
ELI 4	Earth loop impedance PVC & XLPE aluminium strip armoured cables	402
ELI 5	Earth loop impedance and sheath area of micc cables	403

Fuse characteristics

1	WYLEX rewirable fuse and I^2t Characteristcs .5 A TO 30 A	415
2	GEC BS 88 Type NIT Time /Current Characteristics 2A to 20A & 20M32	416
3	GEC BS88 Type T Time/Current Characteristics 2A to 63A & 32M63	417
4	GEC BS 88 Type T Time/Current Characteristics 80A to 1250A and 630M670	418
5	GEC BS 88 Type T Cut-off Characteristcs 2A to 1250A	420
6	GEC BS 88 Type T Cut-off Characteristics 2A to 1250A	421
7	GEC BS 88 Type T and NIT I^2t Characteristcs 2A to 1250A	422

K factors

K 1A	Values of 'k' for different conductor materials and temperatures	397
K 1B	Values of 'k' for different conductor materials and temperatures	398

Miniature circuit breaker characteristics

1	Merlin Gerin C60H M9 Type 2 mcb, Time/Current Characteristics (Maximum)	411
2	Merlin Gerin Compact mccb characteristics (D type trip only), Ref: temperature 40 °C	413
3	Merlin Gerin I^2t energy let-through characteristcs for mcbs and mccbs	414

Motor protection and comparative fuse sizes

HRC 1	Industrial HRC fuse comparison chart	392
MC 1	Full load currents for three phase motors and recommended fuse ratings	391

Miscellaneous formula and tables

MISC 4	Factors for converting conductor temperature into design temperature	407
MISC 5	Resistor colour code, Conversion factors	408
MISC 6	Lighting formula ; Heating formula	409
MISC 7	Temperature rise of a conductor;Water heating formula; Three phase formula	410

Resistance and impedance

RA1	Resistance and impedance of PVC and XLPE solid aluminium conductor cables	393
RC 1	Resistance and impedance of PVC insulated copper cables up to 300 mm^2	394
RXLPE 1	Resistance and impedance of XLPE insulated copper cables up to 300 mm^2	395
ZCT 1	Impedance of steel conduit and trunking	396
MISC 3	Impedance of transformers	406

Temperature conversion

MISC 3	Temperature conversion: degrees F to degrees C	406

I_t GROUPED CABLES CRG 1 ENCLOSED OR BUNCHED

Required current ratings for three-phase & single-phase circuits enclosed in conduit, trunking, or bunched and clipped direct to a non-metallic surface.

Number of circuits or multicore cables	Size of I_n or I_b in amperes, excluding BS 3036 fuses when they are used for overload protection																				
	2	4	5	6	10	15	16	20	25	30	32	35	40	45	50	60	63	80	100	125	160	200	250
	Current-carrying capacity required for conductors in amperes																						
2	2.5	5.0	6.3	7.5	12.5	18.8	20.0	25.0	31.3	37.5	40.0	43.8	50.0	56.3	62.5	75.0	78.8	100.0	125.0	156.3	200.0	250.0	312.5
3	2.9	5.7	7.1	8.6	14.3	21.4	22.9	28.6	35.7	42.9	45.7	50.0	57.1	64.3	71.4	85.7	90.0	114.3	142.9	178.6	228.6	285.7	357.1
4	3.1	6.2	7.7	9.2	15.4	23.1	24.6	30.8	38.5	46.2	49.2	53.8	61.5	69.2	76.9	92.3	96.9	123.1	153.8	192.3	246.2	307.7	384.6
5	3.3	6.7	8.3	10.0	16.7	25.0	26.7	33.3	41.7	50.0	53.3	58.3	66.7	75.0	83.3	100.0	105.0	133.3	166.7	208.3	266.7	333.3	416.7
6	3.5	7.0	8.8	10.5	17.5	26.3	28.1	35.1	43.9	52.6	56.1	61.4	70.2	78.9	87.7	105.3	110.5	140.4	175.4	219.3	280.7	350.9	438.6
7	3.7	7.4	9.3	11.1	18.5	27.8	29.6	37.0	46.3	55.6	59.3	64.8	74.1	83.3	92.6	111.1	116.7	148.1	185.2	231.5	296.3	370.4	463.0
8	3.8	7.7	9.6	11.5	19.2	28.8	30.8	38.5	48.1	57.7	61.5	67.3	76.9	86.5	96.2	115.4	121.2	153.8	192.3	240.4	307.7	384.6	480.8
9	4.0	8.0	10.0	12.0	20.0	30.0	32.0	40.0	50.0	60.0	64.0	70.0	80.0	90.0	100.0	120.0	126.0	160.0	200.0	250.0	320.0	400.0	500.0
10	4.2	8.3	10.4	12.5	20.8	31.3	33.3	41.7	52.1	62.5	66.7	72.9	83.3	93.8	104.2	125.0	131.3	166.7	208.3	260.4	333.3	416.7	520.8
11	4.3	8.6	10.8	12.9	21.5	32.3	34.4	43.0	53.8	64.5	68.8	75.3	86.0	96.8	107.5	129.0	135.5	172.0	215.1	268.8	344.1	430.1	537.6
12	4.4	8.9	11.1	13.3	22.2	33.3	35.6	44.4	55.6	66.7	71.1	77.8	88.9	100.0	111.1	133.3	140.0	177.8	222.2	277.8	355.6	444.4	555.6
13	4.5	9.1	11.4	13.6	22.7	34.1	36.4	45.5	56.8	68.2	72.7	79.5	90.9	102.3	113.6	136.4	143.2	181.8	227.3	284.1	363.6	454.5	568.2
14	4.7	9.3	11.6	14.0	23.3	34.9	37.2	46.5	58.1	69.8	74.4	81.4	93.0	104.7	116.3	139.5	146.5	186.0	232.6	290.7	372.1	465.1	581.4
15	4.8	9.5	11.9	14.3	23.8	35.7	38.1	47.6	59.5	71.4	76.2	83.3	95.2	107.1	119.0	142.9	150.0	190.5	238.1	297.6	381.0	476.2	595.2
16	4.9	9.8	12.2	14.6	24.4	36.6	39.0	48.8	61.0	73.2	78.0	85.4	97.6	109.8	122.0	146.3	153.7	195.1	243.9	304.9	390.2	487.8	609.8
17	5.0	10.0	12.5	15.0	25.0	37.5	40.0	50.0	62.5	75.0	80.0	87.5	100.0	112.5	125.0	150.0	157.5	200.0	250.0	312.5	400.0	500.0	625.0
18	5.1	10.3	12.8	15.4	25.6	38.5	41.0	51.3	64.1	76.9	82.1	89.7	102.6	115.4	128.2	153.8	161.5	205.1	256.4	320.5	410.3	512.8	641.0
19	5.2	10.4	13.0	15.6	26.0	39.0	41.6	51.9	64.9	77.9	83.1	90.9	103.9	116.9	129.9	155.8	163.6	207.8	259.7	324.7	415.6	519.5	649.4
20	5.3	10.5	13.2	15.8	26.3	39.5	42.1	52.6	65.8	78.9	84.2	92.1	105.3	118.4	131.6	157.9	165.8	210.5	263.2	328.9	421.1	526.3	657.9

Note: Where a cable or circuit will not carry more than 30% of the tabulated current rating of its grouped cable size, it can be ignored when counting the number of cables grouped together for the remaining circuits. Where the distance between conductors or cables is more than twice the diameter of the larger cable, no derating for grouping need be applied. Circuit groupings can be a mixture of three-phase and single-phase circuits. The current rating of conductors should be based on the lowest operating temperature of any conductor in the same group. Where the circuit is not likely to carry overload current, use the full load current I_b instead of the protective device rating I_n.

I_t GROUPED CABLES   CRG 2   ENCLOSED OR BUNCHED

Required current ratings for three-phase & single-phase circuits enclosed in conduit, trunking, or bunched and clipped direct to a non-metallic surface. Overload protection by BS 3036 fuse.

Number of circuits or multicore cables	Size of BS 3036 protective device in amperes, when BS 3036 fuse used for overload protection						
	5	10	15	20	30	45	60	80	100
	Current-carrying capacity required for conductors in amperes						
2	8.6	17.2	25.9	34.5	51.7	77.6	103.4	137.9	172.4
3	9.9	19.7	29.6	39.4	59.1	88.7	118.2	157.6	197.0
4	10.6	21.2	31.8	42.4	63.7	95.5	127.3	169.8	212.2
5	11.5	23.0	34.5	46.0	69.0	103.4	137.9	183.9	229.9
6	12.1	24.2	36.3	48.4	72.6	108.9	145.2	193.6	242.0
7	12.8	25.5	38.3	51.1	76.6	114.9	153.3	204.3	255.4
8	13.3	26.5	39.8	53.1	79.6	119.4	159.2	212.2	265.3
9	13.8	27.6	41.4	55.2	82.8	124.1	165.5	220.7	275.9
10	14.4	28.7	43.1	57.5	86.2	129.3	172.4	229.9	287.4
11	14.8	29.7	44.5	59.3	89.0	133.5	178.0	237.3	296.6
12	15.3	30.7	46.0	61.3	92.0	137.9	183.9	245.2	306.5
13	15.7	31.3	47.0	62.7	94.0	141.1	188.1	250.8	313.5
14	16.0	32.1	48.1	64.2	96.2	144.3	192.5	256.6	320.8
15	16.4	32.8	49.3	65.7	98.5	147.8	197.0	262.7	328.4
16	16.8	33.6	50.5	67.3	100.9	151.4	201.9	269.1	336.4
17	17.2	34.5	51.7	69.0	103.4	155.2	206.9	275.9	344.8
18	17.7	35.4	53.1	70.7	106.1	159.2	212.2	282.9	353.7
19	17.9	35.8	53.7	71.7	107.5	161.2	215.0	286.6	358.3
20	18.1	36.3	54.4	72.6	108.9	163.3	217.8	290.4	363.0

Note: Where a cable or circuit will not carry more than 30% of the tabulated current rating of its grouped cable size, it can be ignored when counting the number of cables grouped together for the remaining circuits. Where the distance between conductors or cables is more than twice the diameter of the larger cable, no derating for grouping need be applied. Circuit groupings can be a mixture of three-phase and single-phase circuits. The current rating of conductors should be based on the lowest operating temperature of any conductor in the same group.

I_t GROUPED CABLES   CRG 3   GROUPED-CLIPPED DIRECT

Required current ratings for three-phase & single-phase circuits, single layer of cable clipped direct to, or lying on a non-metallic surface with cable sheaths touching.

Number of Circuits or Multicore cables	Size of I_n or I_b in amperes, excluding BS 3036 fuses when they are used for overload protection																
	2	4	5	6	10	15	16	20	25	30	32	35	40	45	50	60
	Current-carrying capacity required for conductors in amperes																
2	2.4	4.7	5.9	7.1	11.8	17.6	18.8	23.5	29.4	35.3	37.6	41.2	47.1	52.9	58.8	70.6
3	2.5	5.1	6.3	7.6	12.7	19.0	20.3	25.3	31.6	38.0	40.5	44.3	50.6	57.0	63.3	75.9
4	2.7	5.3	6.7	8.0	13.3	20.0	21.3	26.7	33.3	40.0	42.7	46.7	53.3	60.0	66.7	80.0
5	2.7	5.5	6.8	8.2	13.7	20.5	21.9	27.4	34.2	41.1	43.8	47.9	54.8	61.6	68.5	82.2
6	2.8	5.6	6.9	8.3	13.9	20.8	22.2	27.8	34.7	41.7	44.4	48.6	55.6	62.5	69.4	83.3
7	2.8	5.6	6.9	8.3	13.9	20.8	22.2	27.8	34.7	41.7	44.4	48.6	55.6	62.5	69.4	83.3
8	2.8	5.6	7.0	8.5	14.1	21.1	22.5	28.2	35.2	42.3	45.1	49.3	56.3	63.4	70.4	84.5
9	2.9	5.7	7.1	8.6	14.3	21.4	22.9	28.6	35.7	42.9	45.7	50.0	57.1	64.3	71.4	85.7

Number of Circuits	63	80	100	125	160	200	250
2	74.1	94.1	117.6	147.1	188.2	235.3	294.1
3	79.7	101.3	126.6	158.2	202.5	253.2	316.5
4	84.0	106.7	133.3	166.7	213.3	266.7	333.3
5	86.3	109.6	137.0	171.2	219.2	274.0	342.5
6	87.5	111.1	138.9	173.6	222.2	277.8	347.2
7	87.5	111.1	138.9	173.6	222.2	277.8	347.2
8	88.7	112.7	140.8	176.1	225.4	281.7	352.1
9	90.0	114.3	142.9	178.6	228.6	285.7	357.1

Note: Where a cable or circuit will not carry more than 30% of the tabulated current rating of its grouped cable size, it can be ignored when counting the number of cables grouped together for the remaining circuits. Where the distance between conductors or cables is more than twice the diameter of the larger cable, no derating for grouping need be applied. Circuit groupings can be a mixture of three-phase and single-phase circuits. The current rating of conductors should be based on the lowest operating temperature of any conductor in the same group. Where the circuit is not likely to carry overload current, use the full load current I_b instead of the protective device rating I_n.

I_t GROUPED CABLES   CRG 4   GROUPED-CLIPPED DIRECT

Required current ratings for three-phase & single-phase circuits, single layer of cable clipped direct to, or lying on a non-metallic surface with cable sheaths touching.

Overload protection by BS 3036 fuse.

Number of Circuits or Multicore cables	Size of BS 3036 protective device in amperes, when BS 3036 fuse used for overload protection							
	5	10	15	20	30	45	60	80	100
	Current-carrying capacity required for conductors in amperes								
2	8.1	16.2	24.3	32.5	48.7	73.0	97.4	129.8	162.3
3	8.7	17.5	26.2	34.9	52.4	78.6	104.8	139.7	174.6
4	9.2	18.4	27.6	36.8	55.2	82.8	110.3	147.1	183.9
5	9.4	18.9	28.3	37.8	56.7	85.0	113.4	151.2	188.9
6	9.6	19.2	28.7	38.3	57.5	86.2	114.9	153.3	191.6
7	9.6	19.2	28.7	38.3	57.5	86.2	114.9	153.3	191.6
8	9.7	19.4	29.1	38.9	58.3	87.4	116.6	155.4	194.3
9	9.9	19.7	29.6	39.4	59.1	88.7	118.2	157.6	197.0

Note: Where a cable or circuit will not carry more than 30% of the tabulated current rating of its grouped cable size, it can be ignored when counting the number of cables grouped together for the remaining circuits. Where the distance between conductors or cables is more than twice the diameter of the larger cable, no derating for grouping need be applied. Circuit groupings can be a mixture of three-phase and single-phase circuits. The current rating of conductors should be based on the lowest operating temperature of any conductor in the same group.

I_t GROUPED CABLES   CRG 5   GROUPED-CLIPPED DIRECT

Required current ratings for three-phase & single-phase circuits, single layer of cable clipped direct to, or lying on a non-metallic surface with at least one cable diameter between adjacent cable sheaths.

Number of circuits or multicore cables	Size of I_n or I_b in amperes																						
	2	4	5	6	10	15	16	20	25	30	32	35	40	45	50	60	63	80	100	125	160	200	250
	Current-carrying capacity required for conductors in amperes																						
	Use this table for all protective devices when BS 3036 fuse is not being used for overload protection																						
2	2.1	4.3	5.3	6.4	10.6	16.0	17.0	21.3	26.6	31.9	34.0	37.2	42.6	47.9	53.2	63.8	67.0	85.1	106.4	133.0	170.2	212.8	266.0
3 TO 20	2.2	4.4	5.6	6.7	11.1	16.7	17.8	22.2	27.8	33.3	35.6	38.9	44.4	50.0	55.6	66.7	70.0	88.9	111.1	138.9	177.8	222.2	277.8
	Use this table for BS 3036 protective devices when BS 3036 fuse is being used for overload protection																						
2	2.9	5.9	7.3	8.8	14.7	22.0	23.5	29.3	36.7	44.0	47.0	51.4	58.7	66.0	73.4	88.0	92.4	117.4	146.7	183.4	234.8	293.5	366.8
3 to 20	3.1	6.1	7.7	9.2	15.3	23.0	24.5	30.7	38.3	46.0	49.0	53.6	61.3	69.0	76.6	92.0	96.6	122.6	153.3	191.6	245.2	306.5	383.1

Note: Where a cable or circuit will not carry more than 30% of the tabulated current rating of its grouped cable size, it can be ignored when counting the number of cables grouped together for the remaining circuits. Where the distance between conductors or cables is more than twice the diameter of the larger cable, no derating for grouping need be applied. Circuit groupings can be a mixture of three-phase and single-phase circuits. The current rating of conductors should be based on the lowest operating temperature of any conductor in the same group. Where the circuit is not likely to carry overload current, use the full load current I_b instead of the protective device rating I_n.

I_t GROUPED CABLES — CRG 6 — GROUPED ON TRAY

Required current ratings for three-phase & single-phase circuits, single layer of multicore cable on perforated metal cable tray with cable sheaths touching.
(Not applicable to micc cables, see CRG 11.)

Size of I_n or I_b in amperes, excluding BS 3036 fuses when they are used for overload protection

Current-carrying capacity required for conductors in amperes

Number Multicore cables	2	4	5	6	10	15	16	20	25	30	32	35	40	45	50	60	63	80	100	125	160	200	250
2	2.3	4.7	5.8	7.0	11.6	17.4	18.6	23.3	29.1	34.9	37.2	40.7	46.5	52.3	58.1	69.8	73.3	93.0	116.3	145.3	186.0	232.6	290.7
3	2.5	4.9	6.2	7.4	12.3	18.5	19.8	24.7	30.9	37.0	39.5	43.2	49.4	55.6	61.7	74.1	77.8	98.8	123.5	154.3	197.5	246.9	308.6
4	2.6	5.2	6.5	7.8	13.0	19.5	20.8	26.0	32.5	39.0	41.6	45.5	51.9	58.4	64.9	77.9	81.8	103.9	129.9	162.3	207.8	259.7	324.7
5	2.7	5.3	6.7	8.0	13.3	20.0	21.3	26.7	33.3	40.0	42.7	46.7	53.3	60.0	66.7	80.0	84.0	106.7	133.3	166.7	213.3	266.7	333.3
6	2.7	5.4	6.8	8.1	13.5	20.3	21.6	27.0	33.8	40.5	43.2	47.3	54.1	60.8	67.6	81.1	85.1	108.1	135.1	168.9	216.2	270.3	337.8
7	2.7	5.5	6.8	8.2	13.7	20.5	21.9	27.4	34.2	41.1	43.8	47.9	54.8	61.6	68.5	82.2	86.3	109.6	137.0	171.2	219.2	274.0	342.5
8	2.7	5.5	6.8	8.2	13.7	20.5	21.9	27.4	34.2	41.1	43.8	47.9	54.8	61.6	68.5	82.2	86.3	109.6	137.0	171.2	219.2	274.0	342.5
9	2.8	5.6	6.9	8.3	13.9	20.8	22.2	27.8	34.7	41.7	44.4	48.6	55.6	62.5	69.4	83.3	87.5	111.1	138.9	173.6	222.2	277.8	347.2
10	2.8	5.6	7.0	8.5	14.1	21.1	22.5	28.2	35.2	42.3	45.1	49.3	56.3	63.4	70.4	84.5	88.7	112.7	140.8	176.1	225.4	281.7	352.1
11	2.8	5.7	7.1	8.5	14.2	21.3	22.7	28.4	35.5	42.6	45.4	49.6	56.7	63.8	70.9	85.1	89.4	113.5	141.8	177.3	227.0	283.7	354.6
12	2.9	5.7	7.1	8.6	14.3	21.4	22.9	28.6	35.7	42.9	45.7	50.0	57.1	64.3	71.4	85.7	90.0	114.3	142.9	178.6	228.6	285.7	357.1

Note: Where a cable or circuit will not carry more than 30% of the tabulated current rating of its grouped cable size, it can be ignored when counting the number of cables grouped together for the remaining circuits. Where the distance between conductors or cables is more than twice the diameter of the larger cable, no derating for grouping need be applied. Circuit groupings can be a mixture of three-phase and single-phase circuits. The current rating of conductors should be based on the lowest operating temperature of any conductor in the same group. Where the circuit is not likely to carry overload current, use the full load current I_b instead of the protective device rating I_n.

I_t GROUPED CABLES — CRG 7 — GROUPED ON TRAY

Required current ratings for three-phase & single-phase circuits, single layer of multicore cable on perforated metal cable tray with cable sheaths touching. Overload protection by BS 3036 fuse.

(Not applicable to micc cables, see CRG 11)

Number Multicore cables	Size of protective device in amperes, when BS 3036 fuses are used for overload protection									
	5	10	15	20	30	45	60	80	100
	Current-carrying capacity required for conductors in amperes									
2	8.02	16.0	24.1	32.1	48.1	72.2	96	128	160
3	8.51	17.0	25.5	34.1	51.1	76.6	102	136	170
4	8.96	17.9	26.9	35.8	53.7	80.6	107	143	179
5	9.20	18.4	27.6	36.8	55.2	82.8	110	147	184
6	9.32	18.6	28.0	37.3	55.9	83.9	112	149	186
7	9.45	18.9	28.3	37.8	56.7	85.0	113	151	189
8	9.45	18.9	28.3	37.8	56.7	85.0	113	151	189
9	9.58	19.2	28.7	38.3	57.5	86.2	115	153	192
10	9.71	19.4	29.1	38.9	58.3	87.4	117	155	194
11	9.78	19.6	29.3	39.1	58.7	88.0	117	157	196
12	9.85	19.7	29.6	39.4	59.1	88.7	118	158	197

Note: Where a cable or circuit will not carry more than 30% of the tabulated current rating of its grouped cable size, it can be ignored when counting the number of cables grouped together for the remaining circuits. Where the distance between conductors or cables is more than twice the diameter of the larger cable, no derating for grouping need be applied. Circuit groupings can be a mixture of three-phase and single-phase circuits. The current rating of conductors should be based on the lowest operating temperature of any conductor in the same group.

I_t GROUPED CABLES — CRG 8 — GROUPED ON TRAY

Required current ratings for three-phase & single-phase circuits, single layer of multicore cable on perforated metal cable tray with at least one cable diameter between adjacent cable sheaths.

(Not applicable to micc cables, see CRG 11.)

Number of multicore cables	Size of I_n or I_b in amperes																		
	2	4	5	6	10	15	16	20	25	30	32	35	40	45	50	60	63	80	100	125	160	200	250

Current-carrying capacity required for conductors in amperes

Use this table for all protective devices when BS 3036 fuse not being used for overload protection

N	2	4	5	6	10	15	16	20	25	30	32	35	40	45	50	60	63	80	100	125	160	200	250
2	2.20	4.40	5.49	6.59	11.0	16.5	17.6	22.0	27.5	33.0	35.2	38.5	44.0	49.5	54.9	65.9	69.2	87.9	110	137	176	220	275
3	2.25	4.49	5.62	6.74	11.2	16.9	18.0	22.5	28.1	33.7	36.0	39.3	44.9	50.6	56.2	67.4	70.8	89.9	112	140	180	225	281
4	2.27	4.55	5.68	6.82	11.4	17.0	18.2	22.7	28.4	34.1	36.4	39.8	45.5	51.1	56.8	68.2	71.6	90.9	114	142	182	227	284
5	2.30	4.60	5.75	6.90	11.5	17.2	18.4	23.0	28.7	34.5	36.8	40.2	46.0	51.7	57.5	69.0	72.4	92.0	115	144	184	230	287
6	2.30	4.60	5.75	6.90	11.5	17.2	18.4	23.0	28.7	34.5	36.8	40.2	46.0	51.7	57.5	69.0	72.4	92.0	115	144	184	230	287

Use this table for BS 3036 protective devices when BS 3036 fuse is being used for overload protection

N	2	4	5	6	10	15	16	20	25	30	32	35	40	45	50	60	63	80	100	125	160	200	250
2	3.03	6.06	7.58	9.09	15.2	22.7	24.3	30.3	37.9	45.5	48.5	53.1	60.6	68.2	75.8	90.9	95.5	121	152	189	243	303	379
3	3.10	6.20	7.75	9.30	15.5	23.2	24.8	31.0	38.7	46.5	49.6	54.2	62.0	69.7	77.5	93.0	97.6	124	155	194	248	310	387
4	3.13	6.27	7.84	9.40	15.7	23.5	25.1	31.3	39.2	47.0	50.2	54.9	62.7	70.5	78.4	94.0	98.7	125	157	196	251	313	392
5	3.17	6.34	7.93	9.51	15.9	23.8	25.4	31.7	39.6	47.6	50.7	55.5	63.4	71.3	79.3	95.1	99.9	127	159	198	254	317	396
6	3.17	6.34	7.93	9.51	15.9	23.8	25.4	31.7	39.6	47.6	50.7	55.5	63.4	71.3	79.3	95.1	99.9	127	159	198	254	317	396

Note: Where a cable or circuit will not carry more than 30% of the tabulated current rating of its grouped cable size, it can be ignored when counting the number of cables grouped together for the remaining circuits. Where the distance between conductors or cables is more than twice the diameter of the larger cable, no derating for grouping need be applied. Circuit groupings can be a mixture of three-phase and single-phase circuits. The current rating of conductors should be based on the lowest operating temperature of any conductor in the same group. Where the circuit is not likely to carry overload current, use the full load current I_b instead of the protective device rating I_n.

I_t GROUPED CABLES   CRG 9   GROUPED-LADDERS

Required current ratings for three-phase & single-phase circuits, single layer of multicore cable clipped to a ladder support with cable sheaths touching.

Number of Multicore cables	Size of I_n or I_b in amperes, excluding BS 3036 fuses when they are used for overload protection																			
	2	4	5	6	10	15	16	20	25	30	32	35	40	45	50	60	63	80	100	125	160	200	250
	Current-carrying capacity required for conductors in amperes																					
2	2.3	4.7	5.8	7.0	11.6	17.4	18.6	23.3	29.1	34.9	37.2	40.7	46.5	52.3	58.1	69.8	73.3	93.0	116.3	145.3	186.0	232.6	290.7
3	2.4	4.9	6.1	7.3	12.2	18.3	19.5	24.4	30.5	36.6	39.0	42.7	48.8	54.9	61.0	73.2	76.8	97.6	122.0	152.4	195.1	243.9	304.9
4	2.5	5.0	6.3	7.5	12.5	18.8	20.0	25.0	31.3	37.5	40.0	43.8	50.0	56.3	62.5	75.0	78.8	100.0	125.0	156.3	200.0	250.0	312.5
5	2.5	5.1	6.3	7.6	12.7	19.0	20.3	25.3	31.6	38.0	40.5	44.3	50.6	57.0	63.3	75.9	79.7	101.3	126.6	158.2	202.5	253.2	316.5
6	2.6	5.1	6.4	7.7	12.8	19.2	20.5	25.6	32.1	38.5	41.0	44.9	51.3	57.7	64.1	76.9	80.8	102.6	128.2	160.3	205.1	256.4	320.5
7	2.6	5.1	6.4	7.7	12.8	19.2	20.5	25.6	32.1	38.5	41.0	44.9	51.3	57.7	64.1	76.9	80.8	102.6	128.2	160.3	205.1	256.4	320.5
8	2.6	5.1	6.4	7.7	12.8	19.2	20.5	25.6	32.1	38.5	41.0	44.9	51.3	57.7	64.1	76.9	80.8	102.6	128.2	160.3	205.1	256.4	320.5
9	2.6	5.2	6.5	7.8	13.0	19.5	20.8	26.0	32.5	39.0	41.6	45.5	51.9	58.4	64.9	77.9	81.8	103.9	129.9	162.3	207.8	259.7	324.7

Note: Where a cable or circuit will not carry more than 30% of the tabulated current rating of its grouped cable size, it can be ignored when counting the number of cables grouped together for the remaining circuits. Where the distance between conductors or cables is more than twice the diameter of the larger cable, no derating for grouping need be applied. Circuit groupings can be a mixture of three-phase and single-phase circuits. The current rating of conductors should be based on the lowest operating temperature of any conductor in the same group. Where the circuit is not likely to carry overload current, use the full load current I_b instead of the protective device rating I_n.

I_t GROUPED CABLES　　　CRG 10　　　GROUPED - LADDERS

Required current ratings for three-phase & single-phase circuits, single layer of multicore cable clipped to a ladder support with cable sheaths touching. Overload protection by BS 3036 fuse.

Number of multicore cables	Size of protective device in amperes, when BS 3036 fuses are used for overload protection																					
	2	4	5	6	10	15	16	20	25	30	32	35	40	45	50	60	63	80	100	125	160	200	250
	Current-carrying capacity required for conductors in amperes																						
2	3.2	6.4	8.0	9.6	16.0	24.1	25.7	32.1	40.1	48.1	51.3	56.1	64.2	72.2	80.2	96.2	101.0	128.3	160.4	200.5	256.6	320.8	401.0
3	3.4	6.7	8.4	10.1	16.8	25.2	26.9	33.6	42.1	50.5	53.8	58.9	67.3	75.7	84.1	100.9	106.0	134.6	168.2	210.3	269.1	336.4	420.5
4	3.4	6.9	8.6	10.3	17.2	25.9	27.6	34.5	43.1	51.7	55.2	60.3	69.0	77.6	86.2	103.4	108.6	137.9	172.4	215.5	275.9	344.8	431.0
5	3.5	7.0	8.7	10.5	17.5	26.2	27.9	34.9	43.6	52.4	55.9	61.1	69.8	78.6	87.3	104.8	110.0	139.7	174.6	218.2	279.4	349.2	436.5
6	3.5	7.1	8.8	10.6	17.7	26.5	28.3	35.4	44.2	53.1	56.6	61.9	70.7	79.6	88.4	106.1	111.4	141.5	176.8	221.0	282.9	353.7	442.1
7	3.5	7.1	8.8	10.6	17.7	26.5	28.3	35.4	44.2	53.1	56.6	61.9	70.7	79.6	88.4	106.1	111.4	141.5	176.8	221.0	282.9	353.7	442.1
8	3.5	7.1	8.8	10.6	17.7	26.5	28.3	35.4	44.2	53.1	56.6	61.9	70.7	79.6	88.4	106.1	111.4	141.5	176.8	221.0	282.9	353.7	442.1
9	3.6	7.2	9.0	10.7	17.9	26.9	28.7	35.8	44.8	53.7	57.3	62.7	71.7	80.6	89.6	107.5	112.9	143.3	179.1	223.9	286.6	358.3	447.8

Note: Where a cable or circuit will not carry more than 30% of the tabulated current rating of its grouped cable size, it can be ignored when counting the number of cables grouped together for the remaining circuits. Where the distance between conductors or cables is more than twice the diameter of the larger cable, no derating for grouping need be applied. Circuit groupings can be a mixture of three-phase and single-phase circuits. The current rating of conductors should be based on the lowest operating temperature of any conductor in the same group.

I_t GROUPED CABLES — CRG 11 — MICC GROUPED - TRAY

Required current ratings for three-phase & single-phase circuits, multicore micc cable installed on perforated cable tray with cable sheaths touching.

Number of multicore cables	Size of I_n or I_b in amperes, excluding BS 3036 fuses when they are used for overload protection																		
	2	4	5	6	10	15	16	20	25	30	32	35	40	45	50	60	63	80	100	125	160	200

Current-carrying capacity required for conductors in amperes

CURRENT RATING REQUIRED ON HORIZONTAL CABLE TRAY

Cables	2	4	5	6	10	15	16	20	25	30	32	35	40	45	50	60	63	80	100	125	160	200
2	2.22	4.44	5.56	6.67	11.11	16.67	17.78	22.22	27.78	33.33	35.56	38.89	44.44	50.00	55.56	66.67	70.00	88.89	111.11	138.89	177.78	222.22
3	2.50	5.00	6.25	7.50	12.50	18.75	20.00	25.00	31.25	37.50	40.00	43.75	50.00	56.25	62.50	75.00	78.75	100.00	125.00	156.25	200.00	250.00
4	2.50	5.00	6.25	7.50	12.50	18.75	20.00	25.00	31.25	37.50	40.00	43.75	50.00	56.25	62.50	75.00	78.75	100.00	125.00	156.25	200.00	250.00
5	2.58	5.16	6.45	7.74	12.90	19.35	20.65	25.81	32.26	38.71	41.29	45.16	51.61	58.06	64.52	77.42	81.29	103.23	129.03	161.29	206.45	258.06
6	2.67	5.33	6.67	8.00	13.33	20.00	21.33	26.67	33.33	40.00	42.67	46.67	53.33	60.00	66.67	80.00	84.00	106.67	133.33	166.67	213.33	266.67
7	2.67	5.33	6.67	8.00	13.33	20.00	21.33	26.67	33.33	40.00	42.67	46.67	53.33	60.00	66.67	80.00	84.00	106.67	133.33	166.67	213.33	266.67
8	2.67	5.33	6.67	8.00	13.33	20.00	21.33	26.67	33.33	40.00	42.67	46.67	53.33	60.00	66.67	80.00	84.00	106.67	133.33	166.67	213.33	266.67
9	2.67	5.33	6.67	8.00	13.33	20.00	21.33	26.67	33.33	40.00	42.67	46.67	53.33	60.00	66.67	80.00	84.00	106.67	133.33	166.67	213.33	266.67

CURRENT RATING REQUIRED ON VERTICAL CABLE TRAY

Cables	2	4	5	6	10	15	16	20	25	30	32	35	40	45	50	60	63	80	100	125	160	200
2	2.22	4.44	5.56	6.67	11.11	16.67	17.78	22.22	27.78	33.33	35.56	38.89	44.44	50.00	55.56	66.67	70.00	88.89	111.11	138.89	177.78	222.22
3	2.50	5.00	6.25	7.50	12.50	18.75	20.00	25.00	31.25	37.50	40.00	43.75	50.00	56.25	62.50	75.00	78.75	100.00	125.00	156.25	200.00	250.00
4	2.67	5.33	6.67	8.00	13.33	20.00	21.33	26.67	33.33	40.00	42.67	46.67	53.33	60.00	66.67	80.00	84.00	106.67	133.33	166.67	213.33	266.67
5	2.67	5.33	6.67	8.00	13.33	20.00	21.33	26.67	33.33	40.00	42.67	46.67	53.33	60.00	66.67	80.00	84.00	106.67	133.33	166.67	213.33	266.67
6	2.67	5.33	6.67	8.00	13.33	20.00	21.33	26.67	33.33	40.00	42.67	46.67	53.33	60.00	66.67	80.00	84.00	106.67	133.33	166.67	213.33	266.67
7	2.67	5.33	6.67	8.00	13.33	20.00	21.33	26.67	33.33	40.00	42.67	46.67	53.33	60.00	66.67	80.00	84.00	106.67	133.33	166.67	213.33	266.67
8	2.67	5.33	6.67	8.00	13.33	20.00	21.33	26.67	33.33	40.00	42.67	46.67	53.33	60.00	66.67	80.00	84.00	106.67	133.33	166.67	213.33	266.67
9	2.86	5.71	7.14	8.57	14.29	21.43	22.86	28.57	35.71	42.86	45.71	50.00	57.14	64.29	71.43	85.71	90.00	114.29	142.86	178.57	228.57	285.71

Note: Where a cable or circuit will not carry more than 30% of the tabulated current rating of its grouped cable size, it can be ignored when counting the number of cables grouped together for the remaining circuits. Where the distance between conductors or cables is more than twice the diameter of the larger cable, no derating for grouping need be applied. Circuit groupings can be a mixture of three-phase and single-phase circuits. The current rating of conductors should be based on the lowest operating temperature of any conductor in the same group. Where the circuit is not likely to carry overload current, use the full load current I_b instead of the protective device rating I_n.

DERATING FACTORS

GF 1

GROUPED CABLES

Derating factors for two or more circuits using single core cable, or two or more multicore cables, single-phase or three-phase circuits grouped separately or mixed together.

Method of installing circuits or cables		Number of circuits or multicore cables grouped together																		
		2	3	4	5	6	7	8	9	10	11	12	13	14	15	16	17	18	19	20
Installed in trunking or conduit		0.80	0.70	0.65	0.60	0.57	0.54	0.52	0.50	0.48	0.465	0.45	0.44	0.43	0.42	0.41	0.40	0.39	0.385	0.38
Bunched together & clipped direct to a non-metallic surface		0.80	0.70	0.65	0.60	0.57	0.54	0.52	0.50	0.48	0.465	0.45	0.44	0.43	0.42	0.41	0.40	0.39	0.385	0.38
Single layer clipped direct to, or lying on a non-metallic surface, cable sheaths touching		0.85	0.79	0.75	0.73	0.72	0.72	0.71	0.70	0.70	0.70	0.70	0.70	0.70	0.70	0.70	0.70	0.70	0.70	0.70
Single layer clipped direct to, or lying on a non-metallic surface, with one cable diameter between cables		0.94	0.9	0.9	0.9	0.9	0.9	0.9	0.9	0.9	0.9	0.9	0.9	0.9	0.9	0.9	0.9	0.9	0.9	0.9
Single layer multicore cable on perforated metal cable tray, horizontal or vertical with sheaths touching *		0.86	0.81	0.77	0.75	0.74	0.73	0.73	0.72	0.71	0.705	0.70	0.70	0.70	0.70	0.70	0.70	0.70	0.70	0.70
Single layer multicore cable on perforated metal cable tray, horizontal or vertical with one cable diameter between cables *		0.91	0.89	0.88	0.87	0.87	0.87	0.87	0.87	0.87	0.87	0.87	0.87	0.87	0.87	0.87	0.87	0.87	0.87	0.87
Single layer multicore cable on ladder support with cable sheaths touching		0.86	0.82	0.80	0.79	0.78	0.78	0.78	0.77	0.77	0.77	0.77	0.77	0.77	0.77	0.77	0.77	0.77	0.77	0.77
Single layer, single core cable on perforated metal cable tray cable sheaths touching.	Horizontal	0.90	0.85	0.85	0.85	0.85	0.85	0.85	0.85	0.85	0.85	0.85	0.85	0.85	0.85	0.85	0.85	0.85	0.85	0.85
	Vertical	0.85	0.85	0.85	0.775	0.75	0.75	0.75	0.75	0.75	0.75	0.75	0.75	0.75	0.75	0.75	0.75	0.75	0.75	0.75
Micc cables on perforated cable tray cables sheaths touching	Horizontal	0.9	0.8	0.8	0.75	0.75	0.75	0.75	0.75	0.75	0.75	0.75	0.75	0.75	0.75	0.75	0.75	0.75	0.75	0.75
	Vertical	0.9	0.8	0.75	0.75	0.75	0.75	0.70	0.70	0.70	0.70	0.70	0.70	0.70	0.70	0.70	0.70	0.70	0.70	0.70
Multicore cable on ladder sheaths touching		0.86	0.82	0.8	0.79	0.78	0.78	0.78	0.77	0.77	0.77	0.77	0.77	0.77	0.77	0.77	0.77	0.77	0.77	0.77

* Excluding micc cable

Note: Where a cable or circuit will not carry more than 30% of the tabulated current rating of its grouped cable size, it can be ignored when counting the number of cables grouped together for the remaining circuits. Where the distance between adjacent conductors or cables is more than twice the diameter of the larger cable, no derating for grouping need be applied. Circuit groupings can be a mixture of three-phase and single-phase circuits.

FULL LOAD CURRENT (FLC.)   MC1   HRC FUSE RATINGS

Full load currents for 3 phase induction motors at average efficiencies, power factor, with recommended fuse ratings

KW	H.P.	Motor full load current (FLC) in amperes							Direct on line start (DOL) (See notes)				Assisted Start (See notes)			
		220V	240V	380V	415V	440V	550V		Motor FLC amperes		HRC fuse Type 'gG'	Motor fuse Type 'gM'	Motor FLC ampere		HRC fuse Type 'gG'
									From	To			From	To	
0.37	0.50	2.00	1.75	1.15	1.05	1.00	0.80			0.70	2	-		1.40	2
0.55	0.75	2.70	2.60	1.60	1.50	1.40	1.10		0.80	1.40	4	-	1.50	2.10	4
0.75	1.0	3.90	3.30	2.30	2.00	1.90	1.50		1.50	2.00	6	-	2.20	3.10	6
1.10	1.5	4.70	4.35	2.80	2.50	2.40	1.90		2.10	3.00	10	-	3.20	5.50	10
1.50	2.0	6.50	5.90	3.80	3.50	3.30	2.60		3.10	6.10	16	-	5.60	10.0	16
2.20	3.0	9.30	8.30	5.40	5.00	4.70	3.80		6.20	9.00	20	-	10.1	14.0	20
3.00	4.0	12	11.0	7.10	6.50	6.10	4.90		9.10	11.00	25	20M25	14.1	18.0	25
4.00	5.5	15.4	14.0	9.00	8.40	7.90	6.40		11.1	14.4	32	20M32	18.1	22.0	32
5.50	7.5	20.7	20.0	11.9	11	10.3	8.2		14.5	15.4	35	32M35	22.1	28.0	35
7.50	10	28	25.0	16.1	14.4	14	11.2		15.5	18	40	32M40	28.1	32.0	40
11.0	15	39.1	36.5	23	21	19.8	15.8		18.1	22	50	32M50	32.1	40.0	50
15.0	20	52.8	48	30.5	28	26.4	21.1		22.1	28	63	32M63	40.1	51.0	63
18.5	25	66	60	38	35	33	26.4		28.1	45	80	63M80	51.1	80.0	80
22	30	77	72	45	41	39	31		45.1	58	100	63M100	80.1	100	100
30	40	103	96	60	55	52	42		58.1	80	125	100M125	100.1	125	125
37	50	128	120	75	69	65	52		80.1	99	160	100M160	125.1	160	160
45	60	151	144	87	80	75	60		99.1	128	200	-	160.1	200	200
55	75	185	171	107	98	92	74		128.1	180	250	200M250	200.1	250	250
75	100	257	235	148	136	128	102		180.1	216	315	200M315	250.1	315	315
90	120	308	278	180	164	154	123		216.1	270	355	315M355	315.1	355	355
110	150	370	345	214	196	185	148		270.1	328	400	-	355.1	400	400
132	175	426	403	247	226	213	170		328.1	385	450	400M450	400.1	450	450
150	200	500	454	292	268	252	202		385.1	430	500	-	450.1	500	500

Assumed starting conditions D.O.L up to 1 kW = 5 x FLC for 5 secs: 1.1 kW to 7.5 kW = 6 x FLC for 10 secs: 7.6 kW to 75 kW = 7 x FLC for 10 secs: over 75 kW = 6 x FLC for 15 secs. Assisted start up to 1 kW = 2.5 x FLC for 20 secs: over 1 kW = 3.5 x FLC for 20 secs.

391

H.R.C. FUSES

HRC 1

COMPARATIVE TYPES

Industrial HRC fuse comparative types

BS 88 size reference N°	Hawker	GEC	MEM	Lawson	Fluvent	Reyrolle Belmos	Old GEC
-	D19L	LST	LS	LST	-	RLST	-
A1	F21	NIT	SA2	NIT	FOBX	RIT	Y-PA/Y-QA
A2	H07	TIA	SB3	TIA	FOMX	GPE-6E/TIA	Y-RA
A3	K07	TIS	SB4	TIS	FOSX	GPE-6/RIS	Y-SA
A4	L14	TCP	SD5	TCP	FO1X	GPE-6F/RCP	Y-TA
-	M14	TFP	SD6	TFP	F763/2X	GPG-6F/RFP	Y-VAD
-	K08	TB	SE4	TB	754/M-754/S	GPE-5E/RC	Y-SB
-	K09	TBC	SF4	TBC	755/M-755/S	GPE-5F/RC	Y-TB
B1	L09	TC	SF5	TC	FB1	GPE-5F/RC	Y-TB
B2	M09	TF	SF6	TF	FB2	GPG-G/RF	Y-VB
B3	N09	TKF	SF7	TFK	FB3	GPH-5G/RKF	Y-WB
B4	P09	TMF	SF8	TMF	FB4	GPH-5G/RMF	Y-XBD
-	N11	TKM	SG7	TKM	F772/3	RKM	Y-WBL
C1	P11	TM	SH8	TM	F773/4	GPH-5H/RM	Y-XB
-	P20	85TM	SX8	85TM	-	-	Y-XA
-	R20	86TT	SX9	86TT	-	-	Y-YA
C2	R11	TTM	SH9	TTM	TB6	GPJ-5H/RTM	Y-YB
-	R12	TT	SY9	TT	F781-6	GPH5-5J	Y-YBL
C3	S11	TLM	SH10	TLM	FB6-F851/8	GPK-5H/RLM	Y-ZBD
-	S12	TLT	SY10	TLT	F781-6/F781-8	GPH-SJ	Y-Z8
D1	U44	TXU	SJ11	-	-	-	-
E1	DO4	SS	SS	SS	-	RSS	-
E1	FO6	NS	SN2/SN21	NS	F986B	RS	Y-PF/Y-QF
F2	ESD	ES	-	MES	-	-	-
-	K07R	OS	-	-	-	-	-

Note: This list is given for guidance only, since identical performance for the comparative types shown cannot be guaranteed. Under normal circumstances satisfactory interchangeability should be achieved.

RESISTANCE & IMPEDANCE — RA 1 — ALUMINIUM

Resistance, reactance and impedance of solid aluminium conductors for single core cables or two, three and four core armoured cables, ohms per 1000 metres

Aluminium conductors PVC insulated to BS 6346 :1989

Cable size sq.mm.	Temperature 20 °C			Temperature 95 °C			Temperature 107.8 °C			Temperaure 115 °C		
	R	X	Z	R	X	Z	R	X	Z	R	X	Z
16	1.910	-	1.910	2.483	-	2.483	2.581	-	2.581	2.636	-	2.636
25	1.200	-	1.200	1.560	-	1.560	1.621	-	1.621	1.656	-	1.656
35	0.868	-	0.868	1.128	-	1.128	1.173	-	1.173	1.198	-	1.198
50	0.641	0.082	0.646	0.833	0.082	0.837	0.866	0.082	0.870	0.885	0.082	0.888
70	0.443	0.079	0.450	0.576	0.079	0.581	0.599	0.079	0.604	0.611	0.079	0.616
95	0.320	0.078	0.329	0.416	0.078	0.423	0.432	0.078	0.439	0.442	0.078	0.448
120	0.253	0.077	0.264	0.329	0.077	0.338	0.342	0.077	0.350	0.349	0.077	0.358
150	0.206	0.077	0.220	0.268	0.077	0.279	0.278	0.077	0.289	0.284	0.077	0.295
185	0.164	0.077	0.181	0.213	0.077	0.227	0.222	0.077	0.235	0.226	0.077	0.239
240	0.125	0.076	0.146	0.163	0.076	0.179	0.169	0.076	0.185	0.173	0.076	0.189
300	0.100	0.076	0.126	0.130	0.076	0.151	0.135	0.076	0.155	0.138	0.076	0.158

Aluminium conductors XLPE insulated to BS 5467 :1989

Cable size sq.mm.	Temperature 20 °C			Temperature 140 °C			Temperature 159.2 °C			Temperaure 170 °C		
	R	X	Z	R	X	Z	R	X	Z	R	X	Z
16	1.910	-	1.910	2.827	-	2.827	2.973	-	2.973	3.056	-	3.056
25	1.200	-	1.200	1.776	-	1.776	1.868	-	1.868	1.920	-	1.920
35	0.868	-	0.868	1.285	-	1.285	1.351	-	1.351	1.389	-	1.389
50	0.641	0.077	0.646	0.949	0.077	0.952	0.998	0.077	1.001	1.026	0.077	1.028
70	0.443	0.075	0.449	0.656	0.075	0.660	0.690	0.075	0.694	0.709	0.075	0.713
95	0.320	0.073	0.328	0.474	0.073	0.479	0.498	0.073	0.503	0.512	0.073	0.517
120	0.253	0.073	0.263	0.374	0.073	0.381	0.394	0.073	0.401	0.405	0.073	0.411
150	0.206	0.074	0.219	0.305	0.074	0.314	0.321	0.074	0.329	0.330	0.074	0.338
185	0.164	0.074	0.180	0.243	0.074	0.254	0.255	0.074	0.266	0.262	0.074	0.273
240	0.125	0.073	0.145	0.185	0.073	0.199	0.195	0.073	0.208	0.200	0.073	0.213
300	0.100	0.072	0.123	0.148	0.072	0.165	0.156	0.072	0.172	0.160	0.072	0.175

Reactance can be ignored for cables less than 35 mm². The temperatures of 107.8 °C and 159.2 °C are the average of a conductor carrying 80% of its tabulated current-carrying capacity plus the limit temperature for the conductors insulation. The temperature of 20 °C is used for calculating the maximum fault current for the withstand capacity of equipment. The temperatures of 115 °C and 170 °C are for calculating the minimum fault current when determining the withstand capacity of cables.

RESISTANCE & IMPEDANCE RC 1 PVC CABLES

Resistance and impedance of PVC insulated copper conductors for single core cables or two, three and four core armoured cables

Resistance, reactance and impedance per 1000 metres for copper cable at different temperatures

Cable size sq.mm.	Temperature 20 °C			Temperature 95 °C			Temperature 107.8 °C			Temperaure 115 °C		
	R	X	Z	R	X	Z	R	X	Z	R	X	Z
1.0	18.10	-	18.100	23.530	-	23.530	24.457	-	24.457	24.978	-	24.978
1.5	12.10	-	12.100	15.730	-	15.730	16.350	-	16.350	16.698	-	16.698
2.5	7.410	-	7.410	9.633	-	9.633	10.012	-	10.012	10.226	-	10.226
4.0	4.610	-	4.610	5.993	-	5.993	6.229	-	6.229	6.362	-	6.362
6.0	3.080	-	3.080	4.004	-	4.004	4.162	-	4.162	4.250	-	4.250
10	1.830	-	1.830	2.379	-	2.379	2.473	-	2.473	2.525	-	2.525
16	1.150	-	1.150	1.495	-	1.495	1.554	-	1.554	1.587	-	1.587
25	0.727	-	0.727	0.945	-	0.945	0.982	-	0.982	1.003	-	1.003
35	0.524	-	0.524	0.681	-	0.681	0.708	-	0.708	0.723	-	0.723
50	0.387	0.081	0.395	0.503	0.081	0.510	0.523	0.081	0.529	0.534	0.081	0.540
70	0.268	0.079	0.279	0.348	0.079	0.357	0.362	0.079	0.371	0.370	0.079	0.378
95	0.193	0.077	0.208	0.251	0.077	0.262	0.261	0.077	0.272	0.266	0.077	0.277
120	0.153	0.076	0.171	0.199	0.076	0.213	0.207	0.076	0.220	0.211	0.076	0.224
150	0.124	0.076	0.145	0.161	0.076	0.178	0.168	0.076	0.184	0.171	0.076	0.187
185	0.099	0.076	0.1248	0.1287	0.076	0.1495	0.1338	0.076	0.1539	0.1366	0.076	0.1563
240	0.075	0.075	0.1061	0.0975	0.075	0.1230	0.1013	0.075	0.1261	0.1035	0.075	0.1278
300	0.060	0.075	0.0960	0.0780	0.075	0.1082	0.0811	0.075	0.1104	0.0828	0.075	0.1117

Notes: Reactance only has to be taken into account on cables larger than 35 mm^2. In the table 107.8 °C is the average of the temperature of a conductor carrying 80% of its tabulated current rating plus the insulation limit temperature of 160 °C. The temperature of 95 °C = (30 °C + 160 °C) ÷ 2 is used when the conductor is used as a cpc, or earthing conductor, and the conductor temperature is 30 °C. The temperature 115 °C = (70 °C + 160 °C) ÷ 2 is used when the conductor is carrying its full rated current, or the conductor's temperature is 70 °C.

RESISTANCE & IMPEDANCE

R XLPE 1

XLPE CABLES

Resistance and impedance of XLPE insulated copper conductors for single core cable or two, three, and four core armoured cables

Resistance, reactance and impedance per 1000 metres at different temperatures

Cable size in sq. mm.	Temperature 20 °C			Temperature 140 °C			Temperature 159.2 °C			Temperaure 170 °C		
	R	X	Z	R	X	Z	R	X	Z	R	X	Z
1	18.1	-	18.1	26.788	-	26.788	28.178	-	28.178	28.960	-	28.960
1.5	12.1	-	12.1	17.908	-	17.908	18.837	-	18.837	19.360	-	19.360
2.5	7.41	-	7.41	10.967	-	10.967	11.536	-	11.536	11.856	-	11.856
4	4.61	-	4.61	6.823	-	6.823	7.177	-	7.177	7.376	-	7.376
6	3.08	-	3.08	4.558	-	4.558	4.795	-	4.795	4.928	-	4.928
10	1.83	-	1.83	2.708	-	2.708	2.849	-	2.849	2.928	-	2.928
16	1.15	-	1.15	1.702	-	1.702	1.790	-	1.790	1.840	-	1.840
25	0.727	-	0.727	1.076	-	1.076	1.132	-	1.132	1.163	-	1.163
35	0.524	-	0.524	0.776	-	0.776	0.816	-	0.816	0.838	-	0.838
50	0.387	0.076	0.394	0.573	0.076	0.578	0.602	0.076	0.607	0.619	0.076	0.624
70	0.268	0.075	0.278	0.397	0.075	0.404	0.417	0.075	0.424	0.429	0.075	0.435
95	0.193	0.073	0.206	0.286	0.073	0.295	0.300	0.073	0.309	0.309	0.073	0.317
120	0.153	0.072	0.169	0.226	0.072	0.238	0.238	0.072	0.249	0.245	0.072	0.255
150	0.124	0.073	0.144	0.184	0.073	0.198	0.193	0.073	0.206	0.198	0.073	0.211
185	0.0991	0.073	0.1231	0.1467	0.073	0.1638	0.1543	0.073	0.1707	0.1586	0.073	0.1746
240	0.0754	0.072	0.1043	0.1116	0.072	0.1328	0.1174	0.072	0.1377	0.1206	0.072	0.1405
300	0.0601	0.072	0.0938	0.0889	0.072	0.1144	0.0936	0.072	0.1181	0.0962	0.072	0.1201

Notes: Reactance only needs to be taken into consideration with cables larger than 35 mm^2. The temperature 159.2 °C is the average of the temperature of a conductor carrying 80% of its tabulated current rating plus the insulation limit temperature of 250 °C. The temperature 140 °C = (30 °C + 250 °C) ÷ 2 is used when the conductor is used as a cpc, or earthing conductor, and the conductor temperature is 30 °C. The temperature 170 °C = (90 °C + 250 °C) ÷ 2 is used when the conductor is carrying its full rated current, or the conductor's temperature is 90 °C.

CONDUIT - TRUNKING   ZCT 1   IMPEDANCES

Values of impedance to be used when calculating phase earth loop impedance Z_S when conduit or trunking is used as the circuit protective conductor

IMPEDANCE OF CONDUIT in Ω per metre

Conduit size in mm	For fault currents up to 100A				For fault currents over 100A			
	Light gauge		Heavy gauge		Light gauge		Heavy gauge	
	Conduit	With joints	Conduit	With joints	Conduit	With joints	Conduit	With joints
16	0.0078	0.00975	0.0076	0.0095	0.005	0.00625	0.0038	0.00475
20	0.0054	0.00675	0.0047	0.005875	0.004	0.005	0.0025	0.003125
25	0.0035	0.004375	0.0032	0.004	0.0022	0.00275	0.0017	0.002125
32	0.0024	0.003	0.002	0.0025	0.0015	0.001875	0.0011	0.001375

IMPEDANCE OF HEAVY GAUGE TRUNKING WITHOUT LID in Ω per metre

Trunking size			
50 mm × 50 mm	0.00345 Ω	100 mm × 100 mm	0.00123 Ω
75 mm × 75 mm	0.00194 Ω	150 mm × 150 mm	0.00074 Ω

Notes: The impedance of trunking behaves in a similar manner to conduit with a fault current flowing, except the reduction in impedance does not occur until the fault current exceeds 200 A; for practical purposes only the worst conditions are required for calculations, so it was decided that the lower impedances were not necessary. As far as conduit is concerned, joints should not contribute any additional impedance to the circuit; the above values being given in case the designer wishes to make an allowance for deterioration.

'K' FACTORS K 1A DIFFERENT MATERIALS

Values of 'k' for live and protective conductors with various types of insulation and with various initial and final conductor temperatures, for conductors up to 300 mm^2

Type of conductor insulation	Initial temperature at start of fault °C	Limit temperature of conductor insulation °C	Live conductors, or protective conductors in a cable or bunched with live conductors		Protective conductors installed separately and not grouped or bunched with other conductors		Protective conductor as the sheath or armour of a cable			Protective conductor is conduit, trunking, or steel ducting
			Copper	Aluminium	Copper	Aluminium	Aluminium	Steel	Lead	Steel
70 °C P.V.C.	30	160	-	-	143	95	-	-	-	-
	*47.66	200	-	-	-	-	98	54	27	-
	50	160	-	-	-	-	-	-	-	47
	*55.6	160	125	83	-	-	-	-	-	-
	60	200	-	-	-	-	93	51	26	-
	70	160	115	76	-	-	-	-	-	-
85 °C P.V.C.	30	160	-	-	143	95	-	-	-	-
	*57.53	200	-	-	-	-	94	52	26	-
	58	160	-	-	-	-	-	-	-	45
	*65.2	160	118	78	-	-	-	-	-	-
	75	200	-	-	-	-	87	48	24	-
	85	160	104	69	-	-	-	-	-	-
60 °C Rubber	30	200	-	-	159	105	-	-	-	-
	60	200	141	93	-	-	-	-	-	-

Notes: The initial temperatures indicated by * are applicable only when the live conductors are carrying 80% of their tabulated current-carrying capacity given in the CR Tables. The 'k' factor should be chosen by selecting the initial temperature of the conductor at the start of the fault and the limit temperature of the conductor's insulation. Except for those temperatures indicated by * the initial temperatures given in the above table are when the live conductors are carrying their tabulated current-carrying capacity as given in the CR Tables. Where a protective conductor is not installed with live conductors, its initial temperature will be 30 °C. The initial temperature of the cable armour will be approximately 10 °C lower than the live conductors when they are carrying their tabulated current : i.e., for pvc cable, the initial temperature of the armour will be 60 °C

'K' FACTORS

K 1B DIFFERENT MATERIALS

Values of 'k' for live and protective conductors with various types of insulation and with various initial and final conductor temperatures, for conductors up to 300 mm²

Type of conductor insulation	Initial temperature at start of fault °C	Limit temperature of conductor insulation °C	Live conductors, or protective conductors in a cable or bunched with live conductors		Protective conductors installed separately and not grouped or bunched with other conductors		Protective conductor as the sheath or armour of a cable			Protective conductor conduit, trunking, or steel ducting
			Copper	Aluminium	Copper	Aluminium	Aluminium	Steel	Lead	Steel
85 °C Rubber	30	220	-	-	166	110	-	-	-	-
	*57.53	220	-	-	-	-	99	54	27	-
	58	220	-	-	-	-	-	-	-	54
	*65.2	220	146	96	-	-	-	-	-	-
	75	220	-	-	-	-	93	51	26	-
	85	220	134	89	-	-	-	-	-	-
90 °C Thermo-setting	30	250	-	-	176	116	-	-	-	-
	60	250	-	-	-	-	-	-	-	58
	*60.8	200	-	-	-	-	93	51	26	-
	*68.4	250	155	102	-	-	-	-	-	-
	80	200	-	-	-	-	85	46	23	-
	90	250	143	94	-	-	-	-	-	-
Paper cable	70	160	-	-	-	-	76	42	21	-
	80	160	108	71	-	-	-	-	-	-
Micc pvc	70	160	115	-	-	-	-	-	-	-
Micc bare	105	250	135	-	-	-	-	-	-	-

Notes: See notes at bottom of K1A.

ELI 1

Earth loop impedance for PVC and XLPE insulated copper conductor steel wire armoured cables to BS 6346, BS 5467 & BS 6742

Cable type		PVC insulated cables to BS 6346 copper conductors			XLPE insulated cables to BS 5467 and BS6742 copper conductors		
		Maximum resistance of armour Ω / 1000 m	Earth fault loop impedance per 1000 metres	Gross area of armour	Maximum resistance of armour Ω / 1000 m	Earth fault loop impedance per 1000 metres	Gross area of armour
Cores	Size sq.mm.	Design Ω	Design Ω	Sq. mm.	Design Ω	Design Ω	Sq.mm.
2	1.5	14.26	30.96	15	14.35	33.71	16.0
3	1.5	12.91	29.61	16	12.87	32.23	17.0
4	1.5	11.72	28.42	17	11.60	30.96	18.0
5	1.5	10.37	27.06	19	-	-	-
7	1.5	9.75	26.45	20	-	-	-
10	1.5	5.43	22.13	36	-	-	-
2	2.5	12.13	22.36	17	13.43	25.29	17.0
3	2.5	11.14	21.36	19	11.59	23.45	19.0
4	2.5	9.75	19.97	20	10.51	22.36	20.0
5	2.5	8.88	19.11	22	-	-	-
7	2.5	8.39	18.62	24	-	-	-
10	2.5	4.57	14.79	44	-	-	-
2	4	10.00	16.36	21	12.06	19.44	19.0
3	4	8.86	15.22	23	10.61	17.98	21.0
4	4	5.68	12.04	35	9.28	16.65	23.0
5	4	5.06	11.42	39	-	-	-
7	4	4.81	11.17	42	-	-	-
10	4	2.71	9.08	72	-	-	-
2	6	9.06	13.31	24	10.69	15.61	22.0
3	6	5.82	10.07	36	9.33	14.26	23.0
4	6	5.06	9.31	40	5.87	10.80	36.0
2	10	5.20	7.72	41	9.16	12.09	26.0
3	10	4.68	7.21	44	5.66	8.58	39.0
4	10	4.20	6.72	49	5.05	7.98	43.0
2	16	4.67	6.25	46	5.80	7.64	41.0
3	16	4.05	5.64	50	5.09	6.93	44.0
4	16	2.71	4.30	72	4.37	6.21	49.0

Note: Impedance of armour is based on the final temperature which the armour can be expected to reach with fault current flowing. Earth fault loop impedance includes the impedance of one phase conductor and the armour impedance.

ELI 2

Earth loop impedance for BS 6346 : 1969 steel wire armoured p.v.c. insulated cables with copper and aluminium conductors

Cable		Steel wire armouring stranded copper conductors			Steel wire armouring solid aluminium conductors		
Number of cores	Cable size in mm^2.	Design value armour impedance $\Omega/1000$ m	Gross area of armour in mm^2.	Design value earth loop impedance $\Omega/1000$ m.	Design value armour impedance $\Omega/1000$ m	Gross area of armour in mm^2.	Design value earth loop impedance $\Omega/1000$ m.
1	2	3	4	5	6	7	8
2	25	3.466	60	4.47	3.866	54	5.52
3	25	3.037	66	4.04	3.164	62	4.82
4	25	2.591	76	3.59	2.838	70	4.49
2	35	3.213	66	3.94	3.612	58	4.81
3	35	2.674	74	3.40	2.926	68	4.12
4	35	2.364	84	3.09	2.486	78	3.68
2	50	2.815	74	3.36	3.213	66	4.10
3	50	2.423	84	2.96	2.549	78	3.44
4	50	1.632	122	2.17	1.753	113	2.64
2	70	2.550	84	2.93	2.815	74	3.43
3	70	1.797	119	2.17	1.797	113	2.41
4	70	1.511	138	1.89	1.632	128	2.25
2	95	1.759	122	2.04	2.022	109	2.47
3	95	1.548	138	1.82	1.672	128	2.12
4	95	1.246	160	1.52	1.390	147	1.84
2	120	1.627	131	1.85	-	-	-
3	120	1.424	150	1.65	1.548	138	1.91
4	120	0.926	220	1.15	1.008	201	1.37
2	150	1.497	144	1.68	-	-	-
3	150	0.983	211	1.17	1.080	191	1.38
4	150	0.856	240	1.04	0.926	220	1.22
2	185	1.082	201	1.24	-	-	-
3	185	0.911	230	1.07	0.971	215	1.21
4	185	0.787	265	0.94	0.845	245	1.08
2	240	0.967	225	1.10	-	-	-
3	240	0.816	260	0.94	0.876	240	1.06
4	240	0.708	299	0.84	0.765	274	0.95
2	300	0.892	250	1.00	-	-	-
3	300	0.746	289	0.86	0.805	265	0.96
4	300	0.653	333	0.76	0.708	304	0.87

Note: Impedance of armour is based on the final temperature which the armour can be expected to reach with fault current flowing. Earth fault loop impedance includes the impedance of one phase conductor and the armour impedance.

ELI 3

Earth loop impedance for BS 5467 : 1989 steel wire armoured XLPE insulated cables with copper and aluminium conductors

Cable		Steel wire armouring stranded copper conductors			Steel wire armouring solid aluminium conductors		
Number of cores	Cable size in mm^2.	Design value armour impedance Ω / 1000 m.	Gross area of armour in mm^2	Design value earth loop impedance Ω / 1000 m.	Design value armour impedance Ω / 1000 m.	Gross area of armour in mm^2.	Design value earth loop impedance Ω / 1000 m.
1	2	3	4	5	6	7	8
2	25	5.432	42	6.595	6.019	38	7.939
3	25	3.420	62	4.583	3.694	58	5.614
4	25	3.045	70	4.208	3.178	66	5.098
2	35	3.726	62	4.565	4.268	54	5.657
3	35	3.161	70	3.999	3.433	64	4.822
4	35	2.665	80	3.503	2.928	72	4.317
2	50	3.390	68	4.014	3.829	60	4.857
3	50	2.752	78	3.376	3.025	72	4.053
4	50	2.402	90	3.026	2.533	82	3.561
2	70	2.951	80	3.386	3.390	70	4.103
3	70	2.481	90	2.916	2.616	84	3.329
4	70	1.617	131	2.052	1.747	122	2.460
2	95	2.077	113	2.394	2.368	100	2.885
3	95	1.804	128	2.121	1.939	119	2.456
4	95	1.487	147	1.804	1.617	135	2.134
2	120	1.932	125	2.187	-	-	-
3	120	1.669	141	1.924	1.669	131	2.080
4	120	1.050	206	1.305	1.126	191	1.537
2	150	1.787	138	1.998	-	-	-
3	150	1.108	201	1.319	1.214	181	1.552
4	150	0.949	230	1.160	1.025	211	1.363
2	185	1.241	191	1.415	-	-	-
3	185	1.017	220	1.191	1.082	206	1.355
4	185	0.862	255	1.036	0.936	235	1.209
2	240	1.113	215	1.253	-	-	-
3	240	0.913	250	1.053	0.977	230	1.190
4	240	0.775	289	0.916	0.837	265	1.050
2	300	1.028	235	1.148	-	-	-
3	300	0.848	269	0.968	0.913	250	1.088
4	300	0.715	319	0.835	0.775	289	0.950

Note: Impedance of armour is based on the final temperature which the armour can be expected to reach with fault current flowing. Earth fault loop impedance includes the impedance of one phase conductor and the armour impedance.

ELI 4
Earth loop impedance for aluminium strip armoured solidal aluminium PVC insulated cables to BS 6346 and XLPE insulated cables to BS 5467

Number of cores	Cable size in mm^2.	PVC insulated solidal aluminium strip aluminium armoured cables			XLPE insulated solidal aluminium strip aluminium armoured cables		
		Design value armour impedance Ω/1000 m.	Gross area of armour in mm^2.	Design value earth loop impedance Ω/1000 m.	Design value armour impedance Ω/1000 m.	Gross area of armour in mm^2.	Design value earth loop impedance Ω/1000 m.
1	2	3	4	5	6	7	8
2	16	1.944	23	9.800	2.349	22	5.405
3	16	1.730	26	4.366	2.052	23	5.108
4	16	1.450	29	4.086	1.854	26	4.910
2	25	2.074	22	3.730	2.496	20	4.416
3	25	1.607	27	3.263	1.915	24	3.835
4	25	1.450	30	3.106	1.589	29	3.509
2	35	1.816	24	3.014	2.203	23	3.592
3	35	1.486	30	2.684	1.780	27	3.169
4	35	1.332	34	2.530	1.459	31	2.848
2	50	1.557	29	2.446	2.056	26	3.084
3	50	0.709	60	1.598	1.507	31	2.535
4	50	0.623	67	1.512	0.721	63	1.749
2	70	0.781	56	1.397	0.942	53	1.655
3	70	0.635	67	1.251	0.743	63	1.456
4	70	0.539	78	1.155	0.615	74	1.328
2	95	0.703	63	1.152	0.839	60	1.356
3	95	0.415	104	0.864	0.661	70	1.178
4	95	0.349	123	0.798	0.394	117	0.911
3	120	0.366	117	0.723	0.431	110	0.842
4	120	0.314	136	0.671	0.367	130	0.778
3	150	0.342	130	0.636	0.391	123	0.729
4	150	0.290	149	0.584	0.329	143	0.667
3	185	0.306	143	0.545	0.350	136	0.623
4	185	0.211	214	0.450	0.290	162	0.563
3	240	0.211	214	0.400	0.310	156	0.523
4	240	0.188	248	0.377	0.215	237	0.428
3	300	0.200	237	0.358	0.218	226	0.393
4	300	0.177	271	0.335	0.191	260	0.366

Note: Impedance of armour is based on the final temperature which the armour can be expected to reach with fault current flowing. Earth fault loop impedance includes the impedance of one phase conductor and the armour impedance.

ELI 5

Characteristics of mineral insulated copper sheathed cables giving effective sheath area and earth loop impedance when carrying fault current

Cable		Resistance in ohms per 1000 metres					
		Light duty 500 V cable			Heavy duty 750 V cable		
		Earth loop impedance		Effective sheath area in sq.mm.	Earth loop impedance		Effective sheath area in sq.mm.
Number of cores	Size sq. mm.	Exposed to touch, bare or covered	NOT exposed to touch and bare		Exposed to touch, bare or covered	NOT exposed to touch and bare	
1	2	3	4	5	6	7	8
4 x 1	1H6	-	-	-	5.1	5.33	31
4 x 1	1H10	-	-	-	3.2	3.34	38
4 x 1	1H16	-	-	-	2.11	2.21	46
4 x 1	1H25	-	-	-	1.4	1.46	60
4 x 1	1H35	-	-	-	1.05	1.10	71
4 x 1	1H50	-	-	-	0.77	0.807	88
4 x 1	1H70	-	-	-	0.57	0.598	109
4 x 1	1H95	-	-	-	0.44	0.465	130
4 x 1	1H120	-	-	-	0.36	0.382	150
4 x 1	1H150	-	-	-	0.30	0.316	175
2	1.0	30.7	32.0	5.4	-	-	-
2	1.5	21.2	22.1	6.3	19.8	20.7	11
2	2.5	13.3	13.9	8.2	12.4	12.8	13
2	4.0	8.64	9.03	10.7	8.04	8.39	16
2	6.0	-	-	-	5.59	5.86	18
2	10	-	-	-	3.56	3.73	24
2	16	-	-	-	2.36	2.48	30
2	25	-	-	-	1.62	1.70	38
3	1.0	30.0	31.3	6.7	-	-	-
3	1.5	20.6	21.4	7.8	19.6	20.4	12
3	2.5	12.9	13.5	9.5	12.2	12.7	14
3	4.0	-	-	-	7.92	8.25	17
3	6.0	-	-	-	5.49	5.74	20
3	10	-	-	-	3.46	3.63	27
3	16	-	-	-	2.29	2.40	34
3	25	-	-	-	1.57	1.64	42
4	1.0	29.6	30.7	7.7	-	-	-
4	1.5	20.2	21.1	9.1	19.4	20.2	14
4	2.5	12.6	13.1	11.3	12.0	12.6	16
4	4.0	-	-	-	7.74	8.08	20
4	6.0	-	-	-	5.35	5.59	24
4	10	-	-	-	3.36	3.53	30
4	16	-	-	-	2.20	2.31	39
4	25	-	-	-	1.50	1.57	49
7	1.0	28.9	30.1	10.2	-	-	-
7	1.5	19.7	20.5	11.8	19.0	19.8	18
7	2.5	12.1	12.6	15.4	11.7	12.2	22
12	2.5	-	-	-	11.4	11.8	34
19	1.5	-	-	-	18.5	18.9	37

MISC 1

Table M 54 G

Type of insulation cpc conductor in contact with	Type of conductor material used for cpc	Cross-sectional area 'S' of copper phase conductor		
		Up to 16 mm²	16 to 35 mm²	Over 35 mm²
		Cross-sectional area required for cpc		
70 °C P.V.C.	Aluminium sheath or armour Steel wire armour Steel conduit or trunking lead sheath	1.24S 2.25S 2.45S 4.42S	19.78 36.08 39.15 70.77	0.62S 1.13S 1.225S 2.21S
85 °C P.V.C.	Aluminium sheath or armour Steel wire armour Steel conduit or trunking lead sheath	1.2S 2.17S 2.31S 4.33S	19.13 34.67 36.98 69.33	0.6S 1.08S 1.16S 2.17S
85 °C Rubber	Aluminium sheath or armour Steel wire armour Steel conduit or trunking lead sheath	1.44S 2.63S 2.48S 5.15S	23.05 42.04 39.70 82.46	.72S 1.31S 1.24S 2.58S
90 °C Thermosetting	Aluminium sheath or armour Steel wire armour Steel conduit or trunking lead sheath	1.68S 3.11S 2.47S 6.22S	26.92 49.74 39.45 99.48	0.84S 1.55S 1.23S 3.11S

Minimum size of cpc when phase conductor is copper and cpc is a dissimilar material

Area of cpcs, conduit, cable armouring and trunking

Area of c.p.c. in Twin & cpc cable		Cross sectional area steel conduit			Cross sectional area steel trunking to BS 4768: Part 1: 1971	
Cable size sq.mm	C.p.c size sq.mm.	Conduit Size mm	Area sq.mm.		Size in mm	Area sq.mm
			Light gauge	Heavy gauge		
1.0	1.0	16	47	75	50 x 37.5 50 x 50	123.00 148.00
1.5	1.0	20	59	107	75 x 50 75 x 75	207.12 267.12
2.5	1.5	25	89	132		
4.0	1.5	32	116	167	100 x 50 100 x 75 100 x 100	237.12 297.12 416.08
6.0	2.5	38	-	186		
10	4.0				150 x 50 150 x 75 150 x 100 150 x 150	346.08 416.08 554.88 714.88
16	6.0					

Area of cable armour or micc sheath
See tables ELI 1, ELI 2, ELI 3, ELI 4 and ELI 5

MISC 2
Conversion table - Power Factor to Sines

Power factor	Sine	Radians	Degrees	Power factor	Sine	Radians	Degrees
0.10	0.9950	1.4706	84.26	0.55	0.8352	0.9884	56.63
0.11	0.9939	1.4606	83.68	0.56	0.8285	0.9764	55.94
0.12	0.9928	1.4505	83.11	0.57	0.8216	0.9643	55.25
0.13	0.9915	1.4404	82.53	0.58	0.8146	0.9521	54.55
0.14	0.9902	1.4303	81.95	0.59	0.8074	0.9397	53.84
0.15	0.9887	1.4202	81.37	0.60	0.8000	0.9273	53.13
0.16	0.9871	1.4101	80.79	0.61	0.7924	0.9147	52.41
0.17	0.9854	1.4000	80.21	0.62	0.7846	0.9021	51.68
0.18	0.9837	1.3898	79.63	0.63	0.7766	0.8892	50.95
0.19	0.9818	1.3796	79.05	0.64	0.7684	0.8763	50.21
0.20	0.9798	1.3694	78.46	0.65	0.7599	0.8632	49.46
0.21	0.9777	1.3592	77.88	0.66	0.7513	0.8500	48.70
0.22	0.9755	1.3490	77.29	0.67	0.7424	0.8366	47.93
0.23	0.9732	1.3387	76.70	0.68	0.7332	0.8230	47.16
0.24	0.9708	1.3284	76.11	0.69	0.7238	0.8093	46.37
0.25	0.9682	1.3181	75.52	0.70	0.7141	0.7954	45.57
0.26	0.9656	1.3078	74.93	0.71	0.7042	0.7813	44.77
0.27	0.9629	1.2974	74.34	0.72	0.6940	0.7670	43.95
0.28	0.9600	1.2870	73.74	0.73	0.6834	0.7525	43.11
0.29	0.9570	1.2766	73.14	0.74	0.6726	0.7377	42.27
0.30	0.9539	1.2661	72.54	0.75	0.6614	0.7227	41.41
0.31	0.9507	1.2556	71.94	0.76	0.6499	0.7075	40.54
0.32	0.9474	1.2451	71.34	0.77	0.6380	0.6920	39.65
0.33	0.9440	1.2345	70.73	0.78	0.6258	0.6761	38.74
0.34	0.9404	1.2239	70.12	0.79	0.6131	0.6600	37.81
0.35	0.9367	1.2132	69.51	0.80	0.6000	0.6435	36.87
0.36	0.9330	1.2025	68.90	0.81	0.5864	0.6266	35.90
0.37	0.9290	1.1918	68.28	0.82	0.5724	0.6094	34.92
0.38	0.9250	1.1810	67.67	0.83	0.5578	0.5917	33.90
0.39	0.9208	1.1702	67.05	0.84	0.5426	0.5735	32.86
0.40	0.9165	1.1593	66.42	0.85	0.5268	0.5548	31.79
0.41	0.9121	1.1483	65.80	0.86	0.5103	0.5355	30.68
0.42	0.9075	1.1374	65.17	0.87	0.4931	0.5156	29.54
0.43	0.9028	1.1263	64.53	0.88	0.4750	0.4949	28.36
0.44	0.8980	1.1152	63.90	0.89	0.4560	0.4735	27.13
0.45	0.8930	1.1040	63.26	0.90	0.4359	0.4510	25.84
0.46	0.8879	1.0928	62.61	0.91	0.4146	0.4275	24.49
0.47	0.8827	1.0815	61.97	0.92	0.3919	0.4027	23.07
0.48	0.8773	1.0701	61.31	0.93	0.3676	0.3764	21.57
0.49	0.8717	1.0587	60.66	0.94	0.3412	0.3482	19.95
0.50	0.8660	1.0472	60.00	0.95	0.3122	0.3176	18.19
0.51	0.8602	1.0356	59.34	0.96	0.2800	0.2838	16.26
0.52	0.8542	1.0239	58.67	0.97	0.2431	0.2456	14.07
0.53	0.8480	1.0122	57.99	0.98	0.1990	0.2003	11.48
0.54	0.8417	1.0004	57.32	0.99	0.1411	0.1415	8.11

MISC 3

Transformer impedance data

Resistance & reactance in Ω of 11 kV/415V Class T1 transformers referred to the 415V system

kVA rating	Resistance per phase when cold 15 / 20 °C Ω per phase	Resistance per phase fully loaded 75 °C Ω per phase	Reactance Ω per phase
200	0.012	0.014	0.0423
315	0.0071	0.0085	0.027
500	0.0038	0.0045	0.0173
800	0.0022	0.0027	0.011
1000	0.0016	0.0019	0.0087
1500	0.00099	0.0012	0.0068
2000	0.00079	0.001	0.0056

My thanks to John Boulton GEC Transformers for the above typical values of transformer data

Fuse data

Breaking capacity of fuses

Type	Voltage	Breaking capacity kA
BS 3036		
S1 duty	240V	1.0
S2 duty	240V	2.0
S3 duty	240V	4.0
BS 1361		
Type I	240V	16.5
Type II	415V	33.0
BS1362	250V	6.0
BS 88		
Type 2	250V	40.0
Type 2	415V	80.0

Mcb data

Breaking capacity of BS 3871 Part 1 mcb's

Category of duty	Rated breaking capacity kA	Test power factor
M1	1.0	0.85 - 0.9
M1.5	1.5	0.80 - 0.85
M2	2.0	0.75 - 0.8
M3	3.0	0.75 - 0.8
M4	4.0	0.75 - 0.8
M6	6.0	0.75 - 0.8
M9	9.0	0.55 - 0.6

Thermal Insulation factors

Length in mm conductor enclosed in insulation	De-rating factor
50	0.89
100	0.81
200	0.68
400	0.55
500 plus	0.50

Correction factor for BS 3036 fuses when used for overload protection (S)

I_t = Fuse rating ÷ 0.725

Temperature conversion

°C	°F	°C	°F	°C	°F	°C	°F
0	32	35	95	60	140	85	185
10	50	40	104	60.8	141.44	90	194
15	59	47.66	117.79	65.2	149.36	105	221
20	68	50	122	68.4	155.12	115	239
25	77	55.6	132.08	70	158	160	320
30	86	57.53	135.55	80	176	170	338

MISC 4

Factors for converting the resistance of copper or aluminium from 20 °C to the average of conductor final temperature, plus limit temperature of the conductors insulation. For the design of shock, cpc and fault current protection.

PVC insulated copper or aluminium conductor				XLPE insulated copper or aluminium conductors			
Final conductor temperature in °C	Multiply resistance at 20 °C by	Final conductor temperature in °C	Multiply resistance at 20 °C by	Final conductor temperature in °C	Multiply resistance at 20 °C by	Final conductor temperature in °C	Multiply resistance at 20 °C by
30	1.300	61	1.362	30	1.480	61	1.542
31	1.302	62	1.364	31	1.482	62	1.544
32	1.304	63	1.366	32	1.484	63	1.546
33	1.306	64	1.368	33	1.486	64	1.548
34	1.308	65	1.370	34	1.488	65	1.550
35	1.310	66	1.372	35	1.490	66	1.552
36	1.312	67	1.374	36	1.492	67	1.554
37	1.314	68	1.376	37	1.494	68	1.556
38	1.316	69	1.378	38	1.496	69	1.558
39	1.318	70	1.380	39	1.498	70	1.560
40	1.320	71	1.382	40	1.500	71	1.562
41	1.322	72	1.384	41	1.502	72	1.564
42	1.324	73	1.386	42	1.504	73	1.566
43	1.326	74	1.388	43	1.506	74	1.568
44	1.328	75	1.390	44	1.508	75	1.570
45	1.330	76	1.392	45	1.510	76	1.572
46	1.332	77	1.394	46	1.512	77	1.574
47	1.334	78	1.396	47	1.514	78	1.576
48	1.336	79	1.398	48	1.516	79	1.578
49	1.338	80	1.400	49	1.518	80	1.580
50	1.340	81	1.402	50	1.520	81	1.582
51	1.342	82	1.404	51	1.522	82	1.584
52	1.344	83	1.406	52	1.524	83	1.586
53	1.346	84	1.408	53	1.526	84	1.588
54	1.348	85	1.410	54	1.528	85	1.590
55	1.350	86	1.412	55	1.530	86	1.592
56	1.352	87	1.414	56	1.532	87	1.594
57	1.354	88	1.416	57	1.534	88	1.596
58	1.356	89	1.418	58	1.536	89	1.598
59	1.358	90	1.420	59	1.538	90	1.600
60	1.360	91	1.422	60	1.540	91	1.602

Note: The above factors will give the resistance at the average of: conductor operating temperature plus the limit temperature for the conductor's insulation.; e.g PVC cable operating temperature 70 °C: factor from table gives resistance at 115 °C.

MISC 5

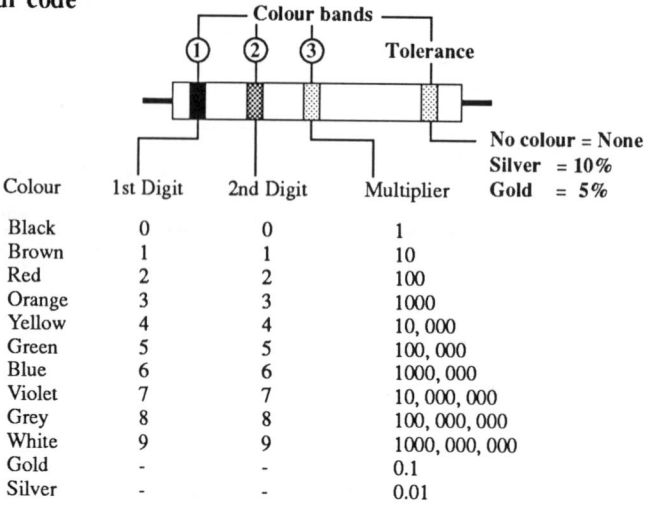

Resistor colour code

No colour = None
Silver = 10%
Gold = 5%

Colour	1st Digit	2nd Digit	Multiplier
Black	0	0	1
Brown	1	1	10
Red	2	2	100
Orange	3	3	1000
Yellow	4	4	10,000
Green	5	5	100,000
Blue	6	6	1000,000
Violet	7	7	10,000,000
Grey	8	8	100,000,000
White	9	9	1000,000,000
Gold	-	-	0.1
Silver	-	-	0.01

Conversion Factors

Length

1 inch	=	1000 mils	1 mm	=	39.37 mils
1 inch	=	25.4 mm	1 mm	=	0.03937 inches
1 foot	=	304.8 mm	1 centimetre	=	0.3937 inches
1 yard	=	914.4 mm	1 metre	=	1.094 yards
1 yard	=	0.9144 m	1 metre	=	3.282 ft
1 mile	=	1 609.34 m	1 kilometre	=	0.621373 miles

Area

1 sq.inch	=	645.16 sq.mm	1 sq.mm	=	0.00155 sq.inches
1 sq.ft	=	0.092903 sq.m	1 sq.m	=	10.764 sq. ft
1 sq.yard	=	0.8361 sq.m	1 sq.m	=	1.196 sq.yards
1 acre	=	4 840 sq.yards	1 acre	=	4046.86 sq.m
1 acre	=	2.471 Hectares	1 Hectares	=	0.4046.86 acres

Liquid

1 lb water	=	0.454 litres	1 ltres	=	2.202 lbs
1 pint	=	0.568 litres	1 litre	=	1.76 pints
1 gallon	=	4.546 litres	1 litre	=	0.22 gallons

Horse-power

1 h.p.	=	746 watts	1 watt	=	0.00134 h.p.
1 h.p	=	0.746 kW	1 kW	=	1.34 h.p.
1 h.p.	=	33 000 ft.lb./min	1 h.p.	=	0.759 kg.m/s

Weight

1 lb	=	0.4536 kg	1 kg	=	2.205 lbs
1 Cwt	=	112 lbs	20 Cwt	=	1 Ton (English)
1 Cwt	=	50.8 kg	1 kg	=	0.01968 Cwt
1 Ton	=	1 016.1 kg	1 kg	=	0.000984 Ton

MISC 6

Lighting
The lumen method is usually used for determining the amount of illumination required for a given area.

$$\text{Total lumens required [F]} = \frac{A \times E_{av}}{CU \times M}$$

Where A is the Area to be illuminated
E_{av} is the average illumination required on the working plane
CU is the coefficient of utilisation
M is the maintenance factor, usually taken as 0.8, but reduced to 0.6 in dirty areas

The average illumination required on the working plane is taken from a table giving the recommended values of illumination for different conditions. Some typical values are :-

Area	lux	Approximate lm/ft²
General office	500	50
Drawing office boards	750	75
Store rooms, corridors	300	30
Shop counters	500	50
Business machine operation	750	75
Watch repair, and fine soldering	3000	300
Proof reading- print works	750	75
Living rooms, halls and landings	100	10
Bedrooms	50	5
Reading	150/300	15/30

To select the coefficient of utilisation from the manufacturer's data, the room index is required:

$$\text{Room index} = \frac{L \times W}{H_m (L \times W)}$$

where L is the length of the room, W the width and H_m is the mounting height of the lighting fitting above the working plane. Where the working plane is a desk top or work bench, the working plane is taken to be 0.85m above the floor. The coefficient of utilisation is then selected from the manufacturer's design data. Having worked out the total illumination required, the design lumens of the lamp is chosen, and the number of lighting points calculated.

$$\text{Total number of lamps} = \frac{\text{Illumination required [F]}}{\text{Design lumens per lamp}}$$

To ensure even illumination, fittings are arranged so that the space to mounting height ratio does not exceed 1.5:1; this means that the spacing between fittings shall not exceed $1.5H_m$. Glare index calculations may also be required.

Conversion factors
1 lumen/ sq.ft (Foot-lambert) = 10.76 Lux or 10.76 apostilb, or 3.426 cd/m², or 0.00221 cd/in²

Heating calculations
Heat lost from a building is the product of the area of the surface, its thermal transmission coefficient, and the temperature difference between the inside and the outside temperatures.

$$Q_f = AU(t_1 - t_2)$$

where Q_f is the heat lost through the fabric of the building in Watts, A is the area of the surface in square metres, U is the thermal transmission coefficient W/m² °C, t_1 is the inside temperature required and t_2 is the outside temperature. Allowance must also be made for the heat lost through floors and ceilings. The power required can be determined approximately by allowing 15 watts per square foot to give a temperature of 20 °C (70 °F), with an outside temperature of 0 °C (32 F).

MISC 7

Final temperature of a conductor

To determine the final temperature of a conductor, when it's not carrying its tabulated current-carrying capacity, the following formula can be used:

$$t_f = t_p - \left(G^2 A^2 - \frac{I_b^2}{I_{tab}^2}\right)(t_p - t_a)$$

where t_f is the final operating temperature of the conductor, t_p is the maximum conductor temperature allowed, t_a is the ambient temperature (taken as 30 °C in the regulations), I_b is the design or full load current carried by the conductor, I_{tab} is the tabulated current-carrying capacity for the conductor in the regulations, G is grouping factor, and A is ambient temperature factor, both being 1 if they are not applicable.

Water

Capacity of containers

Rectangular container capacity in gallons $= \dfrac{L \times B \times H \text{ (cm)}}{4546}$

Rectangular container capacity in litres $= \dfrac{L \times B \times H \text{ (cm)}}{1000}$

Cylindrical container capacity in gallons $= \dfrac{D^2 \times H \text{ (cm)}}{5788}$

Cylindrical container capacity in litres $= \dfrac{D^2 \times H \text{ (cm)}}{1273}$

Power required and time taken

Time taken in minutes to heat water $= \dfrac{\text{Litres} \times \text{Temperature rise °C}}{14.33 \times \text{kW} \times \text{Efficiency}}$

Loading required in kW to heat water $= \dfrac{\text{Litres} \times \text{Temperature rise °C}}{14.33 \times \text{minutes} \times \text{Efficiency}}$

Efficiency is approximately 0.9 for a lagged tank, and 0.8 for an unlagged tank.

Conversion factors

1 kWh	= 3415 BthU	= 860 kilogramme calories	= 3.6 Mega Joules
1 BthU	= 252 calories	= 778 ft lbf	= 1055 joules
1 joule	= 0.239 calories	= 1 watt second	= 0.737 ft lbf
1 gallon	= 4.546 Litres	= 276.9 cu inches	= 10 lb pure water
1 cu ft	= 28.3 litres	= 6.23 gallons	= 62.23 lb pure water
1 cu metre	= 1.308 cu yards	1 kWh	= 2.6552 x 10^6 ft.lbf
1 Newton	= 9.807 kgf	1 litre pure water	= 1 kg

Three phase formula

$$\text{h.p.} = \frac{\text{kW} \times \text{Efficiency}}{0.746} = \frac{\text{kVA} \times \text{Efficiency} \times \text{p.f.}}{0.746} = \frac{I_L \times V_L \times \text{Efficiency} \times 1.732 \times \text{p.f.}}{746}$$

$$I_L = \frac{\text{kW} \times 1000}{V_L \times 1.732 \times \text{p.f.}} = \frac{\text{kVA} \times 1000}{V_L \times 1.732} = \frac{\text{h.p.} \times 746}{V_L \times 1.732 \times \text{Efficiency} \times \text{p.f.}}$$

$$\text{kW} = \frac{\text{h.p.} \times 746}{1000 \times \text{Efficiency}} = \frac{I_L \times V_L \times 1.732 \times \text{p.f.}}{1000} = \text{kVA} \times \text{p.f.}$$

where V_L = Line volts, I_L = Line amperes, p.f. = Power factor and h.p. = Horse power.

MERLIN GERIN C60H M9 Type 2 mcb Characteristics
Rated current. Tripping from no-load (Ref: Calibration temperature 40 °C)

TIME/CURRENT CHARACTERISTICS (MAXIMUM)

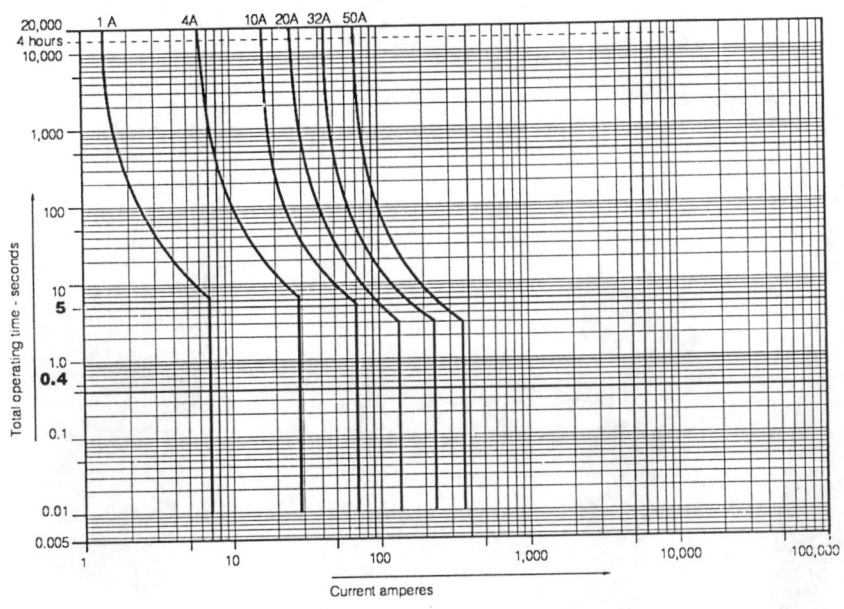

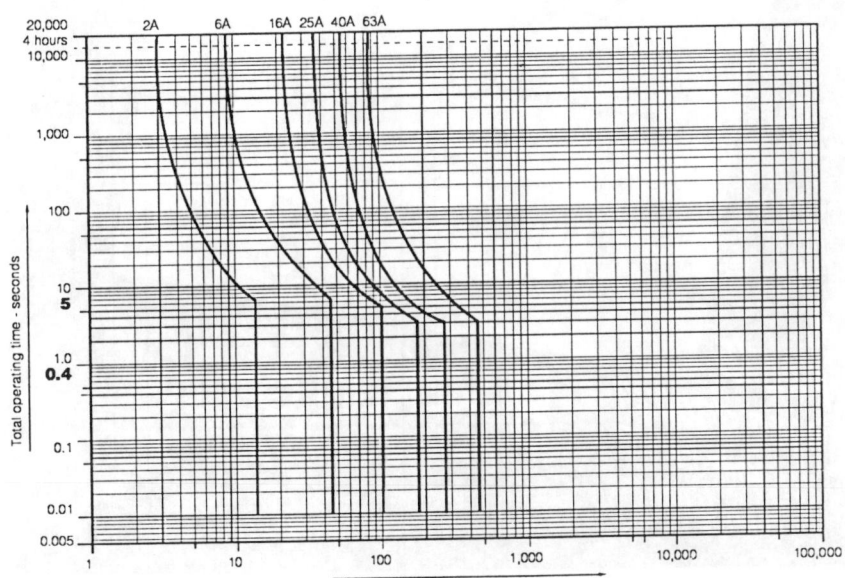

FREE
COMPREHENSIVE GUIDE TO CIRCUIT BREAKER APPLICATION

Over 100 pages of tables, charts and in-depth information on the choice and application of circuit breaker protection and control of circuits

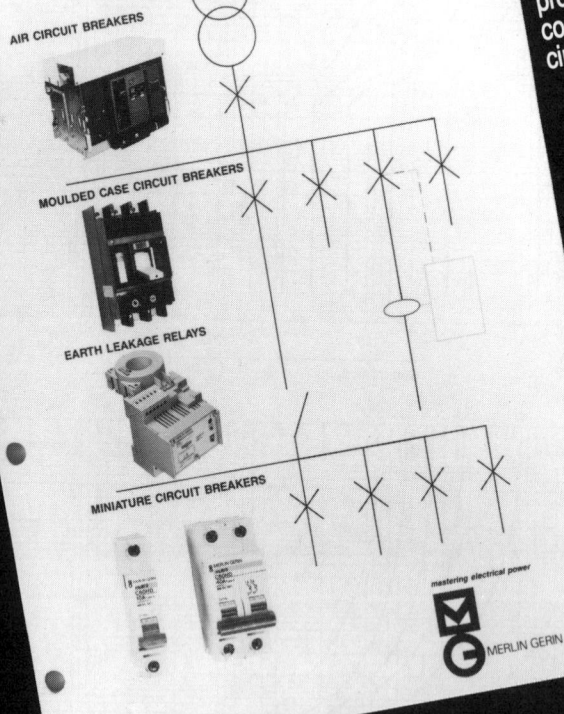

low voltage circuit breaker application guide 1Amp to 6300Amps

AIR CIRCUIT BREAKERS

MOULDED CASE CIRCUIT BREAKERS

EARTH LEAKAGE RELAYS

MINIATURE CIRCUIT BREAKERS

mastering electrical power

MERLIN GERIN

PHONE 0952 290029 for your FREE copy today

MERLIN GERIN
mastering electrical power

MERLIN GERIN LIMITED Stafford Park 5, Telford, Shropshire, TF3 3BL.
Telephone: (0952) 290029 Telex: 35433 Fax: (0952) 290534

MERLIN GERIN Compact mccb characteristics - (D type trip only)
Rated current. Tripping from no-load (Ref: Calibration temperature 40 °C)

TIME/CURRENT CHARACTERISTICS (MAXIMUM)

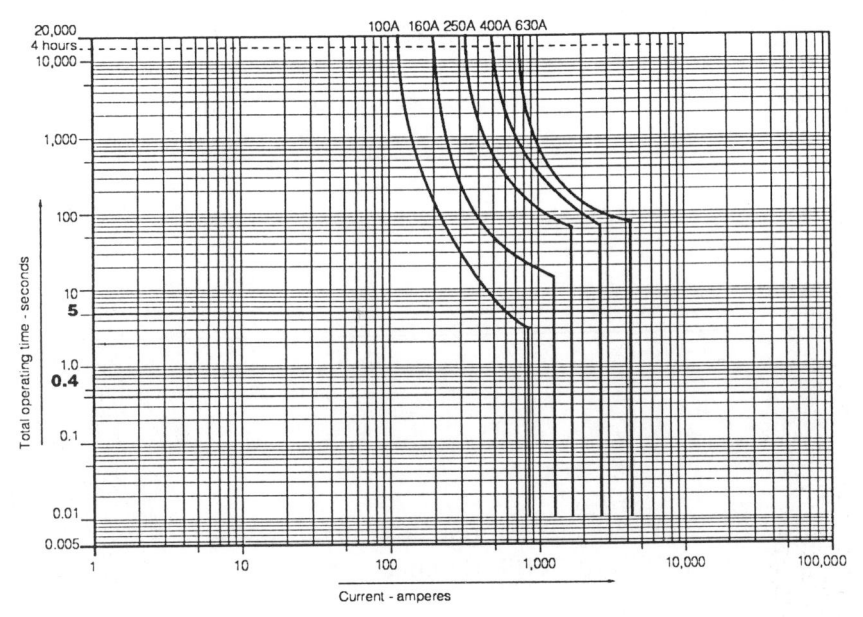

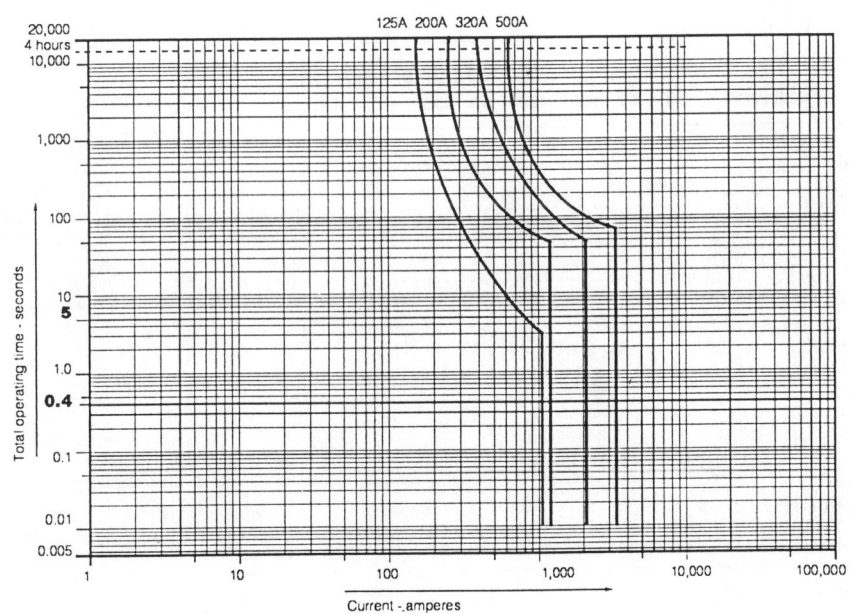

MERLIN GERIN - ENERGY LET-THROUGH MCBS & MCCBS

I^2t Energy let through (values x 10^4) - 240V, 1 pole; 415V, 2, 3 4 pole for MERLIN GERIN C60H mcbs

Circuit breaker Type	Rating (A)	Short circuit prospective current (kA rms)									
		3	4	5	6	7	8	9	11	13	15
C60H	≤ 6	0.55	0.65	0.75	0.8	0.89	0.9	0.95	1.0	1.1	1.2
	≤ 10	0.9	1.1	1.3	1.5	1.7	1.8	2.0	2.2	2.4	2.6
	≤ 25	1.0	1.4	1.8	2.1	2.4	2.8	3.0	3.7	4.5	5.0
	≤ 40	1.6	2.2	2.8	3.3	3.7	4.2	4.5	5.5	6.0	7.0
	≤ 63	1.8	2.6	3.4	4.0	4.5	5.2	6.0	7.0	8.0	9.0

I^2t Energy let through at 415V (values x 10^6) for MERLIN GERIN Compact mccbs N/H/L types

Circuit Breaker	Short circuit prospective current (kA rms)													
	5	10	15	20	25	30	35	40	45	50	70	90	110	130	150
C125N	0.30	0.60	0.90	1.0											
C125H	0.30	0.60	0.90	1.0	1.3	1.4	1.5	1.6	1.7	1.8					
C125L	0.16	0.20	0.24	0.25	0.26	0.27	0.28	0.29	0.29	0.30	0.30	0.30	0.30	0.30	0.30
C161N	0.30	0.60	0.90	1.0	1.3										
C161H	0.30	0.60	0.90	1.0	1.3	1.4	1.5	1.6	1.7	1.8					
C161L	0.16	0.20	0.24	0.25	0.26	0.27	0.28	0.29	0.29	0.30	0.30	0.30	0.30	0.30	0.30
C250N	0.35	1.0	1.7	2.1	2.5	3.0	3.3								
C250H	0.35	1.0	1.7	2.1	2.5	3.0	3.3	3.8	4.0	4.4					
C250L	0.35	0.80	0.90	1.0	1.2	1.2	1.3	1.3	1.3	1.3	1.3	1.3	1.3	1.3	1.3
C401N	0.4	1.50	2.8	4.5	4.6	7.0	8.5								
C401H	0.4	1.50	2.8	4.5	4.6	7.0	8.5	9.5	11	12					
C401L	0.4	1.0	1.5	1.9	2.0	2.4	2.6	2.8	2.9	3.0	3.2	3.2	3.2	3.2	3.2
C630N		2.4	3.6	5.0	6.0	7.0	7.5								
C630H		2.4	3.6	5.0	6.0	7.0	7.5	8.0	8.9	9.0					
C630L		1.6	2.0	2.4	2.7	2.9	3.0	3.2	3.2	3.3	3.5	3.5	3.6	3.7	3.7

WYLEX - BS 3036 Rewirable fuses

Rated Braking capacity of each fuse is 2 kA

Fuse rating	Total I²t energy let-through fuse at 2 kA
5A	12,300 A² sec
15A	18,100 A² sec
20A	23,300 A² sec
30A	36,300 A² sec

Fuse wire size		
	SWG	mm
	34	0.234
	25	0.508
	23	0.609
	22	0.711

Time / Current Characteristics

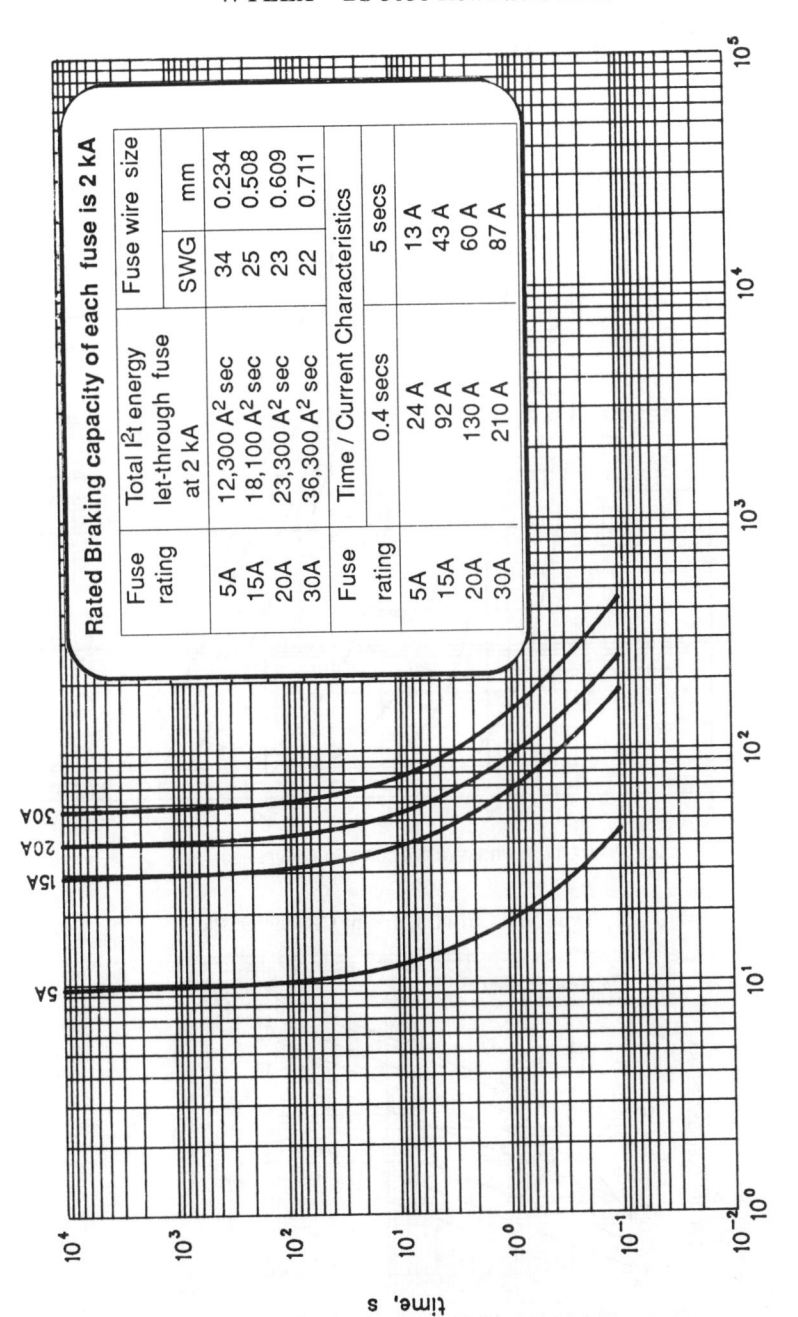

Fuse rating	0.4 secs	5 secs
5A	24 A	13 A
15A	92 A	43 A
20A	130 A	60 A
30A	210 A	87 A

x-axis: prospective current, r.m.s. amperes
y-axis: time, s

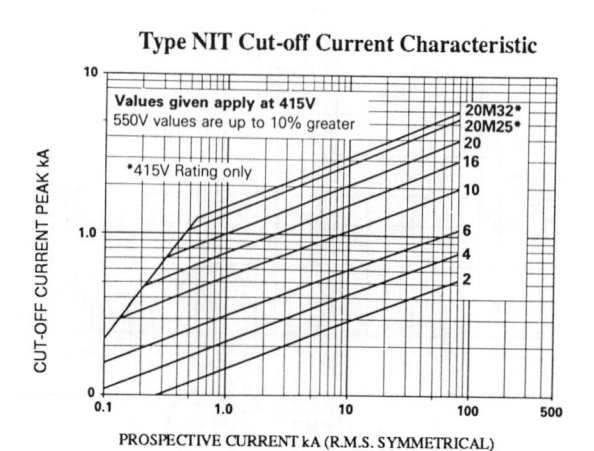

GEC ALSTHOM 2 - 20 Amp BS 88 Type NIT Time/ Current Characteristics

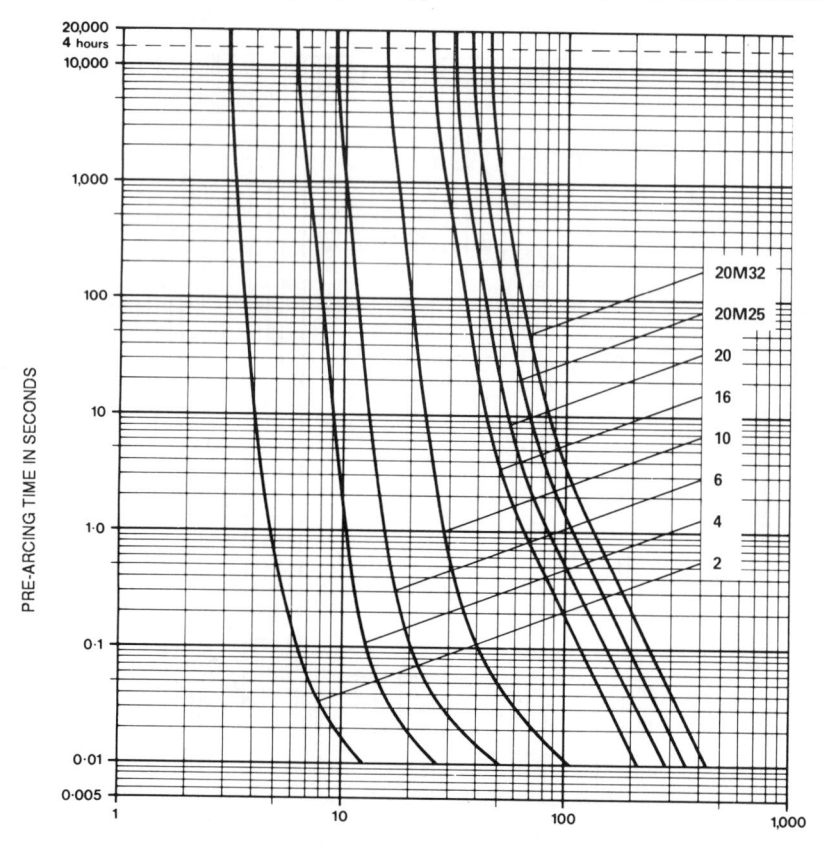

Type NIT Cut-off Current Characteristic

417

GEC ALSTHOM 2 - 63 Amp BS 88 Type T Time/Current Characteristics

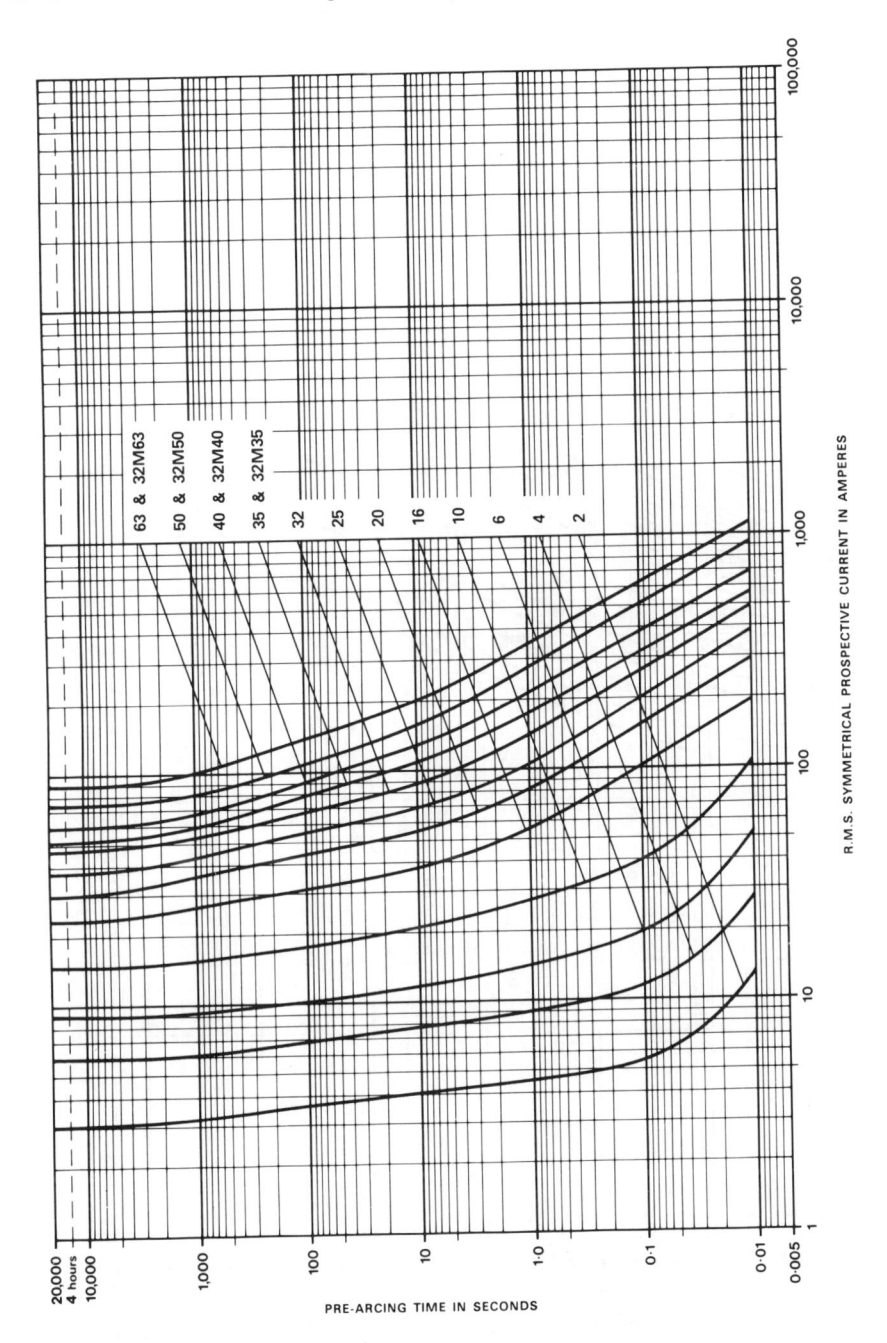

GEC ALSTHOM 80 - 1250 Amp BS 88 Type T Time/Current Characteristics

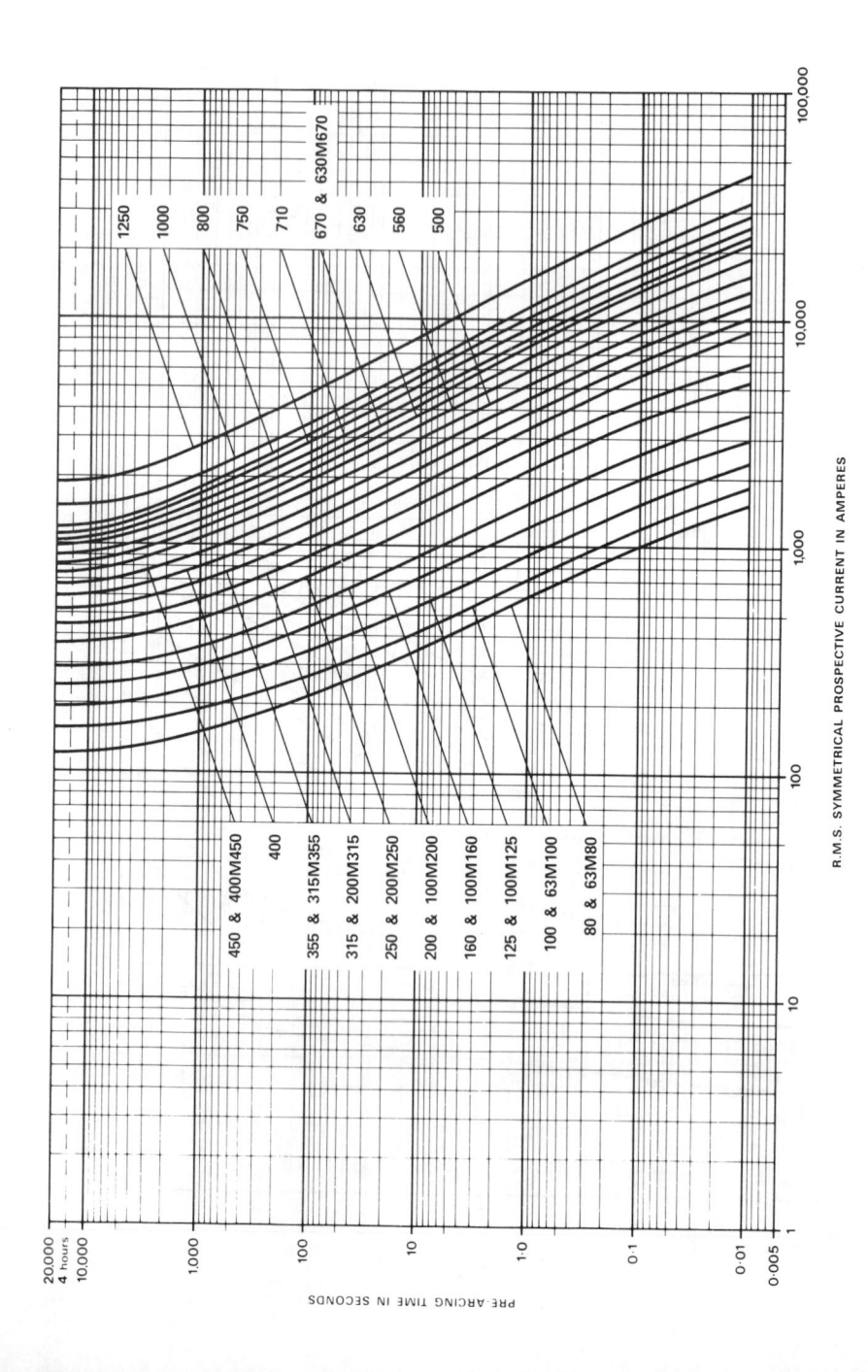

Tried, Tested, Proven

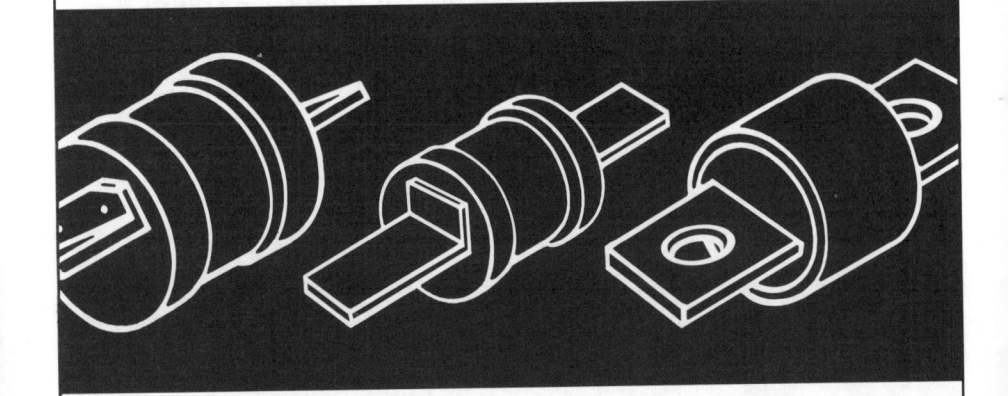

Since the introduction of the HRC fuse link 70 years ago, it has always stayed ahead of its time. Today it's breaking capacity, energy and current-limiting ability are still superior to any other protective device.

setting international standards

▼
GEC ALSTHOM

GEC ALSTHOM INSTALLATION EQUIPMENT LIMITED
East Lancashire Road, Liverpool, England, L10 5HB
Telephone: 051 525 8371   Telex: 627324 FGEAR G   Fax: 051 523 7007

GEC ALSTHOM 2 - 1250 Amp BS 88 Type T Cut-off Characteristics

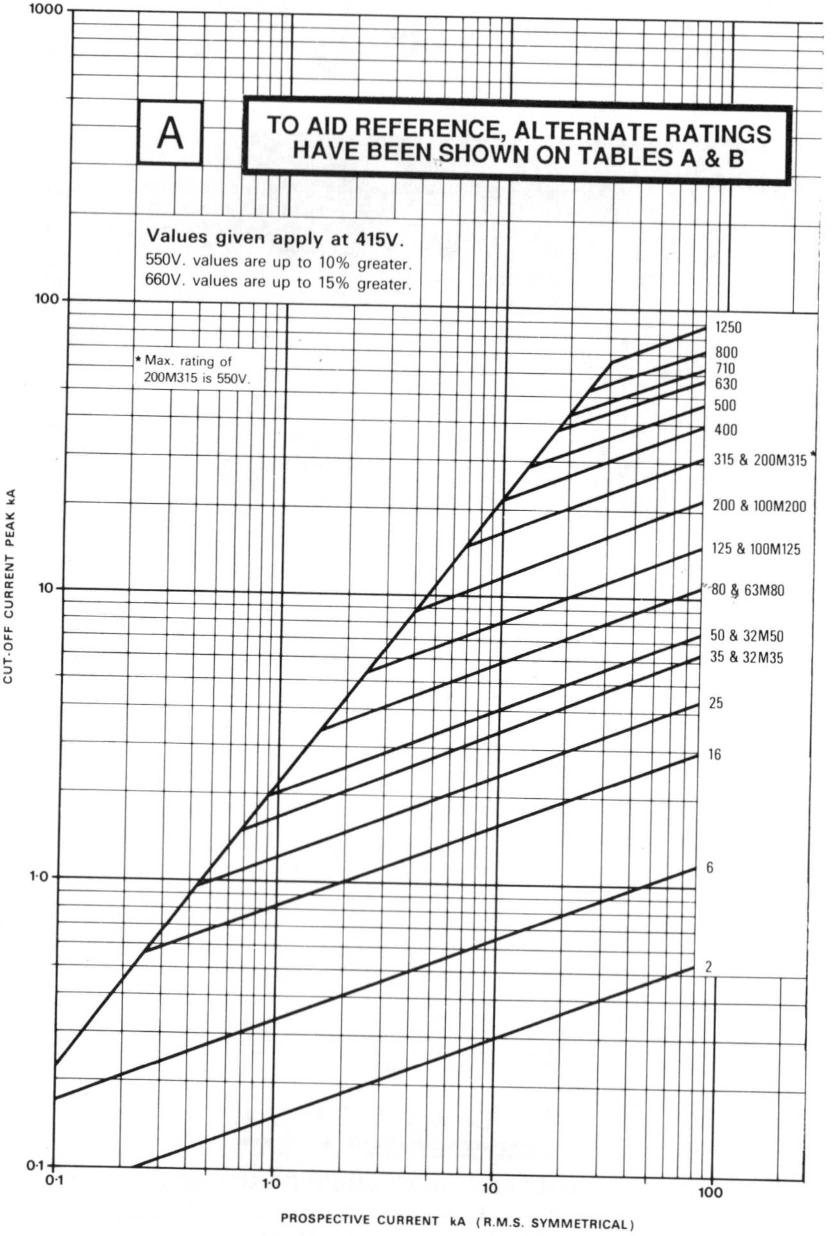
A
TO AID REFERENCE, ALTERNATE RATINGS HAVE BEEN SHOWN ON TABLES A & B
Values given apply at 415V.
550V. values are up to 10% greater.
660V. values are up to 15% greater.
* Max. rating of 200M315 is 550V.
1250
800
710
630
500
400
315 & 200M315*
200 & 100M200
125 & 100M125
80 & 63M80
50 & 32M50
35 & 32M35
25
16
6
2
1000
100
10
1·0
0·1
0·1
1·0
10
100
CUT-OFF CURRENT PEAK kA
PROSPECTIVE CURRENT kA (R.M.S. SYMMETRICAL)

GEC ALSTHOM 2 - 1250 Amp BS 88 Type T Cut-off Characteristics

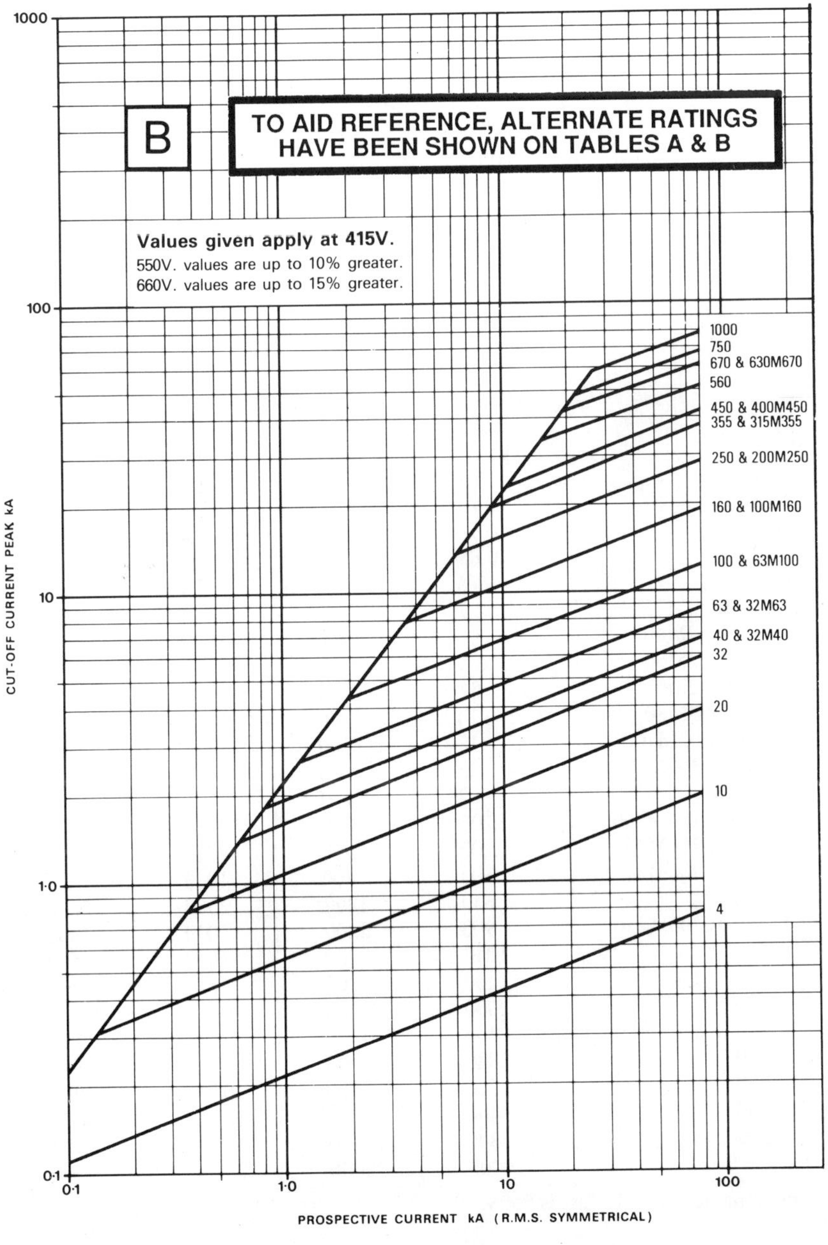
B
TO AID REFERENCE, ALTERNATE RATINGS HAVE BEEN SHOWN ON TABLES A & B
Values given apply at 415V.
550V. values are up to 10% greater.
660V. values are up to 15% greater.
1000
750
670 & 630M670
560
450 & 400M450
355 & 315M355
250 & 200M250
160 & 100M160
100 & 63M100
63 & 32M63
40 & 32M40
32
20
10
4
1000
100
10
1·0
0·1
0·1
1·0
10
100
CUT-OFF CURRENT PEAK kA
PROSPECTIVE CURRENT kA (R.M.S. SYMMETRICAL)

ENERGY LET THROUGH HRC FUSE

GEC ALSTHOM 2 to 1250 Amp Type NIT & T I^2t Fuse Characteristics

Type of fuse	Current rating of fuse in amps	Pre-Arcing I^2t ($A^2\text{sec} \times 1{,}000$)	Total I^2t ($A^2 \text{sec} \times 1{,}000$)		
			415V	550V	660V
		kA	kA	kA	kA
NIT	2	0.0022	0.0054	0.031	-
NIT	4	0.0072	0.018	0.07	-
NIT	6	0.021	0.06	0.4	-
NIT	10	0.10	0.28	1.00	-
NIT	16	0.30	0.85	2.00	-
NIT	20	0.54	1.00	2.50	-
NIT	20M25	0.90	3.00	-	-
NIT	20M32	1.10	4.00	-	-
T	2	0.0022	0.0055	0.0074	0.015
T	4	0.007	0.0185	0.023	0.05
T	6	0.021	0.06	0.08	0.15
T	10	0.1	0.28	0.37	0.7
T	16	0.25	0.55	0.74	1.8
T	20	0.54	1.1	1.4	2.5
T	25	0.85	1.85	2.3	3.7
T	32	1.6	3.4	5.4	8.7
T	35 & 32M35	2.7	5.3	8	15
T	40 & 32M40	4	8.5	11	20.5
T	50 & 32M50	6.3	13.5	18.5	28
T	63 & 32M63	11	24	36	50
T	80 & 63M80	14	40	52	66
T	100 & 63M100	17	60	80	100
T	125 & 100M125	25	85	110	140
T	160 & 100M160	62	160	210	270
T	200 & 100M200	105	260	330	430
T	250 & 200M250	200	550	700	870
T	315 & 200M315*	300	800	1050	1350
T	400	640	1800	2500	3000
T	450 & 400M450	800	2200	3000	3800
T	500	1050	3000	3800	4500
T	560	1400	3800	4250	5400
T	630	2000	5200	6000	7500
T	670 & 630M670	2400	6400	7400	9000
T	710	2800	7000	8000	9700
T	750	3700	7500	10000	12000
T	800	4400	9600	12500	15000
T	1000	5300	12000	14500	17500
T	1250	10000	20000	24000	29000

*Maximum rating of 200M315 fuse is 550V.